Advanced Courses in Mathematics
CRM Barcelona

Centre de Recerca Matemàtica

Managing Editor:
Manuel Castellet

Spiros A. Argyros
Stevo Todorcevic

Ramsey Methods in Analysis

Birkhäuser Verlag
Basel · Boston · Berlin

Authors:

Spiros A. Argyros
National Technical University
Department of Mathematics
Zografou Campus
157 80 Athens
Greece
sargyros@math.ntua.gr

Stevo Todorcevic
Université Paris 7 – C.N.R.S.
U.M.R. 7057
2, Place Jussieu – Case 7012
75251 Paris Cedex 05
France
and
Department of Mathematics
University of Toronto
Toronto, Ontario
M5S 3G3, Canada
stevo@math.toronto.edu

2000 Mathematical Subject Classification 46B20, 05D10, 03E75

A CIP catalogue record for this book is available from the
Library of Congress, Washington D.C., USA

Bibliografische Information Der Deutschen Bibliothek
Die Deutsche Bibliothek verzeichnet diese Publikation in der Deutschen Nationalbibliografie; detaillierte
bibliografische Daten sind im Internet über <http://dnb.ddb.de> abrufbar.

ISBN 3-7643-7264-8 Birkhäuser Verlag, Basel – Boston – Berlin

© 2005 Birkhäuser Verlag, P.O. Box 133, CH-4010 Basel, Switzerland
Part of Springer Science+Business Media
Cover design: Micha Lotrovsky, 4106 Therwil, Switzerland
Printed on acid-free paper produced from chlorine-free pulp. TCF∞

ISBN-10: 3-7643-7264-8
ISBN-13: 978-3-7643-7264-4

9 8 7 6 5 4 3 2 1   www.birkhauser.ch

Contents

B   High-Dimensional Ramsey Theory and Banach Space Geometry
Stevo Todorcevic   **121**

Foreword

This book contains two sets of notes prepared for the Advanced Course on Ramsey Methods in Analysis given at the Centre de Recerca Matemàtica in January 2004, as part of its year-long research programme on Set Theory and its Applications. The common goal of the two sets of notes is to help young mathematicians enter a very active area of research lying on the borderline between analysis and combinatorics. The solution of the distortion problem for the Hilbert space, the unconditional basic sequence problem for Banach spaces, and the Banach homogeneous space problem are samples of the most important recent advances in this area, and our two sets of notes will give some account of this. But our main goal was to try to expose the general principles and methods that lie hidden behind and are most likely useful for further developments. The goal of the first set of notes is to describe a general method of building norms with desired properties, a method that is clearly relevant when testing any sort of intuition about the infinite-dimensional geometry of Banach spaces. The goal of the second set of notes is to expose Ramsey-theoretic methods relevant for describing the rough structure present in this sort of geometry.

We would like to thank the coordinator of the Advanced Course, Joan Bagaria, and the director of the CRM, Manuel Castellet, for giving us this challenging but rewarding opportunity.

Part A

Saturated and Conditional Structures in Banach Spaces

Spiros A. Argyros

Introduction

In 1991, W.T. Gowers and B. Maurey [33], independently, constructed examples of reflexive Banach spaces with no unconditional basis. It is remarkable that, working completely separately, they arrived at the same space. This result is a fundamental discovery with important consequences. Thus, today we have a theory, or at least the beginning of a theory, initializing from Gowers–Maurey paper. Let us recall the basic ingredients of this theory. As it was noticed by W. Johnson the Gowers–Maurey example is a Hereditarily Indecomposable (HI) space. This means that no infinite dimensional closed subspace is the topological direct sum of two further infinite dimensional closed subspaces of it. This is equivalent to the following remarkable geometric property: For any two infinite dimensional subspaces the distance between their unit spheres is zero (sometimes this property is called "the angle zero property"). This is a new concept, defining a new class of Banach spaces, which has good stability properties; for example it is closed in taking subspaces of its members. Also HI spaces have some contradictory properties, resulting from the tightness of their structure. Thus no HI space is isomorphic to any proper subspace, answering in negative the long standing hyperplane problem [31], [33]. On the other side every two closed infinite dimensional subspaces have further subspaces which are almost isometric. The new concept stands in the opposite of the unconditional basic sequence and W.T. Gowers' famous dichotomy ([30], [32]) has provided the first sufficient classification of Banach spaces. Namely, every Banach space either is unconditionally saturated or contains an HI space. Unexpectedly this classification provides a positive solution of the homogenous problem that yields a new characterization of Hilbert spaces.

Another important discovery of Gowers and Maurey concerns the Banach spaces with few operators. As they have shown [33] every bounded linear operator on a complex HI space is of the form $\lambda I + S$ with S strictly singular. This has as consequence that every Fredholm operator on a HI (real or complex) space is of index zero, which yields the aforementioned property that such a space is not isomorphic to any proper subspace. The structure of $\mathcal{L}(X)$ for real HI spaces is studied in [25]. The problem of the existence of a Banach space with very few operators (i.e. every $T \in \mathcal{L}(X)$ is of the form $\lambda I + K$ with K a compact operator) remains open and if such a space existed it should be connected to the methods of HI constructions. Recently it has been shown that there exist spaces with few operators not containing any HI space ([14], [15]).

The third important contribution of Gowers–Maurey discovery concerns the generic character of their method. This is a powerful method of constructing Banach spaces using saturated norms and delicate codings. This method is not related exclusively to HI spaces. As Gowers has shown [31] one can also obtain new spaces with an unconditional basis resulting from variants of Gowers–Maurey method. The recent paper of J. Lopez-Abad, S. Todorcevic and the present author [12], [13] extends this method to reflexive Banach spaces with transfinite Schauder basis where new phenomena concerning the operator spaces occur. I must admit that constructing such spaces requires several steps and although each one is not so difficult putting everything together makes the understanding not easy. In particular dealing with these spaces, one faces three main parts. The first is how we define such spaces. The definition of saturated norms uses induction and additional one has also to involve the special functionals resulting from the coding. The second concerns the form of vectors of which we compute the norms. In order to show the non-existence of unconditional basic sequences we have to compute the norms of certain vectors. To locate these vectors we have to follow several steps. Finally to compute the norms requires some new techniques which have been developed for this purpose.

The goal of the present notes is, on the one hand, to develop the method of strictly singular HI extensions of a ground norm and, on the other hand, to apply this method in specific constructions of HI Banach spaces. In the later part we are mainly interested in two non separable constructions. The first concerns a non separable HI space, such an example appeared in [16], and the second a non separable reflexive space containing no unconditional basic sequence [12], [13]. The method of strictly singular extensions is related to the following problem. Given Y a Banach space with a Schauder basis $(y_n)_n$ and not containing ℓ_1, we are interested in finding a HI space $\mathfrak{X}$ with a Schauder basis $(e_n)_n$ such that the correspondence $e_n \to y_n$ is extended to a bounded linear operator $T : \mathfrak{X} \to Y$. This problem has been answered affirmatively in [16] with the use of a transfinite hierarchy of saturation methods. In the same paper the strictly singular extensions were introduced and studied although not named there. In these notes we discuss a part missing from [16], namely the extensions with low complexity saturation methods. This does not answer the above stated problem in its complete generality, however it permits some of the main constructions to be carried out and moreover, extending with the methods $(\mathcal{A}_{n_j}, \frac{1}{m_j})_j$, we deal exclusively with averages, as in [33], making our approach easier.

The interest for the HI extensions arises from the fact that in some cases specific features of the space Y are preserved in the HI extension of it. For example if Y is $c_0(\mathbb{N})$ and the basis $(y_n)_n$ is the summing basis of $c_0(\mathbb{N})$, then independently of the saturation method we use, the resulting space is a quasi-reflexive HI space. A remarkable consequence of the extension method is a dichotomy contained in

[16], not discussed here, that every separable Banach space either contains ℓ_1 or it is a quotient of a HI space. To some extend the HI extensions, thinking of the specific Y as a model of Banach space theory, attempt to do a work similar to the forcing method in set theory. Namely, they create a new model with a desired property (HI) and in some cases phenomena of absoluteness also occur, preserving properties of the initial space. The HI extensions are presented in Chapters II–IV. Chapter I is devoted to an introduction to Tsirelson type and Mixed Tsirelson spaces. The content of this chapter is not directly related to the HI extensions, thus the familiar reader could skip it. However we recommend to study the proof of Theorem I.4, where important ingredients, like the tree analysis and their usage, appear in a simple setting.

Chapter V contains examples of HI extensions. Thus it is shown the existence of a quasi-reflexive HI space, that every ℓ_p, $1 < p < +\infty$, is a quotient of a HI space and a non separable HI space $\mathfrak{X}_{ns}$ which is the dual as well as the second dual of a separable HI space. Chapter VI concerns the construction of $\mathfrak{X}_{\omega_1}$, a non separable reflexive space with no unconditional basic sequence, or according to Gowers' dichotomy, HI saturated (i.e. every infinite dimensional subspace contains a further HI subspace). The dimension of $\mathfrak{X}_{\omega_1}$ is ω_1 and it is unknown if there exists a space with the same properties and higher dimension. The main obstacle for non separable reflexive constructions with no unconditional basic sequence, concerns the definition of the conditional structure realized by the special functionals. The key ingredient for this is the coding function υ used by Gowers and Maurey and having its roots in the classical Maurey–Rosenthal construction [48]. The basic characteristic of σ, in the separable constructions, is that it is injective (one-to-one), a property not extendable to the non separable setting. Thus the really new ingredient in the definition of $\mathfrak{X}_{\omega_1}$ is the coding function σ_ρ defined with the use of Todorcevic's ρ function [60],[61] . The ρ function is acting on the doubletons of ω_1, taking values in $\mathbb{N}$ and, although not injective, permits a definition of the special functionals sharing similar properties with the corresponding in the separable case. Chapter VII is devoted to the study of the space of diagonal non strictly singular operators of $\mathfrak{X}_{\omega_1}$. Let us mention that $\mathfrak{X}_{\omega_1}$, being a non separable reflexive space, admits many non trivial projections [40]. Therefore there exists a large portion of operators which are different from the identity and non strictly singular. The space $D(\mathfrak{X}_\gamma)$ has a very precise representation, as a Banach space which is related to a long James like space. Chapter VIII is a continuation of Chapter VII in the study of the spaces of operators for subspaces of $\mathfrak{X}_{\omega_1}$.

We have included two appendices. The first concerns transfinite basic sequences and the second is devoted to a unified approach of the basic inequality and the finite block representability of the James like space J_{T_0}.

The present notes are based on a series of lectures delivered by the author in CRM Barcelona. The principal aim was to introduce the audience to the theory of HI Banach spaces. Thus there are directions of the theory not discussed here. One of them is the method developed by W.T. Gowers and B. Maurey [34] for constructing Banach spaces with selected a priori algebra of non strictly singular operators. As consequence they provide a separable Banach space X_s with a Schauder basis $(x_n)_n$ such that the space X_s does not contain an unconditional basic sequence and also the shift operator is an isometry. A second direction concerns the use of interpolation methods in HI constructions, as appeared in [10, 3, 27]. Recently has been studied the problem of the existence of Banach spaces with few operators (i.e. every $T \in \mathcal{L}(X)$ is of the form $T = \lambda I + S$ with S strictly singular) and not containing a HI subspace. The existence of few operators yields that the space X is indecomposable. Related to this type of questions are the papers [14], [15], [17]. Finally we recommend the two survey papers by W.T. Gowers [32] and B. Maurey [46] for results related to HI Banach spaces. In particular Maurey's survey includes a beautiful presentation of some of the aforementioned methods and results not included in these notes. We extend our warm thanks to A. Arvanitakis, P. Dodos, A. Manoussakis, and A. Tolias for their valuable help during the preparation of the notes.

Chapter I

Tsirelson and Mixed Tsirelson Spaces

The Gowers–Maurey example and all the subsequent constructions have their roots in B.S. Tsirelson's fundamental discovery of a reflexive space with an unconditional basis not containing any ℓ_p with $1 < p < \infty$. Tsirelson's space appeared in 1972. Almost twenty years later (1991) Th. Schlumprecht presented his space as an example of an arbitrarily distortable space. This is the unconditional frame for the Gowers–Maurey construction. The two norms share common features and our aim is to explain their generic character, their relation with the classical ℓ_p norms and their differences.

Glossary

For κ an ordinal we denote by $c_{00}(\kappa)$ the vector space of all $x : \kappa \to \mathbb{R}$ such that the set $\operatorname{supp} x = \{\alpha < \kappa : x(\alpha) \neq 0\}$ is finite. For $x \in c_{00}(\kappa)$ we denote by $\operatorname{ran} x$ the minimal interval of κ containing $\operatorname{supp} x$. We also denote by $(e_\alpha)_{\alpha < \kappa}$ the natural basis of $c_{00}(\kappa)$. For the special case of $c_{00}(\omega)$ we shall use the notation c_{00}. For $E_1, E_2 \subset \kappa$ we denote by $E_1 < E_2$ the property that $\sup E_1 < \min E_2$. Further for x_1, x_2 elements of $c_{00}(\kappa)$ we denote by $x_1 < x_2$ the property that $\operatorname{supp} x_1 < \operatorname{supp} x_2$. In most cases we shall use $c_{00}(\kappa)$ as a vector space on which we shall define certain norms. These norms will be induced by an appropriate subset W of $c_{00}(\kappa)$ and will be defined as

$$\|x\|_W = \sup \left\{ |f(x)| = \Big| \sum_{\alpha < \kappa} f(\alpha) x(\alpha) \Big| : f \in W \right\}.$$

Thus in some cases we shall denote the elements of $c_{00}(\kappa)$ as f, g, etc. while its Hamel basis as $(e_\alpha^*)_{\alpha < \kappa}$ meaning that we concern ourselves with the functionals of the norming set. Finally for $x \in c_{00}(\kappa)$ and $E \subset \kappa$ we denote by Ex the restriction of x on E or equivalently the function $x\chi_E$. We refer the reader to [24] and [41] for the fundamental background of the theory of Banach spaces.

Tsirelson space

Tsirelson's norm is the first norm defined by induction. It satisfies a fixed point property and it is implicitly defined as follows.

For $x \in c_{00}$,

$$\|x\|_{\mathcal{T}} = \max\left\{\|x\|_0, \frac{1}{2} \sup_{n \leq E_1 < E_2 < \cdots < E_n} \sum_{i=1}^{n} \|E_i x\|_{\mathcal{T}}\right\}. \tag{I.1}$$

The space $\mathcal{T}$ is the completion of c_{00} endowed with $\|\cdot\|_{\mathcal{T}}$. The above definition is due to T. Figiel and W. Johnson [28]. Tsirelson's original definition [63] was concerning the norm of the dual space. Notice that the norm of $\mathcal{T}^*$ does not admit any explicit or implicit description. Thus Tsirelson actually defined the unit ball of $\mathcal{T}^*$.

Remarks I.1. (a) The existence of a norm satisfying the above implicit formula is established by induction. Namely we inductively define an increasing sequence of norms $(\|\cdot\|_n)_{n\geq 0}$ such that $\|x\|_0 = \|x\|_\infty$ and

$$\|x\|_{n+1} = \max\left\{\|x\|_0, \frac{1}{2} \sup_{n \leq E_1 < E_2 < \cdots < E_n} \sum_{i=1}^{n} \|E_i x\|_n\right\}.$$

Then $\|x\|_{\mathcal{T}} = \lim_n \|x\|_n$.

(b) An important feature of Tsirelson's norm is that it provides a method to saturate the structure of a Banach space with a property (P). For example the space $\mathcal{T}$ has the property that every normalized weakly null sequence contains no Cesaro summable subsequence, which actually shows that no ℓ_p, $1 < p < \infty$ is embedded in $\mathcal{T}$. The existence of a weakly null normalized basic sequence with no Cesaro summable subsequence was a fundamental discovery of J. Schreier in the early 1930s.

(c) Observe that in the definition of (I.1) appears the number $1/2$ and the constrain $n \leq E_1 < E_2 < \cdots < E_n$. The number $1/2$ could be substituted by any $0 < \theta < 1$ and the space retains all its properties. If $\theta = 1$, then the space becomes isomorphic to ℓ_1. The same will happen if we allow arbitrary families $E_1 < E_2 < \cdots < E_n$.

Tsirelson type norms

In the sequel we shall denote by $\mathcal{M}$ a compact family (in the topology of pointwise convergence) of finite subsets of $\mathbb{N}$ which includes all singletons.

Definition I.2. A finite family $E_1 < E_2 < \cdots < E_n$ of subsets of $\mathbb{N}$ is said to be $\mathcal{M}$-*admissible* if there exists $M = \{m_i\}_{i=1}^n$ in $\mathcal{M}$ such that $m_1 \le E_1 < m_2 \le E_2 < \cdots < m_n \le E_n$. Next for $\mathcal{M}$ a compact family of subsets of $\mathbb{N}$ and $0 < \theta < 1$ we define the $\|\cdot\|_{(\mathcal{M},\theta)}$ as follows. For $x \in c_{00}$,

$$\|x\|_{(\mathcal{M},\theta)} = \max\left\{\|x\|_0, \theta \sup \sum_{i=1}^n \|E_i x\|_{(\mathcal{M},\theta)}\right\}$$

where the above sup is taken over all $\mathcal{M}$-admissible families $E_1 < E_2 < \cdots < E_n$. Finally we denote by $T(\mathcal{M}, \theta)$ the completion of $(c_{00}, \|\cdot\|_{(\mathcal{M},\theta)})$.

It is easy to see that Tsirelson's space T is the space $T(\mathcal{S}, 1/2)$, where $\mathcal{S} = \{F \subseteq \mathbb{N} : \#F \le \min F\}$ is the Schreier family used by J. Schreier in the definition of his space [56]. The structure of $T(\mathcal{M}, \theta)$ depends on the complexity of the underlying compact family $\mathcal{M}$. In particular the structure of $T(\mathcal{M}, \theta)$ is strongly related to the Cantor–Bendixson index of the family $\mathcal{M}$ denoted by $i(\mathcal{M})$.

We shall present later some results explaining the interference between $i(\mathcal{M})$ and $T(\mathcal{M}, \theta)$. Now we present an alternative description of the norm of $T(\mathcal{M}, \theta)$ closer to the spirit of Tsirelson's original definition.

Let us denote by $W(\mathcal{M}, \theta)$ the minimal subset of c_{00} containing $\pm e_n^*$, $n \in \mathbb{N}$, and which is closed under the $(\mathcal{M}, \theta)$-operation; i.e. for every $f_1 < f_2 < \cdots < f_n$ in $W(\mathcal{M}, \theta)$ with $\mathrm{supp}\, f_1, \mathrm{supp}\, f_2, \ldots, \mathrm{supp}\, f_n$ $\mathcal{M}$-admissible, then $\theta(f_1 + \cdots + f_n) \in W(\mathcal{M}, \theta)$ (we shall call such $f_1 < f_2 < \cdots < f_n$ an $\mathcal{M}$-admissible family of functionals). The norm induced by $W(\mathcal{M}, \theta)$ (i.e. for $x \in c_{00}$, $\|x\|_{W(\mathcal{M},\theta)} = \sup\{f(x) : f \in W(\mathcal{M}, \theta)\}$) is exactly the norm $\|x\|_{(\mathcal{M},\theta)}$ defined above.

Definition I.3. Let $f \in W(\mathcal{M}, \theta)$. For a finite tree $\mathcal{T}$, a family $(f_t)_{t\in\mathcal{T}}$ is said to be a *tree analysis* of f if the following are satisfied:

(i) $\mathcal{T}$ has a unique root denoted by 0 and $f_0 = f$.

(ii) For every $t \in \mathcal{T}$ maximal (or terminal) $f_t = \varepsilon_t e_k^*$ where $\varepsilon_t = \pm 1$ and $k \in \mathbb{N}$.

(iii) For every $t \in \mathcal{T}$ which is not maximal we have that $\{f_s\}_{s\in S_t}$ is $\mathcal{M}$-admissible and $f_t = \theta \sum_{s\in S_t} f_s$ (here S_t denotes the immediate successors of t).

The tree analysis is a key ingredient which will follow us throughout these notes. Later we shall see some variants of it. Most of the estimations will be based on the tree analysis. It is not difficult to see that every $f \in W(\mathcal{M}, \theta)$ admits a tree analysis. Indeed we could consider the set $W'(\mathcal{M}, \theta)$ that contains $\pm e_k^*$, $k \in \mathbb{N}$, and every $f \in W'(\mathcal{M}, \theta)$ satisfies $f \in W(\mathcal{M}, \theta)$ and f admits a tree analysis. Then we show that $W'(\mathcal{M}, \theta)$ is closed under the $(\mathcal{M}, \theta)$-operation and from the minimality of $W(\mathcal{M}, \theta)$ we obtain that $W'(\mathcal{M}, \theta) = W(\mathcal{M}, \theta)$.

Among all possible compact families $\mathcal{M}$ there are two hierarchies that we are mainly concerned about. The first is the low complexity hierarchy $\{\mathcal{A}_n\}_n$ with $\mathcal{A}_n = \{F \subseteq \mathbb{N} : \#F \le n\}$ and the second is the family $\{\mathcal{S}_\xi\}_{\xi<\omega_1}$ of Schreier families. Each $\mathcal{S}_\xi$ is defined recursively as follows. We set $\mathcal{S}_0 = \{\{n\} : n \in \mathbb{N}\} \cup \{\emptyset\}$. For $\xi = \zeta + 1$ we set

$$\mathcal{S}_\xi = \{F \subseteq \mathbb{N} : \text{ there exists } k \le F_1 < F_2 < \cdots < F_k$$
$$\text{with each } F_i \in \mathcal{S}_\zeta \text{ and } F = \bigcup_{i=1}^{k} F_i\} \cup \{\emptyset\}.$$

For ξ limit we choose $\xi_n \nearrow \xi$ and we set

$$\mathcal{S}_\xi = \{F : \exists n \le F \text{ with } F \in \mathcal{S}_{\xi_n}\} \cup \{\emptyset\}.$$

It is quite possible, and partially proved, that the spaces $\mathcal{T}(\mathcal{M}, \theta)$ where $\mathcal{M} \in \{\mathcal{A}_n\}_n$ or $\mathcal{M} \in \{\mathcal{S}_\xi\}_{\xi<\omega_1}$, describe the structure of $\mathcal{T}(\mathcal{M}', \theta)$ where $\mathcal{M}'$ is an arbitrary family.

The following theorem is quite remarkable showing that the classical ℓ_p spaces admit an equivalent Tsirelson type norm.

Theorem I.4. *For every $0 < \theta < 1$ and every $n \in \mathbb{N}$ the space $\mathcal{T}(\mathcal{A}_n, \theta)$ is isomorphic to c_0 or to some ℓ_p with $1 < p < \infty$. In particular:*

(i) *If $\frac{1}{n} \ge \theta$, then $\mathcal{T}(\mathcal{A}_n, \theta)$ is isomorphic to c_0.*

(ii) *If $\frac{1}{n} < \theta$, then $\mathcal{T}(\mathcal{A}_n, \theta) \cong \ell_p$, where $\theta = \frac{1}{n^{1/q}}$ and $\frac{1}{p} + \frac{1}{q} = 1$.*

Proof. (i) Let $\frac{1}{n} \ge \theta$. It is enough to show that for all $f \in W$ it holds

$$f(\sum_n a_n e_n) \le \max_n |a_n| \text{ for all coefficients } (a_n)_n.$$

Let $f \in W$ and $(f_t)_{t \in \mathcal{T}}$ be a tree analysis of f. An easy inductive argument gives us that

$$f_t(\sum_k a_k e_k) \le \max_k |a_k|,$$

for all $t \in \mathcal{T}$, and this inequality yields that the space $\mathcal{T}(\mathcal{A}_n, \theta)$ is isomorphic to c_0.

(ii) Let $\frac{1}{n} < \theta$. Let p, q be as in the assumption. We prove that the basis $(e_n)_n$ of $\mathcal{T}(\mathcal{A}_n, \theta)$ is equivalent to the standard basis of ℓ_p. The proof goes through the following four steps.

Step 1.   For every $x \in c_{00}$, $\|x\| \le \|x\|_p$.

Proof. It is enough to show that for every $f \in W(\mathcal{M}, \theta)$ it holds

$$|f(x)| \le \|x\|_p. \tag{I.2}$$

Let $(f_t)_{t \in \mathcal{T}}$ be a tree analysis of f. It is trivial that for a terminal node $t \in \mathcal{T}$, f_t satisfies (I.2).

Let $t \in \mathcal{T}$, $f_t = \theta \sum_{s \in S_t} f_s$ where $\theta = \frac{1}{n^{1/q}}$ and $(f_s)_{s \in S_t}$ is $\mathcal{A}_n$-admissible. Assume that for every $s \in S_t$ we have that for all $y \in c_{00}$, $|f_s(y)| \leq \|y\|_p$.

Let $x \in c_{00}$. Then, setting $x_i = (\text{supp}(f_i))(x)$ we have

$$|f_t(x)| \leq \theta \sum_{s \in S_t} |f_s(x)| = \frac{1}{n^{\frac{1}{q}}} \sum_{s \in S_t} |f_s(x_i)|$$

$$\leq \frac{1}{n^{\frac{1}{q}}} \sum_{s \in S_t} \|x_i\|_p \leq (\frac{\#S_t}{n})^{\frac{1}{q}} (\sum_{i=i}^{d} \|x_i\|_p^p)^{\frac{1}{p}} \leq \|x\|_p,$$

where in the previous inequality we use the inductive assumption and Hölder's inequality. $\qquad \square$

Step 2. For all $m \in \mathbb{N}$, $\frac{1}{n^{\frac{1}{p}}} m^{\frac{1}{p}} \leq \|\sum_{i=1}^{m} e_i\|$.

Proof. Suppose first that $m = n^s$ for some $s \in \mathbb{N}$. The functional $f = \frac{1}{n^{s/q}} (\sum_{i=1}^{n^s} e_i)$ clearly belongs to $W(\mathcal{M}, \theta)$. So

$$\|\sum_{i=1}^{n^s} e_i\| \geq f(\sum_{i=1}^{n^s} e_i) = \frac{1}{n^{s/q}} n^s = n^{s(1-1/q)} = n^{s/p} = m^{1/p}.$$

Now let $m \in \mathbf{N}$ and find s such that $n^s \leq m \leq n^{s+1}$. Then

$$\|\sum_{i=1}^{m} e_i\| \geq \|\sum_{i=1}^{n^s} e_i\| = n^{s/p} = \frac{1}{n^{1/p}} n^{(s+1)/p} \geq \frac{1}{n^{1/p}} m^{1/p}. \qquad \square$$

Step 3. For every normalized block sequence $(x_k)_{k=1}^{\infty}$ of the basis $(e_n)_{n=1}^{\infty}$ we have

$$\|\sum a_k x_k\| \leq \frac{2}{\theta} \|\sum a_k e_k\|$$

for all coefficients (a_k).

Proof. It is enough to show that for every $f \in W$ one gets

$$f(\sum a_k x_k) \leq \frac{2}{\theta} \|\sum a_k e_k\|.$$

For the proof we shall need the following definition.

Definition I.5. Let $f \in W(\mathcal{M}, \theta)$, $(f_t)_{t \in \mathcal{T}}$ be a tree analysis of f and let $(x_k)_k$ be a finite block sequence.

a) For every $k \in \mathbb{N}$ we consider the set of nodes

$$\mathcal{T}_k = \{t \in \mathcal{T} \quad : \quad \text{(i)} \;\; \text{ran} f_t \cap \text{ran} x_k \neq \emptyset \text{ for all } s \preceq t \text{ if } s \in S_u$$

$$\text{(ii)} \;\; \text{ran} f_s \cap \text{ran} x_k = \text{ran} f_u \cap \text{ran} x_k$$

$$\text{(iii)} \;\; \text{for all } s \in S_t \quad \text{ran} f_s \cap \text{ran} x_k \subsetneqq \text{ran} f_t \cap \text{ran} x_k\}$$

b) For every $t \in \mathcal{T}$ we set $D_t = \bigcup_{s \succeq t} \{k : s \in \mathcal{T}_k\}$.

It is easy to see that for every $k \in \mathbb{N}$, $\#\mathcal{T}_k \leq 1$. Also a $t \in \mathcal{T}$ could belong to more than one $\mathcal{T}_k$.

The proof of *Step* 3 will be an immediate consequence of the following two lemmas.

Lemma I.6. *Let $f \in W(\mathcal{M}, \theta)$ with a tree analysis $(f_t)_{t \in \mathcal{T}}$ and $(x_k)_k$ be a finite normalized block sequence. Assume that for all $t \in \mathcal{T}$ it holds that*

$$\# \left(\{s \in S_t : D_s \neq \emptyset\} \cup \{k : \mathcal{T}_k = \{t\}\} \right) \leq n .$$

Then there exists $g \in W(\mathcal{M}, \theta)$ such that for all k it holds that

$$f(x_k) \leq \frac{1}{\theta} g(e_k) .$$

Proof. For every $t \in \mathcal{T}$ we set $K = D_t \setminus \bigcup_{s \in S_t} D_s = \{k : \mathcal{T}_k = \{t\}\}$. We also set $E = \{s \in S_t : D_s \neq \emptyset\}$. It follows from the hypothesis that $\#(K \cup E) \leq n$. We shall prove by induction that for every $t \in \mathcal{T}$ there exists $g_t \in W(\mathcal{M}, \theta)$ such that

$$\operatorname{supp} g_t \subset D_t \quad \text{and} \quad f_t(x_k) \leq \frac{1}{\theta} g_t(e_k) \quad \text{for every } k \in D_t .$$

Indeed let $t \in \mathcal{T}$ and $f_t = \theta \sum_{s \in S_t} f_s$. We have that

$$f_t(\sum_k a_k x_k) = \theta \left(\sum_{s \in S_t} f_s(\sum_{k \in D_s} a_k x_k) + \sum_{k \in K} (\sum_{s \in S_t} f_s)(a_k x_k) \right) .$$

We set

$$g_t = \theta (\sum_{s \in E} g_s + \sum_{k \in K} e_k^*) .$$

Since $e_k^*, g_s \in W(\mathcal{M}, \theta)$ for every $k \in K$, $s \in E$ and $\#(K \cup E) \leq n$, it follows that $g_t \in W(\mathcal{M}, \theta)$. From the fact that for every k

$$\sum_{s \in S_t} f_s(x_k) \leq \frac{1}{\theta} \|x_k\| = \frac{1}{\theta} e_k^*(e_k),$$

we get, $f_t(x_k) \leq \frac{1}{\theta} g_t(e_k)$ for every $k \in K$.

If $k \in D_s$ for some $s \in E$, from the inductive hypothesis we get

$$f_t(x_k) = \theta f_s(x_k) \leq g_s(e_k) = \frac{1}{\theta} g_t(e_k) . \qquad \square$$

In order to complete the proof of *Step* 3, we show that there exists a partition of x_k such that the assumptions of the previous lemma are satisfied.

Lemma I.7. *Let* $f \in W$ *with a tree analysis* $(f_t)_{t \in \mathcal{T}}$ *and* $(x_k)_k$ *be a finite block sequence. Then there exists a partition of* x_k, $x_k = x'_k + x''_k$ *such that* $(f_t)_{t \in \mathcal{T}}$, $(x'_k)_k$ *and* $(f_t)_{t \in \mathcal{T}}$, $(x''_k)_k$ *satisfy the assumptions of the previous lemma.*

Proof. Let f, $(x_k)_{k=1}^{\ell}$ be as above. Let $(f_t)_{t \in \mathcal{T}}$ be a fixed analysis of f. For $k = 1, \ldots, \ell$ we set s_k to be the minimum of $n \in \mathbb{N} \cup \{0\}$ such that there exist two $t_1, t_2 \in \mathcal{T}$ with

(i) $|t_1| = |t_2| = n$,

(ii) $\operatorname{ran} f_{t_j} \cap \operatorname{ran} x_k \neq \emptyset$ for $j = 1, 2$

if such an n exists. Otherwise we set s_k to be the maximum of all $n \in \mathbb{N}$ such that there exists $t \in \mathcal{T}$ with

(i) $|t| = n$,

(ii) if $\operatorname{supp}(x_k)$ is a singleton $\operatorname{ran} f_t \cap \operatorname{ran} x_k \neq \emptyset$.

For every k, let $\{f_t : t \in \mathcal{T}, |t| = s_k$ and $\operatorname{ran} f_t \cap \operatorname{ran} x_k \neq \emptyset\} = \{f_1 < \cdots < f_d\}$. We define

$$x'_k = x_k|_{\operatorname{supp} f_1} \quad \text{and} \quad x''_k = x_k|_{\bigcup_{i=2}^{d} \operatorname{supp} f_i}.$$

Then the pairs $(f_t)_{t \in \mathcal{T}}$, $(x'_k)_k$ and $(f_t)_{t \in \mathcal{T}}$, $(x''_k)_k$ satisfies the conclusion.

Indeed, for $t \in \mathcal{T}$ let $K = \{k : \mathcal{T}_k = \{t\}\}$ and $E = \{s \in S_t : D_s \neq \emptyset\}$. For $k \in K$ there exists $t_k \in S_t$ with

$$\operatorname{ran} f_{t_k} \cap \operatorname{ran} x'_k \neq \emptyset \quad \text{and} \quad \max \operatorname{supp} f_{t_k} = \max \operatorname{supp} x'_k.$$

It follows that $t_k \notin E$. Therefore we can define a one-to-one map $G : K \to S_t \setminus E$, hence $\#K + \#E \leq \#S_t = n$.

The proof for the pair $(f_t)_{t \in \mathcal{T}}$, $(x''_k)_k$ is similar. $\qquad\square$

The previous two lemmas completes the proof of *Step 3*. $\qquad\square$

Step 4. For all ℓ and all rational non-negative $(r_j)_{j=1}^{\ell}$,

$$\|\sum_{j=1}^{\ell} r_j^{1/p} e_j\| \geq \frac{1}{2n} (\sum_{j=1}^{\ell} r_j)^{1/p}.$$

Proof. Write $r_j = \frac{k_j}{k}$, $k_j, k \in \mathbb{N}$. Set $s_0 = 0$, $s_j = k_1 + \cdots + k_j$ and $u_j = \sum_{i=s_{j-1}+1}^{s_j} e_i$,
$j = 1, \ldots, \ell$. By Step 1, $\|u_j\| \leq k_j^{1/p}$. So

$$\|\sum_{j=1}^{\ell} r_j^{1/p} e_j\| = \frac{1}{k^{1/p}} \|\sum_{j=1}^{\ell} k_j^{1/p} e_j\| \geq \frac{1}{k^{1/p}} \|\sum_{j=1}^{\ell} \|u_j\| e_j\|$$

by unconditionality.

By Step 3,

$$\frac{1}{k^{1/p}}\|\sum_{j=1}^{\ell}\|u_j\|e_j\| \geq \frac{\theta}{2}\frac{1}{k^{1/p}}\|\sum_{j=1}^{\ell}\|u_j\|\frac{u_j}{\|u_j\|}\|$$

$$= \frac{\theta}{2}\frac{1}{k^{1/p}}\|\sum_{j=1}^{\ell}\sum_{i=s_{j-1}+1}^{s_j}e_i\| = \frac{\theta}{2}\frac{1}{k^{1/p}}\|\sum_{i=1}^{s_\ell}e_i\|.$$

By Step 2, $\|\sum_{i=1}^{s_\ell}e_i\| \geq \frac{1}{n^{1/p}}s_\ell^{1/p}$; so using that $\theta = \frac{1}{n^{1/q}}$ we get

$$\frac{\theta}{2}\frac{1}{k^{1/p}}\|\sum_{i=1}^{s_\ell}e_i\| \geq \frac{1}{2n}\frac{s_\ell^{1/p}}{k^{1/p}} = \frac{1}{2n}\Big(\frac{\sum_{j=1}^{\ell}k_j}{k}\Big)^{1/p} = \frac{1}{2n}\Big(\sum_{j=1}^{\ell}r_j\Big)^{1/p}. \qquad \square$$

Step 4 and the unconditionality of $(e_n)_{n\in\mathbf{N}}$ imply that $\|\sum a_k e_k\| \geq \frac{1}{2n}(\sum |a_k|^p)^{1/p}$ for all coefficients (a_k). This fact combined with Step 1 completes the proof of the Theorem. $\qquad \square$

Remark. The result of the Theorem can also be deduced by Steps 1, 2 and 3 using a well known theorem of Zippin [65].

This is a significant result permitting a unified approach of Tsirelson's space and the classical ℓ_p spaces. Hence the saturated norms enable us to extend the class of the classical sequence spaces. Next we state the following result without proof.

Theorem I.8. *Let $\mathcal{M}$ be a compact family of finite subsets of $\mathbb{N}$ containing the subsets of its elements and all singletons. Let also $0 < \theta < 1$. Then the following hold:*

(i) *If $i(\mathcal{M}) < \omega$ and $\frac{1}{i(\mathcal{M})} \geq \theta$, then $T(\mathcal{M},\theta) \cong c_0$.*

(ii) *If $i(\mathcal{M}) < \omega$ and $\frac{1}{i(\mathcal{M})} < \theta$, then $T(\mathcal{M},\theta) \cong \ell_p$, for some $1 < p < \infty$.*

(iii) *If $i(\mathcal{M}) \geq \omega$, then $T(\mathcal{M},\theta)$ is reflexive with an unconditional basis not containing any ℓ_p, $1 < p < \infty$.*

(iv) *If $i(\mathcal{M}) = \omega$, then there exists a normalized sequence $(x_n)_n$ in $T(\mathcal{M},\theta)$ and a subsequence $(e_{n_k})_k$ in Tsirelson's space $T(\mathcal{S},\theta)$ which are equivalent.*

Here $i(\mathcal{M}) = \min\{\alpha < \omega_1 : \mathcal{M}^{(\alpha)} \subset \{\varnothing\}\}$ and $\mathcal{M}^{(\alpha)}$ is the usual α-Cantor–Bendixson derivative of the countable compact space $\{\chi_F : F \in \mathcal{M}\} \cup \{\varnothing\}$. It is open whether or not for $\mathcal{M}$ as in the theorem with $i(\mathcal{M}) > \omega$ there exists some $\xi < \omega_1$ such that for the spaces $T(\mathcal{M},\theta)$ and $T(\mathcal{S}_\xi,\theta)$ the corresponding conclusion (iv) holds. For results related to this problem we refer the reader to [42]. A proof of the above theorem can be found in [20].

Mixed Tsirelson norms

Schlumprecht space $\mathcal{S}$ is the completion of c_{00} endowed with the following norm. For $x \in c_{00}$,

$$\|x\|_{\mathcal{S}} = \max \big\{ \|x\|_0, \sup_n \sup \frac{1}{\log_2(n+1)} \sum_{i=1}^{n} \|E_i x\|_{\mathcal{S}} \big\},$$

where the inside sup is taken over all choices $E_1 < E_2 < \cdots < E_n$ of subsets of $\mathbb{N}$. The motivation for the construction of such a space was to provide an example of an arbitrarily distortable Banach space. The space $\mathcal{S}$ appeared as an ad hoc construction. However the next definition makes the relation of $\mathcal{S}$ with the Tsirelson type spaces more transparent.

Let $(\mathcal{M}_n)_n$, $(\theta_n)_n$ be two sequences with each $\mathcal{M}_n$ a compact family of finite subsets of $\mathbb{N}$, $0 < \theta_n < 1$ and $\lim_n \theta_n = 0$. The mixed Tsirelson space $T[(\mathcal{M}_n, \theta_n)_n]$ is the completion of c_{00} endowed with the norm

$$\|x\|_* = \max \big\{ \|x\|_0, \sup_n \sup \theta_n \sum_{i=1}^{k} \|E_i x\|_* \big\},$$

where the inside sup is taken over all choices $E_1 < E_2 < \cdots < E_k$ of $\mathcal{M}_n$ admissible families.

Remark I.9. (a) In the above notation the Schlumprecht space $\mathcal{S}$ is the mixed Tsirelson space $T[(\mathcal{A}_n, \frac{1}{\log_2(n+1)})_n]$, where $\mathcal{A}_n = \{F \subseteq \mathbb{N} : \#F \le n\}$.

(b) The mixed Tsirelson space $T[(\mathcal{M}_k, \theta_k)_{k=1}^{n}]$ defined by a finite family $(\mathcal{M}_k, \theta_k)_{k=1}^{n}$ is isomorphic to $T(\mathcal{M}_{k_0}, \theta_{k_0})$ for some $1 \le k_0 \le n$. Thus in order to obtain really new spaces we need to use infinite sequences $(\mathcal{M}_n, \theta_n)_n$.

As in the case of Tsirelson spaces there is an alternative definition of the norm of mixed Tsirelson spaces resulting from a norming set $W[(\mathcal{M}_n, \theta_n)_n]$. This set is defined as the minimal subset of c_{00} satisfying the following properties.

(i) $W[(\mathcal{M}_n, \theta_n)_n]$ contains all $\pm e_k^*$, $k \in \mathbb{N}$.

(ii) It is closed under the operations $(\mathcal{M}_n, \theta_n)_n$.

It follows that for an $f \in W[(\mathcal{M}_n, \theta_n)_n]$ the tree analysis $(f_t)_{t \in \mathcal{T}}$ is also defined, taking into account the necessary modifications. We shall discuss this topic more extensively in the next chapter.

The main property of mixed Tsirelson spaces is that for appropriate choices $(\mathcal{M}_n, \theta_n)_n$, they provide countable many equivalent norms $(\|\cdot\|_n)_n$ such that for each $(x_k)_k$ block sequence in $T[(\mathcal{M}_n, \theta_n)_n]$ and every $n \in \mathbb{N}$ there exists a vector $y_n \in \langle (x_k)_k \rangle$ such that $\|y_n\| = \|y_n\|_n$ and for $m \neq n$ $\|y_n\|_m \le \varepsilon_m$ where $\varepsilon_m \to 0$. In other words, the sequence $(\|\cdot\|_n)_n$ has the property that in every block subspace none of them dominates the rest but also it is not dominated by the rest.

Theorem I.10. *Let $X = T[(\mathcal{M}_n, \theta_n)_n]$. If for some $n \in \mathbb{N}$ it holds that $i(\mathcal{M}_n) \geq \omega$ or $i(\mathcal{M}_n) = r < \omega$ and $\theta_n > \frac{1}{r}$, then X is reflexive.*

Moreover if the first alternative holds, then X does not contain isomorphically any of the spaces ℓ_p, $1 \leq p < \infty$, c_0.

Proof. The proof of the reflexivity is similar to the original proof of Tsirelson [T]. According to a classical result of R.C. James [35], it is enough to show that the basis of X is boundedly complete and shrinking.

(i) The basis $(e_n)_n$ of $T[(\mathcal{M}_n, \theta_n)_n]$ is boundedly complete.

On the contrary, assume that there exist $\varepsilon > 0$ and a block sequence $(x_k)_k$ of $(e_n)_n$ such that $\sup_n \|\sum_{k=1}^n x_k\| \leq 1$ and $\|x_k\| \geq \varepsilon$ for all $k = 1, 2, \dots$. Since the basis is unconditional it is enough to find a finite subset A of $\mathbb{N}$ such that $\|\sum_{k \in A} x_k\| > 1$.

According to the assumption there exist $n, r \in \mathbb{N}$ such that $i(\mathcal{M}_n) \geq r$ and $\theta_n > \frac{1}{r}$. Therefore there exists $n_0 \in \mathcal{M}_n^{(r-1)}$. It follows that if $(y_k)_k$ is a block sequence of $(e_n)_n$ there exist a sequence $(n_t)_{t \in \mathbb{N}}$ of positive integers and a subsequence $(y_{k_t})_{t \in \mathbb{N}}$ of $(y_k)_k$ such that $n_t \leq \operatorname{supp} y_{k_{t+1}} < n_{t+1}$ and for every $l = 0, 1, \dots$ it holds $\{n_0, n_{lr+1}, \dots, n_{lr+(r-1)}\} \in \mathcal{M}_n$. It follows that the sequence $(y_{lr+1}, \dots, y_{(l+1)r})$ is $\mathcal{M}_n$-admissible.

Let $s \in \mathbb{N}$ be such that $(\theta_n r)^s > \frac{1}{\varepsilon}$. From the previous observation, we can choose a subsequence $(x'_t)_t$ of $(x_k)_k$ such that for every $l = 0, 1, \dots$ the sequence $(x'_{lr+1}, \dots, x'_{(l+1)r})$ is $\mathcal{M}_n$-admissible. For $l = 0, 1, \dots$, we set $x_{(1,l)} = \sum_{i=1}^r x'_{lr+i}$. It follows that

$$\|x_{(1,l)}\| \geq \theta_n \sum_{i=1}^r \|x'_{lr+i}\| \geq \theta_n r \varepsilon.$$

We repeat the same procedure for the sequence $(x_{(1,l)})_{l \in \mathbb{N}}$ and we get a block sequence $(x_{(2,l)})_{l \in \mathbb{N}}$ of $(x_n)_n$ such that $\|x_{(2,l)}\| \geq (\theta_n r)^2 \varepsilon$. Repeating this procedure s times we get a block sequence $(x_{(s,l)})_l$, where $x_{s,l}$ is sum of terms of the sequence $(x_n)_n$, such that $\|x_{(s,l)}\| \geq (\theta_n r)^s \varepsilon > 1$, a contradiction.

(ii) The basis $\{e_n\}_{n=1}^\infty$ is a shrinking.

Let $\theta = \max_k \theta_k < 1$. For $f \in X^*$ and $m \in \mathbf{N}$, denote by $Q_m(f)$ the restriction of f to the space generated by $\{e_k\}_{k \geq m}$. It suffices to prove the following: For every $f \in B_{X^*}$ there is $m \in \mathbf{N}$ such that $Q_m(f) \in \theta B_{X^*}$. Recall that $B_{X^*} = \overline{\operatorname{co}}(W)$, where the closure is in the topology of pointwise convergence. We shall first prove the following:

Claim. *For every $f \in \overline{W}$ there is m such that $Q_m(f) \in \theta \operatorname{co}(\overline{W})$.*

To prove this, let $f \in \overline{W}$ and let $\{f^n\}_{n=1}^\infty$ be a sequence in W converging pointwise to f.

If $f^n = e^*_{k_n}$ for an infinite number of n, we have nothing to prove. So suppose that for every n there are $k_n \in \mathbb{N}$, a set $\{m_1^n, \dots, m_{d_n}^n\} \in \mathcal{M}_{k_n}$ and vectors $f_i^n \in W$, $i = 1, \dots, d_n$ such that $m_1^n \leq \operatorname{supp} f_1^n < m_2^n \leq \operatorname{supp} f_2^n < \cdots < m_{d_n}^n \leq \operatorname{supp} f_{d_n}^n$ and $f^n = \theta_{k_n}(f_1^n + \cdots + f_{d_n}^n)$. If there is a subsequence of $\{\theta_{k_n}\}$ converging to 0, then $f = 0$. So we may suppose that there is a k such that $k_n = k$ for all n, i.e. $\theta_{k_n} = \theta_k$ and $\{m_1^n, \ldots, m_{d_n}^n\} \in \mathcal{M}_k$.

Since $\mathcal{M}_k$ is compact, if we substitute $\{f^n\}$ with a subsequence we get that there is a set $\{m_1, \ldots, m_d\} \in \mathcal{M}_k$ such that the sequence of indicator functions of the sets $\{m_1^n, \ldots, m_{d_n}^n\}$ converges to the indicator function of $\{m_1, \ldots, m_d\}$. So, for large n, $m_i^n = m_i$, $i = 1, \ldots, d$, and $m_{d+1}^n \to \infty$ as $n \to \infty$.

Passing to a further subsequence of $(f^n)_{n=1}^{\infty}$, we get that there exist $f_i \in \overline{W}$, $i = 1, \ldots, d$, with supp $f_i \subset [m_i, m_{i+1})$, $i = 1, \ldots, d-1$, and supp $f_d \subset [m_d, \infty)$ such that $f_j^n \to f_j$ pointwise for $j = 1, \ldots, d$. We conclude that $f = \theta_k(f_1 + \cdots + f_d)$, so $Q_{m_d}(f) = \theta_k f_d \in \theta \, \mathrm{co}(\overline{W})$.

The proof of the claim is complete. In particular we get that $\overline{W}$ is a weakly compact subset of c_0.

Consider now an $f \in B_{X^*} = \overline{\mathrm{co}(W)}$. In the set $B_{X^*} = \overline{\mathrm{co}(W)}$ the pointwise topology coincides with the restriction of the w^* topology of ℓ_∞ onto $\overline{\mathrm{co}(W)}$. According to Choquet's theorem there exists a measure $\mu_1 \in M_1(\overline{W})$ such that

$$\int_{\overline{W}} G d\mu = G(f_0)$$

for every linear w^*-continuous function G. Consider the sets

$$A_m = \{f \in \overline{W}, Q_m(f) \in \theta^2 \overline{\mathrm{co}(W)}\}, \; m \in \mathbb{N}.$$

The sequence $(A_m)_m$ is an increasing sequence of closed sets converging to $\overline{W}$, and hence there exists m_0 such that $|\mu|(A_{m_0}) > 1 - \theta(1 - \theta)$. We claim that $\frac{1}{\theta} Q_{m_0}(f_0) \in \overline{\mathrm{co}(W)}$. Indeed if not, then by the Hahn–Banach theorem there exists a linear w^*-continuous function G such that

$$\sup\{G(f) : f \in \overline{\mathrm{co}(W)}\} = 1 < \frac{1}{\theta} G(Q_{m_0}(f_0)).$$

On the other hand,

$$\frac{1}{\theta} G(Q_{m_0}(f_0)) = \frac{1}{\theta} \int_{\overline{W}} (G \circ Q_{m_0}) d\mu$$

$$= \frac{1}{\theta} \int_{A_{m_0}} (G \circ Q_{m_0}) d\mu + \frac{1}{\theta} \int_{\overline{W} \setminus A_{m_0}} (G \circ Q_{m_0}) d\mu$$

$$\leq \frac{1}{\theta} \theta^2 |\mu|(A_{m_0}) + \frac{1}{\theta} |\mu|(\overline{W} \setminus A_{m_0}) = \theta + 1 - \theta = 1,$$

a contradiction. Hence $Q_{m_0}(f_0) \in \theta \overline{\mathrm{co}(W)}$. This completes the proof that the basis is shrinking and hence it follows that X is reflexive.

For the remaining part, since X is reflexive it follows that ℓ_1 and c_0 do not embed in X. Assume that for some p, $1 < p < \infty$, ℓ_p, embeds in X. By standard arguments, there exists a block sequence $(x_n)_{n \in \mathbb{N}}$ of $(e_n)_n$ equivalent to the standard basis of ℓ_p. Hence there exists $C > 0$ such that $\|\sum_n a_n x_n\| \leq C(\sum_n |a_n|^p)^{\frac{1}{p}}$ for all coefficients $(a_n)_n$. Choose $m \in \mathbb{N}$ such that $m^{1-\frac{1}{p}} > \frac{C}{\theta_n}$. Since $i(\mathcal{M}_n) \geq \omega$, there exist $n_1, \ldots, n_m \in \mathbb{N}$ such that the sequence $(x_{n_1}, \ldots, x_{n_m})$ is $\mathcal{M}_n$-admissible. It follows

$$\theta_n m \leq \|\sum_{i=1}^m x_i\| \leq Cm^{\frac{1}{p}},$$

a contradiction. $\qquad\qquad\qquad\qquad\qquad\qquad\qquad\qquad\qquad\qquad\qquad\qquad\qquad$ $\square$

Notes and Remarks. Tsirelson space is an important discovery in Banach space theory. It is the "first truly non-classical space" according to E. Odell and Th. Schlumprecht [53]. It is the first space with its norm inductively defined, and, more important, it introduces a fundamental method for saturating the structure of a Banach space with a property (P). To illustrate this we recall that J. Schreier [56] provided the first example of a weakly null sequence $(x_n)_n$ with no norm Cesaro summable subsequence. Tsirelson space retains this property for all seminormalized weakly null sequences and it remains reflexive. Let us mention that Tsirelson's initial construction used the forcing method and he actually defined the dual of what we call Tsirelson space. The implicit form is due to T. Figiel and W.B. Johnson [28]. We recommend the interested reader Tsirelson's web page where it is explained how he discovered his space. A tentative study followed Tsirelson's discovery. Thus T. Fiegel and W.B. Johnson [28], introduced the p-convexification of T and W.B. Johnson [39], defined the modified version of T spaces with remarkable properties. We refer the reader to [21] for a comprehensive presentation of the results concerning Tsirelson space.

Tsirelson-type spaces of the form $T(\mathcal{M}, \theta)$ were introduced in an unpublished paper [5], where also the proof of Theorem I.4 was presented. The later result was initially proved in [18] with a different proof. It is worth noticing that Lemmas I.6, I.7 are the simplest approach of the basic inequality, which will be discussed in the following chapters.

The first mixed Tsirelson space is Schlumprecht's space [55]. This space was the decisive ingredient for Gowers–Maurey example [33]. The concept of Mixed Tsirelson space was introduced in [6]. The modified versions of mixed Tsirelson spaces are discussed in [8] and [9]. It is interesting that while Tsirelson space is isomorphic to its modified version, genuine mixed Tsirelson spaces are totally incomparable to their modified version. A. Manoussakis in [43] introduced and studied a class of mixed Tsirelson defined as p-spaces. These are spaces of the form $T[(\mathcal{A}_{n_j}, \theta_j)_j]$ satisfying the following property:

For $j \in \mathbb{N}$ we denote by $T_j = T[(\mathcal{A}_{n_i}, \theta_i)_{i=1}^j]$. Then T_j is isomorphic to ℓ_{p_j} with $1 < p_j \leq \infty$, [20]. A p-space is a space of the above such that the sequence $(p_j)_j$ strictly decreases to p. Under this consideration Schlumprecht space S is an 1-space.

Theorem I.10 is proved in [6] and Theorem I.8 which is stated with no proof can be found in [20]. We refer the reader to survey papers [11] and [51] for related results to the content of the present section. The hierarchy of Schreier families $(\mathcal{S}_\xi)_{\xi<\omega}$ was introduced in [1]. We refer the reader to S. Todorcevic's part of this book for a systematic presentation of the compact families of finite subsets of $\mathbb{N}$ and their remarkable Ramsey properties.

Chapter II

Tree Complete Extensions of a Ground Norm

In this chapter we start the novel construction of HI extensions of a ground norm. Our approach shares some common metamathematical ideas with the extension of models in set theory. Namely, one may think of a ground norm as an initial model, and the HI extension of it, being a new Banach space, which is HI and at the same time preserves some of the properties of the initial space. In this part we shall discuss the mixed Tsirelson extensions which correspond to the unconditional extensions.

II.1 Mixed Tsirelson Extension of a Ground Norm

We fix two sequences of integers $(m_j)_j$ and $(n_j)_j$ such that $m_1 = 2$, $m_{j+1} = m_j^5$ and $n_1 = 4$, $n_{j+1} = (5n_j)^{s_j}$ where $s_j = \log_2 m_{j+1}^3$. These sequences will follow us throughout these notes.

Definition II.1. Let κ be an ordinal and $G \subset c_{00}(\kappa)$. The set G is said to be a *ground set* provided the following are fulfilled:

(i) For $\alpha < \kappa$, $e_\alpha^* \in G$, the set G is symmetric (i.e. $g \in G$ iff $-g \in G$) and closed in the restriction of its elements on intervals of κ (i.e. for $E \subset \kappa$ interval and $g \in G$, $Eg = g \cdot \chi_E \in G$).

(ii) For $g \in G$, $\|g\|_\infty \leq 1$ and $g(\alpha) \in \mathbb{Q}$ for $\alpha < \kappa$.

A ground norm is the norm induced on $c_{00}(\kappa)$ by a ground set G. Namely for $x \in c_{00}(\kappa)$

$$\|x\|_G = \sup\{g(x) : \ g \in G\}.$$

We shall denote by Y_G the completion of $(c_{00}(\kappa), \|\cdot\|_G)$.

A property of Y_G is that the natural Hamel basis $(e_\alpha)_{\alpha<\kappa}$ of $c_{00}(\kappa)$ defines a normalized bimonotone transfinite Schauder basis of Y_G. This is a consequence of the fact that the norming set is closed in the interval projections and also the property that $\|g\|_\infty \leq 1$ for all $g \in G$. In the opposite, for every Banach space Z with a transfinite Schauder basis $(z_\alpha)_{\alpha<\kappa}$ there exists a ground set G such that the natural correspondence $e_\alpha \to z_\alpha$ from Y_G to Z defines an isomorphism.

The mixed Tsirelson extension of the ground norm $\|\cdot\|_G$ is defined on $c_{00}(\kappa)$ by the next formula. For $x \in c_{00}(\kappa)$ we set

$$\|x\|_* = \max\big\{\|x\|_G, \sup_j \sup\{\frac{1}{m_j}\sum_{i=1}^{d}\|E_i\|_*,\ E_1 < \cdots < E_d,\ d \leq n_j\}\big\}.$$

The space $T_\kappa[G]$ is the completion of $(c_{00}(\kappa), \|\cdot\|_*)$.

Observe that the above defined norm is greater than or equal to $\|\cdot\|_G$ and its difference from the usual mixed Tsirelson norms is that we have substituted $\|x\|_0$ by $\|x\|_G$.

Definition II.2. Let $\|\cdot\|_G$ be a ground norm in $c_{00}(\kappa)$. The mixed Tsirelson extension $T_\kappa[G]$ is said to be a *strictly singular extension* of Y_G if the identity map $I : T_\kappa[G] \to Y_G$ is a strictly singular operator.

We remind the reader that an operator $T : X \to Y$ is strictly singular if its restriction to any infinite dimensional closed subspace of X is not an isomorphism. The following is a well known result from the theory of strictly singular operators [41].

Theorem II.3. *Let $T : X \to Y$ be a strictly singular operator. Then for every infinite dimensional subspace Z of X and every $\varepsilon > 0$ there exists an infinite dimensional subspace W of Z such that $\|T|_W\| < \varepsilon$. If moreover X has a transfinite Schauder basis and Z is a block subspace of X then W may be selected as a block subspace of Z.*

The norming sets $W_\kappa[G]$, $W'_\kappa[G]$

As in the case of mixed Tsirelson spaces there is an alternative definition of the norm of the space $T_\kappa[G]$ through employing a norming set of functionals as follows.

Definition II.4. We shall denote by $W_\kappa[G]$ the minimal subset of $c_{00}(\kappa)$ satisfying the following conditions:

  (i) It contains the ground set G.

 (ii) It is closed in the $(\mathcal{A}_{n_j}, \frac{1}{m_j})$ operations.

(iii) It is rationally convex.

It is easy to check that the set $W_\kappa[G]$ is symmetric and closed in the restriction of its elements on the intervals of κ.

We shall also denote by $W'_\kappa[G]$ the minimal subset of $c_{00}(\kappa)$ satisfying the above (i) and (ii) (i.e. we do not require the set $W'_\kappa[G]$ to be closed in rational convex combinations).

The tree analysis of $f \in W_\kappa[G]$, $f \in W'_\kappa[G]$ and their relation

In the sequel for $f \in W_\kappa[G]$ ($f \in W'_\kappa[G]$) resulting from the operation $(\mathcal{A}_{n_j}, \frac{1}{m_j})$ for some $j \in \mathbb{N}$ we shall denote by $w(f)$ the *weight* of f which is equal to m_j. The weight $w(f)$ is not necessarily uniquely determined. Also for $f \in W_\kappa[G]$ we say that f is of *type* 0 if $f \in G$, f is of *type* I if $w(f)$ exists and f is of *type* II if it is a rational convex combination. It is easy to show that every $f \in W_\kappa[G]$ is of one of the above defined types. For $f \in W'_\kappa[G]$ there are only two possibilities, namely type 0 and type I. As it happens with the weight of f the type of f is not necessarily unique.

Definition II.5. Let $f \in W_\kappa[G]$. A family $(f_t)_{t \in \mathcal{T}}$ with $\mathcal{T}$ a rooted finite tree, is a *tree analysis* of f if the following hold:

(i) The functional f_0 equals to f where 0 denotes the root of $\mathcal{T}$.

(ii) Each f_t belongs to $W_\kappa[G]$ and if $t \prec s$ then $\operatorname{ran}(f_s) \subset \operatorname{ran}(f_t)$.

(iii) If t is a maximal element of $\mathcal{T}$ then $f_t \in G$.

(iv) For $t \in \mathcal{T}$ which is not maximal, denoting by S_t the set of immediate successors of t, either $f_t = \frac{1}{m_j} \sum_{s \in S_t} f_s$ as a result of an $(\mathcal{A}_{n_j}, \frac{1}{m_j})$ operation of the functionals $(f_s)_{s \in S_t}$, or $f_t = \sum_{s \in S_t} r_s f_s$ as a rational convex combination of the functionals $(f_s)_{s \in S_t}$ and each f_s is not a convex combination of its immediate successors.

Proposition II.6. *Every $f \in W_\kappa[G]$ admits a tree analysis.*

The proof is easy. We simply consider all $f \in W_\kappa[G]$ admitting a tree analysis and then we show that this set satisfies (i), (ii) and (iii) of Definition II.5. The result follows from the minimality of the set $W_\kappa[G]$.

Remark II.7. For $f \in W'_\kappa[G]$ the tree analysis $(f_t)_{t \in \mathcal{T}}$ is defined in the same manner. The only difference concerns condition (iv) in the above definition where the second alternative does not occur. The conclusion of the above proposition remains valid for $f \in W'_\kappa[G]$.

The next proposition connects the sets $W_\kappa[G]$, $W'_\kappa[G]$.

Proposition II.8. *Let G be a ground subset of $c_{00}(\kappa)$.*

(i) *If $f \in W_\kappa[G]$ is of type* I *with $w(f) = m_j$ then there exist $(f_i)_{i=1}^n$ in $W'_\kappa[G]$ with $w(f_i) = m_j$ such that f is a rational convex combination of $(f_i)_{i=1}^n$. Moreover, if f admits a tree analysis $(f_t)_{t \in \mathcal{T}}$ such that for every $t \in \mathcal{T}$ with f_t of type* I, $w(f_t) \neq m_{j_0}$ *then each f_i admits a tree analysis with the same property.*

(ii) *If $f \in W_\kappa[G]$ is of type* II *then f is a rational convex combination of a family $(f_i)_{i=1}^n$ of elements of $W'_\kappa[G]$.*

(iii) $W_\kappa[G] = \mathrm{conv}_{\mathbb{Q}}(W'_\kappa[G])$.

(iv) *The sets $W_\kappa[G]$, $W'_\kappa[G]$ induce the same norm on $c_{00}(\kappa)$ which is the mixed Tsirelson extension defined above.*

The proof is easy and we leave it to the reader.

The auxiliary space $T[(\mathcal{A}_{5n_j}, \frac{1}{m_j})_j]$ and the norm of the averages of its basis

We consider the mixed Tsirelson space $T[(\mathcal{A}_{5n_j}, \frac{1}{m_j})_j]$ and we denote by W_0, W'_0 the norming sets corresponding to the norm of this space where W_0 is closed in the $(\mathcal{A}_{5n_j}, \frac{1}{m_j})$ operations and the rational convex combinations while W' is closed only in the corresponding operations. We shall prove the following lemma.

Lemma II.9. *Let $j_0 \in \mathbb{N}$ and $f \in W'_0$. Then for every $k_1 < k_2 < \ldots < k_{n_{j_0}}$ we have that*

$$|f(\frac{1}{n_{j_0}} \sum_{l=1}^{n_{j_0}} e_{k_l})| \leq \begin{cases} \frac{2}{m_i \cdot m_{j_0}}, & \text{if } w(f) = m_i, \ i < j_0 \\ \frac{1}{m_i}, & \text{if } w(f) = m_i, \ i \geq j_0. \end{cases} \tag{II.1}$$

The same estimates hold for $f \in W_0$ with $w(f) = m_i$.

If we additionally assume that the functional f admits a tree $(f_t)_{t \in \mathcal{T}}$ such that $w(f_t) \neq m_{j_0}$ for all $t \in \mathcal{T}$, then we have that

$$|f(\frac{1}{n_{j_0}} \sum_{l=1}^{n_{j_0}} e_{k_l})| \leq \begin{cases} \frac{2}{m_i \cdot m_{j_0}^2}, & \text{if } w(f) = m_i, \ i < j_0 \\ \frac{1}{m_i}, & \text{if } w(f) = m_i, \ i > j_0. \end{cases} \tag{II.2}$$

In particular $|f(\frac{1}{n_{j_0}} \sum_{l=1}^{n_{j_0}} e_{k_l})| \leq \frac{1}{m_{j_0}^2}$. The same estimates remain valid for $f \in W_0$ with $w(f) = m_i$ admitting a tree $(f_t)_{t \in \mathcal{T}}$ such that $w(f_t) \neq m_{j_0}$ for all $t \in \mathcal{T}$.

Proof. We first prove the following claim.

Claim. *Let $h \in W'_0$. Then*

(i) $\#\{k : |h(e_k)| > \frac{1}{m_{j_0}}\} \leq (5n_{j_0-1})^{\log_2(m_{j_0})-1}$.

(ii) *If the functional h admits a tree $(h_a)_{a \in \mathcal{A}}$ with $w(h_a) \neq m_{j_0}$ for each $a \in \mathcal{A}$ then*

$$\#\{k : |h(e_k)| > \tfrac{1}{m_{j_0}^2}\} \leq (5n_{j_0-1})^{2\log_2(m_{j_0})-1}.$$

Proof of the claim. We shall prove only part (i) of the claim, as the proof of (ii) is similar. Let $(h_a)_{a \in \mathcal{A}}$ be a tree of h and let n be its height (i.e. the maximal length of its branches). We may assume that $|h(e_k)| > \frac{1}{m_{j_0}}$ for all $k \in \operatorname{supp} h$. Let $h = (h_0, h_1, \ldots, h_n)$ be a branch (then $h_n = \pm e_p^*$) and let $k \in \operatorname{supp} h_n$. Then

$$\frac{1}{m_{j_0}} < |h(e_k)| = \prod_{l=0}^{n-1} \frac{1}{w(h_l)} \leq \frac{1}{2^n}, \text{ hence } n < \log_2(m_{j_0}).$$

On the other hand, since $|h(e_k)| > \frac{1}{m_{j_0}}$ for all $k \in \operatorname{supp} h$, each h_α with α non maximal node is a result of an $(\mathcal{A}_{5n_j}, \frac{1}{m_j})$ operation for $j \leq j_0 - 1$. An inductive argument yields that for $i \leq n$ the cardinality of the set $\{h_t : |t| = i\}$ is less or equal to $(5n_{j_0-1})^i$. The fact that $n < \log_2(m_{j_0})$ yields that $\#(\operatorname{supp}(h)) \leq (5n_{j_0-1})^{\log_2(m_{j_0})-1}$.

The proof of the claim is complete. $\qquad\square$

We pass to the proof the lemma. The case $w(f) = m_i$, $i \geq j_0$ is straightforward. Let $f \in W_0'$ with $w(f) = m_i$, $i < j$. Then $f = \frac{1}{m_i} \sum_{l=1}^{d} f_l$ where $f_1 < \cdots < f_d$ belong to W_0' and $d \leq n_i$.

For $p = 1, \ldots, d$ we set $H_p = \{k : |f_p(e_k)| > \frac{1}{m_{j_0}}\}$. Part (i) of the claim yields that $\#(H_p) < (5n_{j_0-1})^{\log_2(m_{j_0})}$. Thus, setting $H = \bigcup_{p=1}^{d} H_p$, we get that $\#(H) \leq d(5n_{j_0-1})^{\log_2(m_{j_0})-1} \leq (5n_{j_0-1})^{\log_2(m_{j_0})}$. Therefore

$$|f(\frac{1}{n_{j_0}} \sum_{l=1}^{n_{j_0}} e_{k_l})| \leq \frac{1}{m_i}\Big(|(\sum_{p=1}^{d} f_p)_{|H}(\frac{1}{n_{j_0}} \sum_{l=1}^{n_{j_0}} e_{k_l})| + |(\sum_{p=1}^{d} f_p)_{|(\mathbb{N} \setminus H)}(\frac{1}{n_{j_0}} \sum_{l=1}^{n_{j_0}} e_{k_l})|\Big)$$

$$\leq \frac{1}{m_i}(\#(H)\frac{1}{n_{j_0}} + \frac{1}{m_{j_0}}) < \frac{2}{m_i m_{j_0}}.$$

The result for $f \in W_0$ follows from Proposition II.8(i) and the above estimates.

The second part is proved similarly by using part (ii) of the claim. $\qquad\square$

Notes and Remarks. The space $T_\kappa[G]$ is the frame for the HI extensions which will be presented in the next chapter. As we will see the norming set D_G of the HI extensions will be a subset of the norming set $W[G]$. This actually means that the HI space $\mathfrak{X}[G, \sigma]$ is interpolated between $T[G]$ and Y_G. The results of the next section yield that if $T_\kappa[G]$ is a strictly singular extension of Y_G, then it is reflexive and unconditionally saturated. Moreover $T_\kappa[G]$ does not contain any ℓ_p, $1 \leq p < \infty$. The transfinite basis of $T_\kappa[G]$ is boundedly complete (for the definition we refer to Appendix A) hence $T_\kappa[G]$ is the dual of the space generated by the biorthogonals of the basis $(e_\alpha)_{\alpha < \kappa}$.

The auxiliary space $T[(\mathcal{A}_{5n_j}, \frac{1}{m_j})_j]$ will be used in the estimates of the norm of certain block averages. For this reason we compute the norm of the averages of the basis in this space. The tool to reduce from the block averages to the averages of the basis of the auxiliary space, is the basic inequality stated and proved in the next section and also in Appendix B.

II.2   R.I.S. Sequences and the Basic Inequality

The tree complete extension D_G of a ground set is a subset of $W_k[G]$ which satisfies certain properties. The most important one is that D_G is closed in the even operations $(\mathcal{A}_{n_{2j}}, \frac{1}{m_{2j}})$ while the behavior of D_G with respect to the odd operations $(\mathcal{A}_{n_{2j+1}}, \frac{1}{m_{2j+1}})$ is left open. We shall obtain some properties of the spaces $\mathfrak{X}_k[D_G]$ in this section. In the next chapter we shall be more specific for the behavior of the D_G with respect to the odd operations in order to obtain HI extensions of the ground norm. The main result of this part concerns the basic inequality and its applications.

Definition II.10. Let G be a ground set. A subset D_G of $W_\kappa[G]$ is said to be a *tree complete* extension of G if the following are fulfilled.

 (i) The ground set G is a subset of D_G.

 (ii) D_G is symmetric, closed in the restrictions of its elements on intervals of κ and closed in the $(\mathcal{A}_{n_{2j}}, \frac{1}{m_{2j}})_j$ operations.

(iii) Every $f \in D_G$ admits a tree analysis $(f_t)_{t \in \mathcal{T}}$ with $f_t \in D_G$ for all $t \in \mathcal{T}$.

We shall denote by $\mathfrak{X}_\kappa[D_G]$ the completion of $(c_{00}(\kappa), \|\cdot\|_{D_G})$ and for $\kappa = \omega$ we write $\mathfrak{X}[D_G]$.

Remark II.11. Let's observe that the difference between $W_\kappa[G]$ and D_G concerns that in the later set the behavior of the odd operations $(\mathcal{A}_{n_{2j+1}}, \frac{1}{m_{2j+1}})_j$ is remaining undefined. Actually we shall use the odd operations in order to impose the conditional structure on the space normed by the set D_G. The complete definition of the set D_G will be given in the next chapter.

Definition II.12. The tree complete extension is said to be *strictly singular* if the identity operator $I : \mathfrak{X}_\kappa[D_G] \to Y_G$ is strictly singular.

For $f \in D_G$ we say that is of type I (or II) if it is a result of an $(\mathcal{A}_{n_j}, \frac{1}{m_j})$ operation (or a rational convex combination) of a family $(f_t)_t$ with $f_t \in D_G$.

Definition II.13 (R.I.S.). A block sequence $(x_k)_k$ in $\mathfrak{X}_\kappa[D_G]$ is said to be a (C, ε) *rapidly increasing sequence* (R.I.S.), if $\|x_k\| \leq C$, and there exists a strictly increasing sequence (j_k) of positive integers such that
  (a) $\frac{1}{m_{j_{k+1}}} \cdot \# \operatorname{supp} x_k < \varepsilon$.
  (b) For every $k = 1, 2, \ldots$ and every $f \in D_G$ with $w(f) = m_i$, $i < j_k$ we have that $|f(x_k)| \leq \frac{C}{m_i}$.

The next proposition is the fundamental tool for the computation of the norm for certain vectors in $\mathfrak{X}_\kappa[D_G]$.

Proposition II.14 (The basic inequality). *Let $(x_k)_k$ be a (C, ε) R.I.S. in $\mathfrak{X}_\kappa[D_G]$ such that for every $g \in G$ we have that $|g(x_k)| > \varepsilon$ for at most one k. Let also $(\lambda_k)_k \in c_{00}$ be a sequence of scalars. Then for every $f \in D_G$ of type I we can find g_1, such that either $g_1 = h_1$ or $g_1 = e_t^* + h_1$ with $t \notin \operatorname{supp} h_1$ where $h_1 \in W_0$ with $w(h_1) = w(f)$ and $g_2 \in c_{00}(\mathbb{N})$ with $\|g_2\|_\infty \leq \varepsilon$ with g_1, g_2 having nonnegative coordinates and such that*

$$|f(\sum \lambda_k x_k)| \leq C(g_1 + g_2)(\sum |\lambda_k| e_k). \tag{II.3}$$

If we additionally assume that there exists $j_0 \in \mathbb{N}$ such that for every $h \in D_G$ with $w(h) = m_{j_0}$ and every interval E of the natural numbers we have that

$$|h(\sum \lambda_k x_k)| \leq C(\max_{k \in E} |\lambda_k| + \varepsilon \sum_{k \in E} |\lambda_k|), \tag{II.4}$$

then, if $w(f) \neq m_{j_0}$, we may select h_1 to have a tree analysis $(h_t)_{t \in \mathcal{T}}$ with $w(h_t) \neq m_{j_0}$ for all $t \in \mathcal{T}$ with h_t of type I.

Proof. The proof in the general case (where (II.4) is not assumed) and in the special case (where we assume (II.4)) is actually the same. We shall give the proof only in the special case. The proof in the general case arises by omitting any reference to distinguishing cases whether a functional has weight m_{j_0} or not and treating the functionals with $w(f) = m_{j_0}$ as for any other j.

We fix a tree analysis $(f_t)_{t \in \mathcal{T}}$ of f. Before passing to the proof we adopt some useful notation and state two lemmas.

Definition II.15. For each k we define the set $\mathcal{T}_k$ as follows:

$$\mathcal{T}_k = \Big\{ t \in \mathcal{T} \text{ such that } f_t \text{ is of type 0 or I and}$$

      (i) $\operatorname{ran} f_t \cap \operatorname{ran} x_k \neq \emptyset$

      (ii) $\forall\, u < t$, if f_u is of type I, then $w(f_u) \neq m_{j_0}$

      (iii) $\forall\, s \leq t$ if $s \in S_u$ and f_u is of type I,

            then $\operatorname{ran} f_s \cap \operatorname{ran} x_k = \operatorname{ran} f_u \cap \operatorname{ran} x_k$

      (iv) if $w(f_t) \neq m_{j_0}$, then for all $s \in S_t$

           $\operatorname{ran} f_s \cap \operatorname{ran} x_k \subsetneq \operatorname{ran} f_t \cap \operatorname{ran} x_k \Big\}$

The next lemma describes the properties of the set $\mathcal{T}_k$.

Lemma II.16. *For every k we have the following:*

  (i) *If $t \in \mathcal{T}$ and f_t is of type II then $t \notin \mathcal{T}_k$.*

  (ii) *If $t \in \mathcal{T}_k$, then for every $s < t$: if f_s is of type I, then $w(f_s) \neq m_{j_0}$.*

(iii) *If T_k is not a singleton, then its members are incomparable members of the tree T. Moreover if t_1, t_2 are two different elements of T_k and s is the (necessarily uniquely determined) maximal element of T satisfying $s < t_1$ and $s < t_2$ then f_s is of type II.*

(iv) *If $t \in T$ is such that supp $f_t \cap \operatorname{ran} x_k \neq \emptyset$ and $u \notin T_k$ for all $u < t$, then there exists $s \in T_k$ with $t \leq s$. In particular if supp $f \cap \operatorname{ran} x_k \neq \emptyset$, then $T_k \neq \emptyset$.*

Definition II.17. For every $t \in T$ we define $D_t = \bigcup_{s \geq t} \{k : s \in T_k\}$.

Lemma II.18. *According to the notation above we have the following:*

(i) *If supp $f \cap \operatorname{ran} x_k \neq \emptyset$, then $k \in D_0$ (remember that 0 denotes the unique root of T and $f = f_0$). Hence $f(\sum \lambda_k x_k) = f(\sum_{k \in D_0} \lambda_k x_k)$.*

(ii) *If f_t is of type I with $w(f_t) = m_{j_0}$, then D_t is an interval of $\mathbb{N}$.*

(iii) *If f_t is of type I with $w(f_t) \neq m_{j_0}$, then*

$$
\left\{ \{k\} : k \in D_t \setminus \bigcup_{s \in S_t} D_s \right\} \cup \{D_s : s \in S_t\}
$$

is a family of successive subsets of $\mathbb{N}$. Moreover for every $k \in D_t \setminus \bigcup_{s \in S_t} D_s$ (i.e. for k such that $t \in T_k$) such that supp $f_t \cap \operatorname{ran} x_k \neq \emptyset$ there exists a $s \in S_t$ such that either $\min \operatorname{supp} x_k \leq \max \operatorname{supp} f_s < \max \operatorname{supp} x_k$ or $\min \operatorname{supp} x_k < \min \operatorname{supp} f_s \leq \max \operatorname{supp} x_k$.

(iv) *If f_t is of type II, $s \in S_t$ and $k \in D_t \setminus D_s$, then supp $f_s \cap \operatorname{ran} x_k = \emptyset$ and hence $f_s(x_k) = 0$.*

Recall that we have fixed a tree analysis $(f_t)_{t \in T}$ for the given f. We shall construct two families $(g_t^1)_{t \in T}$ and $(g_t^2)_{t \in T}$ such that the following conditions are fulfilled.

(i) For every $t \in T$ such that f_t is not of type II, $g_t^1 = h_t$ or $g_t^1 = e_{k_t}^* + h_t$ with $k_t \notin \operatorname{supp} h_t$, where $h_t \in W_0$ and $g_t^2 \in c_{00}(\mathbb{N})$ with $\|g_t^2\|_\infty \leq \varepsilon$.

(ii) For every $t \in T$, supp $g_t^1 \subset D_t$ and supp $g_t^2 \subset D_t$ and the functionals g_t^1, g_t^2 have nonnegative coordinates.

(iii) For $t \in T$ with $f_t \in G$ and $D_t \neq \emptyset$ we have that $g_t^1 = e_p^*$.

(iv) For f_t of type II with $f = \sum_{s \in S_t} r_s f_s$ (where $r_s \in \mathbb{Q}^+$ for every $s \in S_t$ and $\sum_{s \in S_t} r_s = 1$) we have $g_t^1 = \sum_{s \in S_t} r_s g_s^1$ and $g_t^2 = \sum_{s \in S_t} r_s g_s^2$.

(v) For f_t of type I with $w(f) = m_{j_0}$ we have $g_t^1 = e_{k_t}^*$ where $k_t \in D_t$ is such that $|\lambda_{k_t}| = \max_{k \in D_t} |\lambda_k|$ and $g_t^2 = \sum_{k \in D_t} \varepsilon e_k^*$.

(vi) For f_t of type I with $w(f) = m_j$ for $j \neq j_0$ we have $g_t^1 = h_t$ or $g_t^1 = e_{k_t}^* + h_t$ where $h_t \in W_0$ with $w(h_t) = m_j$ and $k_t \notin \operatorname{supp} h_t$.

(vii) For every $t \in T$ the following inequality holds:

$$|f_t(\sum_{k \in D_t} \lambda_k x_k)| \leq C(g_t^1 + g_t^2)(\sum_{k \in D_t} |\lambda_k| e_k).$$

When the construction of $(g_t^1)_{t \in T}$ and $(g_t^2)_{t \in T}$ has been accomplished, we set $g_1 = g_0^1$ and $g_2 = g_0^2$ (where 0 is the root of T and $f = f_0$) and we observe that these are the desired functionals. To show that such $(g_t^1)_{t \in T}$ and $(g_t^2)_{t \in T}$ exist we use finite induction starting with those $t \in T$ which are maximal and in the general inductive step we assume that g_s^1, g_s^2 have been defined for all $s > t$ satisfying the inductive assumptions and we define g_t^1 and g_t^2.

1st inductive step
Let $t \in T$ be maximal; then $f_t \in G$. If $D_t = \emptyset$, we define $g_t^1 = 0$ and $g_t^2 = 0$. If $D_t \neq \emptyset$, we set

$$E_t = \{k \in D_t : |f_t(x_k)| > \varepsilon\} \quad \text{and} \quad F_t = D_t \setminus E_t.$$

If $E_t \neq \emptyset$, then from our assumption in the statement of the proposition we have that $E_t = \{k_t\}$. We define

$$g_t^1 = \sum_{k \in E_t} e_k^*. \quad g_t^2 = \sum_{k \in F_t} \varepsilon e_k^*,$$

and $\|g_t^2\|_\infty \leq \varepsilon$. Inequality (vii) is easily checked.

General inductive step
Let $t \in T$ and suppose that g_u^1 and g_u^2 have been defined for every $u > t$ satisfying the inductive assumptions. If $D_t = \emptyset$, we set $g_t^1 = 0$ and $g_t^2 = 0$. In the remainder of the proof we assume that $D_t \neq \emptyset$. We consider the following three cases:

1st **case:** The functional f_t is of type II.
Let $f_t = \sum_{s \in S_t} r_s f_s$ where $r_s \in \mathbb{Q}^+$ are such that $\sum_{s \in S_t} r_s = 1$. In this case, we have that $D_t = \bigcup_{s \in S_t} D_s$. We define

$$g_t^1 = \sum_{s \in S_t} r_s g_s^1 \quad \text{and} \quad g_t^2 = \sum_{s \in S_t} r_s g_s^2.$$

Inequality (vii) is easily verified.

2nd **case:** The functional f_t is of type I with $w(f) = m_{j_0}$.
In this case D_t is an interval of the natural numbers (Lemma II.18(ii)). Let $k_t \in D_t$

be such that $|\lambda_{k_t}| = \max\limits_{k \in D_t} |\lambda_k|$. We define

$$g_t^1 = e_{k_t}^* \quad \text{and} \quad g_t^2 = \sum_{k \in D_t} \varepsilon e_k^*.$$

Inequality (vii) is easily established.

3rd case: The functional f_t is of type I with $w(f) = m_j$ for $j \neq j_0$.

Then $f_t = \frac{1}{m_j} \sum\limits_{s \in S_t} f_s$ and the family $\{f_s : s \in S_t\}$ is a family of successive functionals with $\#(S_t) \leq n_j$. We set

$$
\begin{aligned}
E_t \;\; &= \;\; \{k : t \in \mathcal{T}_k \text{ and } \operatorname{supp} f_t \cap \operatorname{ran} x_k \neq \emptyset\} \\
(&= \;\; \{k \in D_t \setminus \bigcup_{s \in S_t} D_s : \operatorname{supp} f_t \cap \operatorname{ran} x_k \neq \emptyset\}).
\end{aligned}
$$

We consider the following partition of E_t.

$$E_t^2 = \{k \in E_t : m_{j_{k+1}} \leq m_j\} \quad \text{and} \quad E_t^1 = E_t \setminus E_t^2.$$

We define

$$g_t^2 = \sum_{k \in E_t^2} \varepsilon e_k^* + \sum_{s \in S_t} g_s^2.$$

Observe that $\|g_t^2\|_\infty \leq \varepsilon$. Let $E_t^1 = \{k_1 < k_2 < \cdots < k_l\}$. From the definition of E_t^1 we get that $m_j < m_{j_{k_2}} < \cdots < m_{j_{k_l}}$. We set

$$k_t = k_1 \quad \text{and} \quad g_t^1 = e_{k_t}^* + h_t \quad \text{where} \quad h_t = \frac{1}{m_j}\Big(\sum_{i=2}^{l} e_{k_i}^* + \sum_{s \in S_t} g_s^1\Big).$$

(The term $e_{k_t}^*$ does not appear if $E_t^1 = \emptyset$.)

It is easy to verify that inequality (vii) holds.

By the second part of Lemma II.18(iii), for every $k \in E_t$ there exists an element of the set $N = \{\min \operatorname{supp} f_s, \max \operatorname{supp} f_s : s \in S_t\}$ belonging to $\operatorname{ran} x_k$. Hence $\#(E_t^1) \leq \#(E_t) \leq 2n_j$.

We next show that $h_t \in W_0$ with $w(f_t) = m_j$. We first examine the case that for every $s \in S_t$ the functional f_s is not of type II. Then for every $s \in S_t$ one of the following holds:

(i) $f_s \in G$. In this case $g_s^1 = e_{k_s}^* \in W_0$ (by the first inductive step).

(ii) f_s is of type I with $w(f_s) = m_{j_0}$. In this case $g_s^1 = e_{k_s}^* \in W_0$.

(iii) f_s is of type I with $w(f_s) = m_j$ for $j \neq j_0$. In this case $g_s^1 = e_{k_s}^* + h_s$ (or $g_s^1 = h_s$) where $h_s \in W_0$ and $k_s \notin \operatorname{supp} h_s$. We set $E_s^1 = \{n \in \mathbb{N} : n < k_s\}$, $E_s^2 = \{n \in \mathbb{N} : n > k_s\}$ and $h_s^1 = E_s^1 h_s$, $h_s^2 = E_s^2 h_s$. The functionals $h_s^1, e_{t_s}^*$, h_s^2 are successive and belong to W_0.

We set

$$
\begin{aligned}
T_t^1 &= \{s \in S_t : f_s \in G\} \\
T_t^2 &= \{s \in S_t : f_s \text{ of type } I \text{ and } w(f_s) = m_{j_0}\} \\
T_t^3 &= \{s \in S_t : f_s \text{ of type } I \text{ and } w(f_s) \neq m_{j_0}\}.
\end{aligned}
$$

The family of successive (see Lemma II.18(iii)) functionals of W_0,

$$
\{e_{k_i}^* : i = 2, \ldots, l\} \cup \{g_s^1 : s \in T_t^1\} \cup \{g_s^1 : s \in T_t^2\}
$$
$$
\cup \{h_s^1 : s \in T_t^3\} \cup \{e_{k_s}^* : s \in T_t^3\} \cup \{h_s^2 : s \in T_t^3\}
$$

has cardinality $\leq 5n_j$, thus we get that $h_t \in W_0$ with $w(h_t) = m_j$.

For the case that for some $s \in S_t$ the functional f_s is of type II, the conclusion follows from the previous case and Lemma II.8. $\qquad\square$

Proposition II.19. *Let* $(x_k)_{k=1}^{n_{j_0}}$ *be a* (C, ε) *R.I.S. with* $\varepsilon \leq \frac{2}{m_{j_0}^2}$ *such that for every* $g \in G$, $|g(x_k)| > \varepsilon$ *for at most one* k. *Then:*

1) *For every* $f \in D$ *with* $w(f) = m_i$,

$$
|f(\frac{1}{n_{j_0}} \sum_{k=1}^{n_{j_0}} x_k)| \leq
\begin{cases}
\frac{3C}{m_{j_0} m_i}, & \text{if } i < j_0 \\
\frac{C}{n_{j_0}} + \frac{C}{m_i} + C\varepsilon, & \text{if } i \geq j_0.
\end{cases}
$$

In particular $\|\frac{1}{n_{j_0}} \sum_{k=1}^{n_{j_0}} x_k\| \leq \frac{2C}{m_{j_0}}$.

2) *If* $(b_k)_{k=1}^{n_{j_0}}$ *are scalars with* $|b_k| \leq 1$ *such that*

$$
|h(\sum_{k \in E} b_k x_k)| \leq C(\max_{k \in E} |b_k| + \varepsilon \sum_{k \in E} |b_k|) \tag{II.5}
$$

for every interval E *of positive integers and every* $h \in D$ *with* $w(h) = m_{j_0}$, *then*

$$
\|\frac{1}{n_{j_0}} \sum_{k=1}^{n_{j_0}} b_k x_k\| \leq \frac{4C}{m_{j_0}^2}.
$$

Proof. The proof is an application of the basic inequality (Proposition II.14) and Lemma II.9. Indeed, let $f \in D$ with $w(f) = m_i$. Proposition II.14 yields the existence of an $h_1 \in W_0$ with $w(h_1) = m_i$, a $t \in \mathbb{N}$ and an $h_2 \in c_{00}(\mathbb{N})$ with $\|h_2\|_\infty \leq \varepsilon$, such that

$$
|f(\frac{1}{n_{j_0}} \sum_{k=1}^{n_{j_0}} x_k)| \leq C(e_t^* + h_1 + h_2)(\frac{1}{n_{j_0}} \sum_{k=1}^{n_{j_0}} e_k).
$$

If $i \geq j_0$, we get that $|f(\frac{1}{n_{j_0}} \sum_{k=1}^{n_{j_0}} x_k)| \leq C(\frac{1}{n_{j_0}} + \frac{1}{m_i} + \varepsilon) < \frac{C}{n_{j_0}} + \frac{C}{m_i} + C\varepsilon$.

If $i < j_0$, using Lemma II.9 we get that $|f(\frac{1}{n_{j_0}} \sum_{k=1}^{n_{j_0}} x_k)| \leq C(\frac{1}{n_{j_0}} + \frac{2}{m_i \cdot m_{j_0}} + \varepsilon) < \frac{3C}{m_i \cdot m_{j_0}}$.

In order to prove 2) let $(b_k)_{k=1}^{n_{j_0}}$ be scalars with $|b_k| \leq 1$ such that (II.5) is satisfied. Then condition (II.4) of the basic inequality is satisfied for the linear combination $\frac{1}{n_{j_0}} \sum_{k=1}^{n_{j_0}} b_k x_k$. Thus for every $f \in D$ with $w(f) = m_i$, $i \neq j_0$ there exist $t \in \mathbb{N}$, $h_1 \in W_0$, $h_2 \in c_{00}(\mathbb{N})$ with h_1, h_2 having nonnegative coordinates, $\|h_2\|_\infty \leq \varepsilon$ and h_1 admitting a tree $(h_t)_{t \in \mathcal{T}}$ with $w(h_t) \neq m_{j_0}$ for every $t \in \mathcal{T}$ with h_t of type I, such that

$$|f(\frac{1}{n_{j_0}} \sum_{k=1}^{n_{j_0}} b_k x_k)| \leq C(e_t^* + h_1 + h_2)(\frac{1}{n_{j_0}} \sum_{k=1}^{n_{j_0}} |b_k| e_k) \leq C(e_t^* + h_1 + h_2)(\frac{1}{n_{j_0}} \sum_{k=1}^{n_{j_0}} e_k)$$
(II.6)

Using the second part of Lemma II.9 we deduce that

$$|f(\frac{1}{n_{j_0}} \sum_{k=1}^{n_{j_0}} b_k x_k)| \leq C(\frac{1}{n_{j_0}} + \frac{1}{m_{j_0}^2} + \varepsilon) < \frac{4C}{m_{j_0}^2}.$$

For $f \in D$ with $w(f) = m_{j_0}$, from condition (II.5) we get that $|f(\frac{1}{n_{j_0}} \sum_{k=1}^{n_{j_0}} b_k x_k)| \leq \frac{C}{n_{j_0}}(1 + \frac{2}{m_{j_0}^2} n_{j_0}) < \frac{4C}{m_{j_0}^2}$. $\square$

Proposition II.20. *Suppose that there exists a universal constant $C > 0$ such that for every $(x_k)_k$ block sequence in $\mathfrak{X}_\kappa[D_G]$ and every $j \in \mathbb{N}$ there exists a $(C, \frac{1}{m_{2j+1}^2})$ R.I.S. $y_1, y_2, \ldots, y_{n_{2j+1}}$ in $\mathrm{span}\{x_k : k \in \mathbb{N}\}$ such that*

(i) $\|\frac{1}{n_{2j+1}} \sum_{k=1}^{n_{2j+1}} y_k\| \geq \frac{1}{m_{2j+1}}$.

(ii) *For every $f \in D_G$ with $w(f) = m_{2j+1}$ we have $|f(\frac{1}{n_{2j+1}} \sum_{k=1}^{n_{2j+1}} (-1)^{k+1} y_k)| < \frac{C}{m_{2j+1}^2}$.*

Then the space $\mathfrak{X}_\kappa[D_G]$ contains no unconditional basic sequence.

Proof. Proposition II.19 yields that

$$\left\| \frac{1}{n_{2j+1}} \sum_{k=1}^{n_{2j+1}} (-1)^{k+1} y_k \right\| < \frac{4C}{m_{2j+1}^2}.$$

This and assumption (i) of the statement yield that for every block sequence $(x_k)_{k \in \mathbb{N}}$ is not unconditional.

The result follows from the fact that every subspace of $\mathfrak{X}_\kappa[D_G]$ contains a further subspace isomorphic to a subspace generated by a block sequence $(x_k)_{k \in \mathbb{N}}$ (Proposition A.3). $\qquad\square$

Definition II.21 (ℓ_1^k-averages). Let $k \in \mathbb{N}$. A finitely supported vector $x \in \mathfrak{X}_\kappa[D_G]$ is said to be a $C - \ell_1^k$ average if $\|x\| > 1$ and there exist $x_1 < \ldots < x_k$ with $\|x_i\| \leq C$ such that $x = \frac{1}{k} \sum_{i=1}^{k} x_i$.

Lemma II.22. *Let $k \in \mathbb{N}$ and $\varepsilon > 0$. Then every block subspace of $\mathfrak{X}_\kappa[D_G]$ contains a vector x which is a $2 - \ell_1^k$ average. If $\mathfrak{X}_\kappa[D_G]$ is a strictly singular extension of Y_G then we may select x satisfying additionally $\|x\|_G < \varepsilon$.*

Proof. If $\mathfrak{X}_\kappa[D_G]$ is a strictly singular extension of Y_G we may pass to a further subspace Z on which the restriction of the identity map $I : \mathfrak{X}_\kappa[D_G] \to Y_G$ has norm less than $\frac{\varepsilon}{2}$.

We choose $j, s \in \mathbb{N}$ such that $k^s \leq n_{2j}$ and $2^s > m_{2j}$. Such a choice is possible from the definition of the sequences $(m_j)_j$, $(n_j)_j$. Let $(x_i)_i$ be a normalized block sequence in the block subspace Z. We set $x_{p,q} = \sum_{l=k^p(q-1)+1}^{k^p q} x_l$ for $p = 0, 1, 2, \ldots$ and $q = 1, 2, \ldots$.

Assume that the conclusion of the lemma fails. Then an easy inductive argument yields that $\|x_{p,q}\| < (\frac{k}{2})^p$ for all $p, q \geq 1$. In particular $\|x_{s,1}\| < (\frac{k}{2})^s$.

Since $k^s < n_{2j}$, using the fact that the set D_G is closed in the $(\mathcal{A}_{n_{2j}}, \frac{1}{m_{2j}})$ operations we get that $\|x_{s,1}\| \geq \frac{1}{m_{2j}} \sum_{l=1}^{k^s} \|x_l\| = \frac{k^s}{m_{2j}}$. Thus $2^s < m_{2j}$, a contradiction.

Therefore there exists a $2 - \ell_1^k$ average z in Z, while our choice of Z yields that $\|z\|_G < \varepsilon$. $\qquad\square$

Lemma II.23. *Let x be a $C - \ell_1^k$ average. Then for every $n \leq k$ and every sequence of intervals $E_1 < \ldots < E_n$, we have that $\sum_{l=1}^{n} \|E_l x\| \leq C(1 + \frac{2n}{k})$. In particular if x is an $C - \ell_1^{n_j}$ average, then for every $f \in D$ with $w(f) = m_i$, $i < j$, we have $|f(x)| \leq \frac{1}{m_i} C(1 + \frac{2n_{j-1}}{n_j}) \leq \frac{3C}{2} \frac{1}{m_i}$.*

Proof. Let $x = \frac{1}{k} \sum_{i=1}^{k} x_i$ be a $C - \ell_1^k$ average. Let also $E_1 < E_2 < \cdots < E_n$ be a sequence of intervals, where $n \leq k$. For $l = 1, \ldots, n$, let I_l (J_l resp.) be the set of all i such that $\operatorname{supp} x_i$ is contained in (resp. intersects) E_l. Clearly $\sum_{l=1}^{n} \#I_l \leq k$, while for each l we have that $\|E_l x\| \leq \frac{1}{k} \sum_{i \in J_l} \|E_l x_i\| \leq \frac{1}{k} C(\#I_l + 2)$. Therefore

$$\sum_{l=1}^{n} \|E_l x\| \leq C \frac{1}{k} (\sum_{l=1}^{n} \#I_l + 2n) \leq C(1 + \frac{2n}{k}).$$
$\qquad\square$

Lemma II.24. *Let $(x_k)_{k\in\mathbb{N}}$ be a block sequence in $\mathfrak{X}_\kappa[D_G]$ such that each x_k is a $C - \ell_1^{l_k}$ average, where $(l_k)_{k\in\mathbb{N}}$ is a strictly increasing sequence of integers, and let $\varepsilon > 0$. Then there exists a subsequence of $(x_k)_{k\in\mathbb{N}}$ which is a $(\frac{3C}{2}, \varepsilon)$ R.I.S.*

Proof. For each k we set $j_k = \max\{j : n_j \le l_k\}$. There exists a subsequence of $(x_k)_{k\in\mathbb{N}}$ (we denote this subsequence again by $(x_k)_{k\in\mathbb{N}}$) such that $(j_k)_{k\in\mathbb{N}}$ is a strictly increasing sequence and $m_{j_{k+1}} > \frac{1}{\varepsilon}\#\operatorname{supp}(x_k)$ for all k. From Lemma II.23 we also get that for each k and every $f \in D_G$ with $w(f) = m_i$, $i < j_k$, we have that $|f(x_k)| \le \frac{3C}{2}\frac{1}{m_i}$.

Therefore this subsequence is a $(\frac{3C}{2}, \varepsilon)$ R.I.S. $\qquad\square$

Proposition II.25 (Existence of R.I.S.). *If $\mathfrak{X}_\kappa[D_G]$ is a strictly singular extension of Y_G, then for every $\varepsilon > 0$ and every block subspace Z of $\mathfrak{X}_\kappa[D_G]$ there exists a $(3, \varepsilon)$ R.I.S. $(x_k)_k$ in Z with $\|x_k\| > 1$ and $\|x_k\|_G < \varepsilon$.*

Proof. It follows from Lemma II.22 and Lemma II.24. $\qquad\square$

Proposition II.26. *Let G be a ground subset of $c_{00}(\omega)$. If $\mathfrak{X}[D_G]$ is a strictly singular extension of Y_G, then the dual space $\mathfrak{X}[D_G]^*$ of $\mathfrak{X}[D_G]$ is the norm closed linear span of the w^* closure of G.*

$$\mathfrak{X}[D_G]^* = \overline{\operatorname{span}}(\overline{G}^{w^*}).$$

Proof. Assume the contrary. Then setting $Z = \overline{\operatorname{span}}(\overline{G}^{w^*})$ there exists $x^* \in \mathfrak{X}_\kappa[D_G]^* \setminus Z$ with $\|x^*\| = 1$ and $x^{**} \in X_G^{**}$ such that $Z \subset \operatorname{Ker} x^{**}$, $\|x^{**}\| = 2$ and $x^{**}(x^*) = 2$. First we observe that the space $\mathfrak{X}[D_G]$ contains no isomorphic copy of ℓ_1. Indeed, if not, then there exists a normalized block sequence $(z_n)_{n\in\mathbb{N}}$ equivalent to the usual ℓ_1 basis. But from Propositions II.19 and II.25 for every block sequence $(u_n)_{n\in\mathbb{N}}$ and for all j there exist $(y_i)_{i=1}^{n_{2j}}$ normalized block sequence of $(u_n)_{n\in\mathbb{N}}$ such that $\|\frac{1}{n_{2j}}\sum_{i=1}^{n_{2j}} y_i\| \le \frac{6}{m_{2j}}$. This leads to a contradiction.

From Odell–Rosenthal's theorem there exists a sequence $(x_k)_{k\in\mathbb{N}}$ in $\mathfrak{X}[D_G]$ with $\|x_k\| \le 2$ such that $x_k \xrightarrow{w^*} x^{**}$. Since each e_n^* belongs to Z we get that $\lim_k e_n^*(x_k) = 0$ for all n, thus, using a sliding hump argument, we may assume that $(x_k)_{k\in\mathbb{N}}$ is a block sequence. Since also $x^*(x_k) \to x^{**}(x^*) = 2$ we may also assume that $1 < x^*(x_k)$ for all k. Let's observe that every convex combination of $(x_k)_{k\in\mathbb{N}}$ has norm greater than 1.

Considering each x_k as a continuous function $x_k : \overline{G}^{w^*} \to \mathbb{R}$ we have that the sequence $(x_k)_{k\in\mathbb{N}}$ is uniformly bounded and tends pointwise to 0, hence it is a weakly null sequence in $C(\overline{G}^{w^*})$. Since Y_G is isometric to a subspace of $C(\overline{G}^{w^*})$ we get that $x_k \xrightarrow{w} 0$ in Y_G, thus there exists a convex block sequence $(y_k)_{k\in\mathbb{N}}$ of $(x_k)_{k\in\mathbb{N}}$ with $\|y_k\|_G \to 0$. We may thus assume that $\|y_k\|_G < \frac{\varepsilon}{2}$ for all k, where $\varepsilon = \frac{1}{n_4}$. We may construct a block sequence $(z_k)_{k\in\mathbb{N}}$ of $(y_k)_{k\in\mathbb{N}}$ such that $(z_k)_{k\in\mathbb{N}}$ is a $(3, \varepsilon)$ R.I.S. of ℓ^1 averages and each z_k is an average of $(y_k)_{k\in\mathbb{N}}$ with $\|z_k\| < \varepsilon$.

Proposition II.19 yields that the vector $z = \frac{1}{n_4} \sum_{k=1}^{n_4} z_k$ satisfies $\|z\| \leq \frac{2 \cdot 3}{m_4} < 1$. On the other hand, the vector z, being a convex combination of $(x_k)_{k \in \mathbb{N}}$, satisfies $\|z\| < 1$. This contradiction completes the proof of the proposition. $\qquad\square$

Remark II.27. The content of the above proposition is that the strictly singular extension $\mathfrak{X}[D_G]$ of the space Y_G is actually a *reflexive extension*. Namely if $\overline{G}^{w^*}$ is a subset of $c_{00}(\mathbb{N})$ then a consequence of Proposition II.26 is that the space $\mathfrak{X}[D_G]$ is reflexive. Furthermore, if $\mathfrak{X}[D_G]$ is nonreflexive then the quotient space $\mathfrak{X}[D_G]^*/\mathfrak{X}[D_G]_*$ is norm generated by the classes of the elements of the set $\overline{G}^{w^*}$.

We denote by $G_0(\kappa)$ the ground set $\{\pm e_\alpha : \alpha < \kappa\}$.

Proposition II.28. *For every ordinal κ and every $D_{G_0(\kappa)}$ tree complete extension of $G_0(\kappa)$ the bimonotone transfinite basis $(e_\alpha)_{\alpha < \kappa}$ of $\mathfrak{X}[D_{G_0(\kappa)}]$ is boundedly complete and shrinking, hence the space $\mathfrak{X}[D_{G_0(\kappa)}]$ is reflexive.*

We refer the reader to Appendix A for the definitions of boundedly complete and shrinking transfinite bases and the proof that this properties yield the reflexivity of the space.

Proof. We prove it for the case $\kappa = \omega$. The general case is reduced to this one.

The fact that the norming set $D_{G_0(\kappa)}$ is closed in the $(\mathcal{A}_{n_{2j}}, \frac{1}{m_{2j}})$ operations implies that for every $j \in \mathbb{N}$ and every sequence of finite intervals $E_1 < E_2 < \cdots < E_{n_{2j}}$ we have that for every $x \in c_{00}$, $\|x\| \geq \frac{1}{m_{2j}} \sum_{i=1}^{n_{2j}} \|E_i x\|$. Thus, since the sequence $(\frac{n_{2j}}{m_{2j}})_j$ increases to infinity, it follows that the basis $(e_n)_n$ is boundedly complete.

We next show that the basis $(e_n)_n$ is shrinking. Assume the contrary. Then there exists a $x^* \in \mathfrak{X}[D_{G_0(\kappa)}]^* \setminus \overline{\mathrm{span}}\{e_n^* : n \in \mathbb{N}\}$. Let $x^* = w^* - \sum_{n=1}^{\infty} b_n e_n$. We may choose an $\varepsilon > 0$ and a sequence of successive intervals $(E_n)_n$ with $\|E_n x^*\| > \varepsilon$. We choose $j \in \mathbb{N}$ with $m_{2j} > \frac{3}{\varepsilon}$.

Pick $(x_n)_n$ a sequence in $\mathfrak{X}[D_{G_0(\kappa)}]$ such that $\mathrm{supp}\, x_n \subset E_n$, $\|x_n\| = \frac{1}{\varepsilon}$ and $x^*(x_n) > 1$. The action of x^* yields that every convex combination of $(x_n)_n$ has norm greater than 1.

On the other hand we may select a $(\frac{3}{2\varepsilon}, \frac{1}{m_{2j}})$ R.I.S. of ℓ_1 averages $y_1, y_2, \ldots, y_{n_{2j}}$ such that each y_l is an average of $(x_n)_n$. Proposition II.19 yields that the vector $y = \frac{1}{n_{2j}} \sum_{l=1}^{n_{2j}} y_l$ satisfies $\|y\| < \frac{3}{\varepsilon m_{2j}}$. But since y is a convex combination of $(x_n)_n$ we also have that $\|y\| > 1$, a contradiction. $\qquad\square$

Remark II.29. It is easy to see that similar arguments yield that the basis $(e_\alpha)_{\alpha < \kappa}$ of $\mathfrak{X}[D_G]$ is boundedly complete for an arbitrary ground subset G of $c_{00}(\kappa)$ and any tree complete extension D_G.

Proposition II.30. *Let G be a ground subset of $c_{00}(\omega)$. If $\mathfrak{X}[D_G]$ is a strictly singular extension of Y_G, then $\mathfrak{X}[D_G]$ is reflexive saturated (or somewhat reflexive).*

Proof. Let Z be a block subspace of $\mathfrak{X}[D_G]$. From the fact that the identity operator $I : \mathfrak{X}[D_G] \to Y_G$ is strictly singular we may choose a normalized block sequence $(z_n)_{n\in\mathbb{N}}$ in Z, with $\sum_{n=1}^{\infty} \|z_n\|_G < \frac{1}{2}$. We claim that the space $Z' = \overline{\text{span}}\{z_n \ n \in \mathbb{N}\}$ is a reflexive subspace of Z.

It is enough to show that the Schauder basis $(z_n)_{n\in\mathbb{N}}$ of Z' is boundedly complete and shrinking. The first follows from the fact that $(z_n)_{n\in\mathbb{N}}$ is a block sequence of the boundedly complete basis $(e_n)_{n\in\mathbb{N}}$ of $\mathfrak{X}[D_G]$. To see that $(z_n)_{n\in\mathbb{N}}$ is shrinking it is enough to show that $\|f|_{\overline{\text{span}}\{z_i \ i\geq n\}}\| \xrightarrow{n\to\infty} 0$ for every $f \in X_G^*$. From Proposition II.26 it is enough to prove it for $f \in \overline{G}^{w^*}$. Since $\sum_{n=1}^{\infty} \|z_n\|_G < \frac{1}{2}$ the conclusion follows. $\qquad\square$

Definition II.31 (exact pair). A pair (x, ϕ) with $x \in \mathfrak{X}_\kappa[D_G]$ and $\phi \in D_G$ is said to be a (θ, C, j) *exact pair* (where $\theta \in \{0, 1\}$, $C \geq 1$, $j \in \mathbb{N}$) if the following conditions are satisfied:

(i) $1 \leq \|x\| \leq C$, for every $\psi \in D_G$ of type I with $w(\psi) = m_i$, $i \neq j$ we have that $|\psi(x)| \leq \frac{3C}{m_i}$ if $i < j$, while $|\psi(x)| \leq \frac{C}{m_j^2}$ if $i > j$.

(ii) ϕ is of type I with $w(\phi) = m_j$.

(iii) $\phi(x) = \theta$ and $\text{ran } x = \text{ran } \phi$.

Proposition II.32. *If $\mathfrak{X}_\kappa[D_G]$ is a strictly singular extension of Y_G, then for every block subspace Z of $\mathfrak{X}_\kappa[D_G]$, every $\varepsilon > 0$ and $j \in \mathbb{N}$ there exists a $(1, 6, 2j)$ exact pair (x, ϕ) with $x \in Z$ and $\|x\|_G < \varepsilon$.*

Proof. From Proposition II.25 there exists $(x_k)_{k=1}^{n_{2j}}$ a $(3, \varepsilon)$-R.I.S. with $\varepsilon \leq 1/m_{2j}^3$. Choose $x_k^* \in D_G$ with $x_k^*(x_k) = 1$ and $\text{ran } x_k^* \subset \text{ran } x_k$. Then Proposition II.19 yields that

$$\left(\frac{m_{2j}}{n_{2j}} \sum_{k=1}^{n_{2j}} x_k, \frac{1}{m_{2j}} \sum_{k=1}^{n_{2j}} x_k^* \right)$$

is a $(1, 6, 2j)$ exact pair. $\qquad\square$

Definition II.33. Let $(X, \|\cdot\|)$ be a Banach space. X is said to be λ-*distortable*, $\lambda > 1$, if there exists an equivalent norm $|\cdot|$ on X such that,

$$\inf_{Y \subset X} \sup \left\{ \frac{|x|}{|y|} : x, y \in Y \text{ and } \|x\| = \|y\| = 1 \right\} \geq \lambda.$$

X is said to be *arbitrarily distortable* if it is λ-distortable for every $\lambda > 1$.

Theorem II.34. *If $\mathfrak{X}_\omega[D_G]$ is a strictly singular extension of Y_G, then $\mathfrak{X}_\omega[D_G]$ is arbitrarily distortable.*

Proof. Let $\lambda > 1$ and choose $j_0 \in \mathbb{N}$ such that $m_{2j_0}/144 > \lambda$. Define

$$|x| = \frac{1}{m_{2j_0}} \|x\| + \sup\{|\phi(x)| : \phi \in D_G \text{ and } w(\phi) = m_{2j_0}\}.$$

Then $|\cdot|$ is an equivalent norm . Let Y be a subspace of $\mathfrak{X}_\omega[D_G]$. By standard arguments we may assume that Y is a block subspace. By Proposition II.32 there exists a $(1, 6, 2j_0)$ exact pair (x_1, ϕ_1) with $x_1 \in Y$ and $w(\phi_1) = m_{2j_0}$. It follows that

$$|x_1| \geq \phi_1(x_1) = 1. \tag{II.7}$$

Also by Proposition II.32 there exists a $(1, 6, 2j_0 + 2)$ exact pair (x_2, ϕ_2) with $x_2 \in Y$. From the properties of the exact pairs, in particular from Definition II.31(i), it follows that

$$|x_2| \leq \frac{6}{m_{2j_0}} + \frac{18}{m_{2j_0}} = \frac{24}{m_{2j_0}}. \tag{II.8}$$

Setting $x = \frac{x_1}{\|x_1\|}$ and $y = \frac{x_2}{\|x_2\|}$, it follows from (II.7) and (II.8) that

$$\frac{|x|}{|y|} \geq m_{2j_0}/144 > \lambda.$$

Since Y was arbitrary chosen we have that $\mathfrak{X}_\omega[D_G]$ is arbitrarily distortable. □

Remark II.35. For a transfinite ordinal κ and every $\lambda > 1$ it can be shown, using the previous method, that there exists an equivalent norm which λ-distorts every block subspace of $\mathfrak{X}_\kappa[D_G]$. Since the block subspaces do not describe all subspaces of $\mathfrak{X}_\kappa[D_G]$, $\kappa > \omega$, it remains unknown if the previous result holds in full generality.

Notes and Remarks. As we have mentioned in the introduction, strictly singular extensions were introduced in [16], for higher complexity saturation methods. In the present form they are contained in [4]. The tree complete extensions and the basic inequality were also introduced and studied in [16]. With the tree complete extensions we attempt to isolate the results obtained from the unconditional part of the definition of the norming set D_G. Thus, independently of the definition of the special functions, resulting from $(\mathcal{A}_{n_{2j+1}}, \frac{1}{m_{2j}})_j$ operations which act on special sequences, one can have seminormalized ℓ_1 averages and from these the exact pairs. As we will see in the next chapter exact pairs are the key component in the n_{2j+1}-dependent sequences.

There is a very interesting interference of the distortion of Banach spaces with the discovery of HI spaces, that we would like to present. R.C. James proved that the spaces c_0 and ℓ_1 are not distortable [37]. In the late 60's V. Milman [49] proved the following. If X is a Banach space with the property that for every equivalent norm on X and every subspace Y of X and $\epsilon > 0$ there exists a subspace Z of Y such that the initial and the new norm are $(1 + \epsilon)$ equivalent, then either some ℓ_p or c_0 is isomorphic to a subspace of X. When Tsirelson discovered his space it became clear that there are Banach spaces not satisfying Milman's condition.

In the early 1990s E. Odell constructed a norm on Tsirelson space which $(2 - \epsilon)$ distorts the original norm. After this, Schlumprecht presented his space as the first example of an arbitrarily distortable Banach space, and based on this, Gowers and Maurey proceeded to the construction of their space. Subsequently Odell and Schlumprecht [52], [53], showed that ℓ_p, $1 < p < \infty$ are arbitrarily distortable and B. Maurey extended this to Banach spaces not containing uniformly ℓ_1^n [45]. Finally N. Tomczack [63], proved that every HI space is arbitrarily distortable.

It remains open whether or not the notions of distortable and arbitrarily distortable are equivalent. In particular we do not know if Tsirelson space is arbitrarily distortable. Notice that the p-convexification $(1 < p < \infty)$ of Tsirelson space is arbitrarily distortable. This is a consequence of the aforementioned Maurey theorem. The key role of the Tsirelson space in the study of the above problem arises from a result of V. Milman and N. Tomczak [50] that asserts the following. If X is a Banach space with no arbitrarily distortable subspace, then X contains an asymptotic ℓ_p (or c_0) space. Results related to the distortion of asymptotic ℓ_1 spaces are contained in [9] and [44].

Chapter III

Hereditarily Indecomposable Extensions with a Schauder Basis

III.1 The HI Property in $\mathfrak{X}[G, \sigma]$

In this chapter we make the final step in the definition of HI extensions for spaces with a Schauder basis. Thus we introduce a coding σ and then we define the n_{2j+1}-special sequences. The tree complete norming set is closed for all $(\mathcal{A}_{n_{2j+1}}, \frac{1}{m_{2j+1}})$ operations acting on n_{2j+1}-special sequences. We also provide sufficient conditions for the HI property of the predual $\mathfrak{X}[G, \sigma]_*$.

Definition III.1. A Banach space X is said to be *Hereditarily Indecomposable* (HI) if for every infinite dimensional closed subspace Y of X there is no nontrivial projection $P : Y \to Y$. (A projection is said to be trivial if either the dimension or the codimension of its kernel is finite).

For equivalent reformulations of the above definition we refer to Proposition IV.4. Throughout this section we shall work in c_{00} (i.e. $\kappa = \omega$).

Given G a ground set we define a tree complete set D_G as follows.

Definition III.2. The set D_G is the minimal subset of c_{00} satisfying the following conditions.

(i) $G \subset D_G$.

(ii) D_G is symmetric (i.e. if $f \in D_G$ then $-f \in D_G$).

(iii) D_G is closed under the restriction of its elements on intervals of $\mathbb{N}$ (i.e. if $f \in D_G$ and E is an interval of $\mathbb{N}$, then $Ef \in D_G$).

(iv) D_G is closed under the $(\mathcal{A}_{n_{2j}}, \frac{1}{m_{2j}})$ operations, i.e. if $f_1 < f_2 < \cdots < f_{n_{2j}}$ belong to D_G, then the functional $f = \frac{1}{m_{2j}}(f_1 + f_2 + \cdots + f_{n_{2j}})$ belongs also to D_G.

(v) D_G is closed under the $(\mathcal{A}_{n_{2j-1}}, \frac{1}{m_{2j-1}})$ operations on special sequences i.e. for every n_{2j-1}-special sequence $(f_1, f_2, \ldots, f_{n_{2j-1}})$ of length n_{2j-1} the functional $f = \frac{1}{m_{2j-1}}(f_1 + f_2 + \cdots + f_{n_{2j-1}})$ belongs to D_G.

(vi) The set D_G is rationally convex.

Next we define the n_{2j-1}-special sequences. For this we shall use a coding function σ defined as follows.

The coding function σ

Let $\mathbb{Q}_s$ denote the set of all finite sequences $(\phi_1, \phi_2, \ldots, \phi_d)$ such that $\phi_i \in c_{00}(\mathbb{N})$, $\phi_i \neq 0$ with $\phi_i(n) \in \mathbb{Q}$ for all i, n and $\phi_1 < \phi_2 < \cdots < \phi_d$. We fix a pair Ω_1, Ω_2 of disjoint infinite subsets of $\mathbb{N}$. From the fact that $\mathbb{Q}_s$ is countable we are able to define an injective coding function $\sigma : \mathbb{Q}_s \to \{2j : j \in \Omega_2\}$ such that $m_{\sigma(\phi_1, \phi_2, \ldots, \phi_d)} > \max\{\frac{1}{|\phi_i(e_l)|} : l \in \operatorname{supp} \phi_i, i = 1, \ldots, d\} \cdot \max \operatorname{supp} \phi_d$.

The n_{2j-1}-special sequences

A finite sequence $(f_i)_{i=1}^{n_{2j-1}}$ is said to be a n_{2j-1}-*special sequence* provided that

(i) $(f_1, f_2, \ldots, f_{n_{2j-1}}) \in \mathbb{Q}_s$ and $f_i \in D_G$ for $i = 1, 2, \ldots, n_{2j-1}$.

(ii) $w(f_1) = m_{2k}$ with $k \in \Omega_1$, $m_{2k}^{1/2} > n_{2j-1}$ and $w(f_{i+1}) = m_{\sigma(f_1, \ldots, f_i)}$ for each $1 \leq i < n_{2j-1}$.

This completes the definition of D_G and it is easy to check that the set D_G, the definition of which depends on the coding function σ, is a tree complete extension of G. We denote by $\mathfrak{X}[G, \sigma]$ the Banach space $\mathfrak{X}[D_G]$.

As we have mentioned the weight $w(f)$ of a functional $f \in D_G$ of type I is not unique. However when we refer to a n_{2j-1}-special sequence $(f_i)_{i=1}^{n_{2j-1}}$ by $w(f_i)$ for $2 \leq i \leq n_{2j-1}$ we shall always mean $w(f_i) = m_{\sigma(f_1, \ldots, f_{i-1})}(\in \Omega_2)$.

Proposition III.3 (The tree-like property of n_{2j-1}-special sequences). *Let $\Phi = (\phi)_{i=1}^{n_{2j-1}}$, $\Psi = (\psi)_{i=1}^{n_{2j-1}}$ be two distinct n_{2j-1}-special sequences. Then*

(i) *For $1 \leq i < j \leq n_{2j-1}$ we have that $w(\phi_i) \neq w(\psi_j)$.*

(ii) *There exists $k_{\Phi,\Psi}$ such that $\phi_i = \psi_i$ for $i < k_{\Phi,\Psi}$ and $w(\phi_i) \neq w(\psi_i)$ for $i > k_{\Phi,\Psi}$.*

We leave the easy proof to the reader.

Definition III.4 (dependent sequences). A double sequence $(x_k, x_k^*)_{k=1}^{n_{2j}-1}$ with $x_k \in \mathfrak{X}[G,\sigma]$ and $x_k^* \in D_G$ is said to be a $(\theta, C, 2j-1)$ dependent sequence (for $\theta \in \{0,1\}$, $C > 1$ and $j \in \mathbb{N}$) if there exists a sequence $(2j_k)_{k=1}^{n_{2j}-1}$ of even integers such that the following conditions are fulfilled:

(i) $(x_k^*)_{k=1}^{n_{2j}-1}$ is a n_{2j-1}-special sequence with $w(x_k^*) = m_{2j_k}$ for all $k \le n_{2j-1}$.

(ii) Each (x_k, x_k^*) is a $(\theta, C, 2j_k)$ exact pair.

Proposition III.5. *Suppose that* $\mathfrak{X}[G,\sigma]$ *is a strictly singular extension of* Y_G *and let* $\varepsilon > 0$, $j \in \mathbb{N}$. *Then for every pair of block subspaces* Z, W *of* $\mathfrak{X}[G,\sigma]$ *there exists a* $(1, 6, 2j-1)$ *dependent sequence* $(x_k, x_k^*)_{k=1}^{n_{2j}-1}$ *with* $\|x_k\|_G < \varepsilon$, $x_{2k-1} \in Z$ *and* $x_{2k} \in W$ *for all* k.

Proof. It follows easily from an inductive application of Proposition II.32. $\square$

Proposition III.6. *Let* $(x_k, x_k^*)_{k=1}^{n_{2j}-1}$ *be a* $(\theta, C, 2j-1)$ *dependent sequence such that* $\|x_k\|_G < \frac{2}{m_{2j-1}^2}$ *for all* k. *Then*

a) *If* $\theta = 1$, *it holds*

$$\|\frac{1}{n_{2j-1}} \sum_{k=1}^{n_{2j-1}} (-1)^{k+1} x_k\| \le \frac{8C}{m_{2j-1}^2}.$$

b) *If* $\theta = 0$, *it holds*

$$\|\frac{1}{n_{2j-1}} \sum_{k=1}^{n_{2j-1}} x_k\| \le \frac{8C}{m_{2j-1}^2}.$$

Proof. a) It is easy to see that the sequence $(x_k)_{k=1}^{n_{2j}-1}$ is a $(2C, \frac{1}{n_{2j-1}^2})$ R.I.S.

The conclusion will follow from Proposition II.19 2) after showing that for every $f \in D$ with $w(f) = m_{2j-1}$ and every interval E of positive integers we have that

$$|f\left(\sum_{k \in E} (-1)^{k+1} x_k\right)| \le 2C(1 + \frac{2}{m_{2j-1}^2} \#(E)).$$

Such an f is of the form $f = \frac{1}{m_{2j-1}}(Fx_{t-1}^* + x_t^* + \cdots + x_r^* + f_{r+1} + \cdots + f_d)$ for some special sequence $(x_1^*, x_2^*, \ldots, x_r^*, f_{r+1}, \ldots, f_{n_{2j-1}})$ of length n_{2j-1} with $x_{r+1}^* \ne f_{r+1}$ and $w(x_{r+1}^*) = w(f_{r+1})$, $d \le n_{2j-1}$ and F an interval of the form $[m, \max \operatorname{supp} x_{t-1}^*]$ (see Proposition III.3).

We estimate the quantity $f(x_k)$ for each k.

- If $k < t-1$, we have that $f(x_k) = 0$.

- If $k = t-1$, we get $|f(x_{t-1})| = \frac{1}{m_{2j-1}}|Fx_{t-1}^*(x_{t-1})| \le \frac{1}{m_{2j-1}}\|x_{t-1}\| \le \frac{C}{m_{2j-1}}$.

- If $k \in \{t, \ldots, r\}$, we have that $f(x_k) = \frac{1}{m_{2j-1}} x_k^*(x_k) = \frac{1}{m_{2j-1}}$.

- If $k > r + 1$, Proposition III.3 yields that $w(f_i) \neq m_{2j_k}$ for all $i > r$. Using the fact that (x_k, x_k^*) is a $(C, 2j_k)$ exact pair and taking in account that $n_{2j-1}^2 < m_{2j_1} \leq m_{2j_k}$ we get

$$|f(x_k)| =$$

$$= \frac{1}{m_{2j-1}}|(f_r + \ldots + f_d)(x_k)|$$

$$\leq \frac{1}{m_{2j-1}}\Big(\sum_{w(f_i)<m_{2j_k}} |f_i(x_k)| + \sum_{w(f_i)>m_{2j_k}} |f_i(x_k)| + \sum_{2r+2\leq 2i\leq d} |f_{2i}(x_{2k-1})| \Big)$$

$$\leq \frac{1}{m_{2j-1}}\Big(\sum_{2j-1<l<2j_k} \frac{3C}{m_l} + n_{2j-1}\frac{C}{m_{2j_k}^2} \Big) \leq \frac{C}{m_{2j-1}^2}.$$

- For $k = r + 1$, the same argument as in the previous case yields that $|f(x_{r+1})| \leq \frac{C}{m_{2j-1}} + \frac{1}{m_{2j-1}^2} < \frac{C+1}{m_{2j-1}}$.

Let now E be an interval. From the previous estimates we get that

$$|f(\sum_{k\in E}(-1)^{k+1}x_k)| \leq |f(x_{t-1})| + |\sum_{k\in E\cap[t,r]} \frac{1}{m_{2j-1}}(-1)^{k+1}|$$

$$+ |f(x_{r+1})| + \sum_{k\in E\cap[r+2,n_{2j-1}]} |f(x_k)|$$

$$\leq \frac{C}{m_{2j-1}} + \frac{1}{m_{2j-1}} + \frac{C+1}{m_{2j-1}} + \frac{C}{m_{2j-1}^2}\#(E)$$

$$< 2C(1 + \frac{2}{m_{2j-1}^2}\#(E)).$$

This completes the proof of the first part.

b) The proof of b) follows using similar arguments. $\qquad\square$

Theorem III.7. *Every $\mathfrak{X}[G, \sigma]$ which is a strictly singular extension of Y_G is a HI space. In other words, strictly singular extensions are HI extensions.*

Proof. Let Z, W be infinite dimensional subspaces of $\mathfrak{X}[G, \sigma]$. We shall show that for every $\varepsilon > 0$ there exist $z \in Z$, $w \in W$ with $\|z - w\| < \varepsilon\|z + w\|$. It is easy to check that this yields the HI property of $\mathfrak{X}[G, \sigma]$. From the well known gliding hump argument we may assume that Z, W are block subspaces. Then for $j \in \mathbb{N}$, using Proposition III.5, we select $(x_k, x_k^*)_{k=1}^{n_{2j-1}}$ a $(6, 2j - 1)$ dependent sequence with $\|x_k\|_G < \frac{2}{m_{2j-1}^2}$, $x_{2k-1} \in Z$ and $x_{2k-1} \in W$ for all k. Observe that

$$\|\frac{1}{n_{2j-1}}\sum_{k=1}^{n_{2j-1}} x_k\| \geq \frac{1}{m_{2j-1}} \text{ and } \|\frac{1}{n_{2j-1}}\sum_{k=1}^{n_{2j-1}} (-1)^{k+1}x_k\| \leq \frac{48}{m_{2j-1}^2} \text{ (Proposition III.6).}$$

Thus setting $z = \sum_{k=1}^{n_{2j-1}/2} x_{2k-1}$ and $w = \sum_{k=1}^{n_{2j-1}/2} x_{2k}$ we get that $z \in Z$, $w \in W$ and $\|z - w\| \leq \frac{48}{m_{2j-1}}\|z + w\|$ which for sufficiently large j yields the desired result.

Therefore the space $\mathfrak{X}[G,\sigma]$ is HI. $\qquad\square$

The space $\mathfrak{X}[G_0,\sigma]$ where $G_0 = \{\pm e_n^* : n < \omega\}$ is an example of a reflexive HI space.

Theorem III.8. *The space* $\mathfrak{X}[G_0,\sigma]$ *where* $G_0 = \{\pm e_n^* : n \in \omega\}$ *is a reflexive HI space.*

Proof. It is easy to check that Y_{G_0} is the space c_0. Hence $\mathfrak{X}[G_0,\sigma]$ is a strictly singular extension of Y_{G_0} and therefore HI. The reflexivity of $\mathfrak{X}[G_0,\sigma]$ follows from Proposition II.28. $\qquad\square$

III.2   The HI Property in $\mathfrak{X}[G,\sigma]_*$

In this section we study the HI property of the predual $\mathfrak{X}[G,\sigma]_*$. We are not able to establish this property for all strictly singular extensions. Actually we require an additional property of the space Y_G defined as uniformly bounded averages (definition III.13). The HI property $\mathfrak{X}[G,\sigma]_*$ is basic for the existence of a non separable HI space.

Notation III.9. We shall denote by $\mathfrak{X}[G,\sigma]_*$ the subspace of the dual space $\mathfrak{X}[G,\sigma]^*$ generated by the biorthogonal functionals $(e_n^*)_{n\in\mathbb{N}}$ of the Schauder basis $(e_n)_{n\in\mathbb{N}}$ of the space $\mathfrak{X}[G,\sigma]$. Notice that the basis $(e_n)_n$ of $\mathfrak{X}[G,\sigma]$ is boundedly complete hence $\mathfrak{X}[G,\sigma]$ is the dual of $\mathfrak{X}[G,\sigma]_*$.

Proposition III.10. *Assume that there exists a constant* $C > 0$ *such that for every infinite dimensional subspace* Z *of* $\mathfrak{X}[G,\sigma]_*$, *every* $\varepsilon > 0$ *and every* $j \in \mathbb{N}$ *there exists a* $(C, 2j)$ *exact pair* (x, ϕ) *with* $\mathrm{dist}(\phi, Z) \leq \frac{5m_{2j}}{n_{2j}}$ *and* $\|x\|_G \leq \frac{5m_{2j}}{n_{2j}}$. *Then the space* $\mathfrak{X}[G,\sigma]_*$ *is HI.*

Proof. Let Z, W be a pair of infinite dimensional subspaces of $\mathfrak{X}[G,\sigma]_*$ and let $\delta > 0$. We shall find $f_Z \in Z$ and $f_W \in W$ such that $\|f_Z + f_W\| < \delta\|f_Z - f_W\|$.

For $j \in \mathbb{N}$, using our assumption, we may inductively select a $(C, 2j-1)$ dependent sequence $(x_k, x_k^*)_{k=1}^{n_{2j-1}}$ such that $\|x_k\|_G < \frac{5m_{2j_k}}{n_{2j_k}} < \frac{m_{2j-1}}{n_{2j-1}}$, $\mathrm{dist}(x_{2k-1}^*, Z)$ $< \frac{5m_{2j_k}}{n_{2j_k}} \frac{m_{2j-1}}{n_{2j-1}}$ and $\mathrm{dist}(x_{2k}^*, W) < \frac{5m_{2j_{2k}}}{n_{2j_{2k}}} \frac{m_{2j-1}}{n_{2j-1}}$ for all k (where $w(x_k^*) = m_{2j_k}$).

From Proposition III.6 the vector $x = \frac{1}{n_{2j-1}} \sum_{k=1}^{n_{2j-1}} (-1)^{k+1} x_k$ satisfies $\|x\| \leq \frac{8C}{m_{2j-1}^2}$.

We set $h_Z = \frac{1}{m_{2j-1}} \sum_{k=1}^{n_{2j-1}} x_{2k-1}^*$ and $h_W = \frac{1}{m_{2j-1}} \sum_{k=1}^{n_{2j-1}} x_{2k}^*$. The functional $h_Z + h_W$ belongs to the norming set D_G, as it is the result of an $(\mathcal{A}_{n_{2j-1}}, \frac{1}{m_{2j-1}})$ operation on the n_{2j-1}-special sequence $(x_1^*, x_2^*, \ldots, x_{n_{2j-1}}^*)$. Hence $\|h_Z + h_W\| \leq 1$. On the other hand $\|h_Z - h_W\| \geq \frac{(h_Z - h_W)(x)}{\|x\|} \geq \frac{\frac{1}{m_{2j-1}}}{\frac{8C}{m_{2j-1}^2}} = \frac{m_{2j-1}}{8C}$.

From our choice of the dependent sequence $(x_k, x_k^*)_{k=1}^{n_{2j-1}}$ we also get that $\text{dist}(h_Z, Z) < \frac{1}{2}$ and $\text{dist}(h_W, W) < \frac{1}{2}$. We may thus select $f_Z \in Z$ and $f_W \in W$ such that $\|f_Z + f_W\| < \|h_Z + h_W\| + \frac{1}{2} + \frac{1}{2} \le 2$ and $\|f_Z - f_W\| > \|h_Z - h_W\| - \frac{1}{2} - \frac{1}{2} \ge \frac{m_{2j-1}}{8C} - 1$.

Hence $\|f_Z - f_W\| > (\frac{m_{2j-1}}{16C} - 2)\|f_Z + f_W\|$ which for large j yields the desired result, therefore the space $\mathfrak{X}[G, \sigma]_*$ is HI. $\square$

Definition III.11. Let $k \in \mathbb{N}$. A finitely supported vector $x^* \in \mathfrak{X}[G, \sigma]_*$ is said to be a $C - c_0^k$ vector if there exist $x_1^* < \cdots < x_k^*$ such that $\|x_i^*\| > C^{-1}$, $x^* = x_1^* + \cdots + x_k^*$ and $\|x^*\| \le 1$.

Lemma III.12. *Let Z be a block subspace of $\mathfrak{X}[G, \sigma]_*$ and let $k \in \mathbb{N}$. Then there exists a block sequence $(z_i^*)_{i \in \mathbb{N}}$ in Z such that for every $i_1 < i_2 < \cdots < i_k$ the vector $z_{i_1}^* + z_{i_2}^* + \cdots + z_{i_k}^*$ is a $2 - c_0^k$ vector.*

Proof. Assume that the conclusion of the lemma fails. We choose $j, s \in \mathbb{N}$ with $k^s \le n_{2j}$ and $2^s > m_{2j}$. Let $(f_i)_{i \in \mathbb{N}}$ be a normalized block sequence in Z. We set

$$\mathcal{B}_1 = \big\{ \{l_1 < l_2 < \cdots < l_k\} : \|f_{l_1} + f_{l_2} + \cdots + f_{l_k}\| > 2 \big\}.$$

The Ramsey theorem yields that there exists $L_1 \in [\mathbb{N}]$ such that either $[L_1]^k \subset \mathcal{B}_1$ or $[L_1]^k \cap \mathcal{B}_1 = \emptyset$. From our assumption on the failure of the lemma the second alternative can not hold thus $[L_1]^k \subset \mathcal{B}_1$. We may assume that $L_1 = \mathbb{N}$. We set $f_{1,l} = \sum_{i=(l-1)k+1}^{lk} f_l$ for $l = 1, 2, \ldots$. As above we may assume, passing to a subsequence, that $\|f_{1,l_1} + f_{1,l_2} + \cdots + f_{1,l_k}\| > 2^2$ for every $l_1 < \cdots < l_k$. After s steps, using the same argument, we arrive at a functional $f \in Z$ which is of the form $f = \sum_{i=1}^{k^s} f_{l_i}$ with $\|f\| > 2^s$. Since $k^s \le n_{2j}$ we get that the functional $\frac{1}{m_{2j}} f$ is the result of a $(\mathcal{A}_{n_{2j}}, \frac{1}{m_{2j}})$ on norm 1 functionals and hence $\|f\| \le m_{2j}$. Therefore $m_{2j} < 2^s$, a contradiction which completes the proof of the lemma. $\square$

Definition III.13. Let G be a ground set. The space Y_G has *uniformly bounded averages* if $\ell_1 \not\hookrightarrow Y_G$ and for every $\varepsilon > 0$ and $n_0 \in \mathbb{N}$ there exists $k \in \mathbb{N}$ such that for every weakly null block sequence $(z_n)_{n \in \mathbb{N}}$ in Y_G with $\|z_n\|_G \le 1$ there exist $i_1 < i_2 < \cdots < i_k$ such that for every $g \in G$ with $\min \text{supp}\, g < n_0$ we have that $|g(\frac{z_{i_1} + z_{i_2} + \cdots + z_{i_k}}{k})| < \varepsilon$.

Proposition III.14. *Assume that Y_G has uniformly bounded averages and $\mathfrak{X}[G, \sigma]$ is a strictly singular extension of Y_G. Let Z be an infinite dimensional subspace of $\mathfrak{X}[G, \sigma]_*$ and $j \in \mathbb{N}$. Then there exists $(1, 12, 2j)$ exact pair (y, ϕ) with $\text{dist}(\phi, Z) < \frac{5m_{2j}}{n_{2j}}$ and $\|y\|_G < \frac{5m_{2j}}{n_{2j}}$.*

Proof. We may assume that Z is a block subspace of $\mathfrak{X}[G, \sigma]_*$. We select $0 < \varepsilon < \frac{1}{n_{2j}}$.

Let $k_1 \in \mathbb{N}$. We choose a $2 - c_0^{k_1}$ vector $f_1 = z_{1,1}^* + \cdots + z_{1,k_1}^*$ in Z and $x_1 = \frac{1}{k_1}(z_{1,1} + \cdots + z_{1,k_1})$ a $2\ell_1^{k_1}$ with $\operatorname{ran} f_1 = \operatorname{ran} x_1$ and $f_1(x_1) > 1$.

We set $t_1 = \max \operatorname{supp} f_1$. Since Y_G has uniformly bounded averages there exists $k_2 = k(\frac{\varepsilon}{4}, t_1)$ satisfying the property of Definition III.13; we may also assume that $k_2 > k_1$. We choose a block sequence $(z_{2,l}^*)_l$ in Z with $\min \operatorname{supp} z_{2,1}^* > \max \operatorname{supp} f_1$ such that $\|z_{2,l}^*\| > \frac{1}{2}$ and $\|\sum_{l=1}^{k_2} z_{2,i_l}^*\| \leq 2$ for every $i_1 < \cdots < i_{k_2}$ in $\mathbb{N}$. For each l we select $z_{2,l} \in \mathfrak{X}[G,\sigma]$ such that $\operatorname{ran} z_{2,l} = \operatorname{ran} z_{2,l}^*$, $\|z_{2,l}\| < 2$ and $z_{2,l}^*(z_{2,l}) > 1$. Since $\ell_1 \nrightarrow Y_G$ and $(z_{2,l})_l$ is a bounded sequence (in $\mathfrak{X}[G,\sigma]$ and thus) in Y_G, we may assume that $(z_{2,l})_l$ is a weakly Cauchy sequence in Y_G. Thus the sequence $(w_{2,l})_l$ defined by $w_{2,l} = z_{2,2l-1} - z_{2,2l}$ is a weakly null sequence in Y_G with $\|w_{2,l}\| \leq 4$. From our choice of k_2 we may assume, passing to a subsequence of $(w_{2,l})_l$ that $|g(\frac{1}{k_2}\sum_{l=1}^{k_2} w_{2,l})| < \varepsilon$ for every $g \in G$ with $\min \operatorname{supp} g \leq t_1$.

We set $x_2 = \frac{1}{k_2} \sum_{l=1}^{k_2} w_{2,l}$ and $f_2 = \sum_{l=1}^{k_2} z_{2l-1}^*$. Observe that $f_2 \in Z$ with $\|f_2\| \leq 1$ while

$$\|x_2\| \geq f_2(x_2) = \frac{1}{k_2} \sum_{l=1}^{k_2} z_{2l-1}^*(z_{2l-1}) > 1$$

which, in particular yields that x_2 is a $4 - \ell_1^{k_2}$ average.

We select $k_3 = k(\varepsilon, t_2)$ with $k_3 > k_2$ satisfying the property of Definition III.13 where $t_2 = \max \operatorname{supp} f_2$ and we define $x_3 \in \mathfrak{X}[G,\sigma]$ and $f_3 \in Z$ similarly to the second step. Following this procedure we may construct a block sequence $(x_r)_{r\in\mathbb{N}}$ in $\mathfrak{X}[G,\sigma]$ and a sequence $(f_r)_{r\in\mathbb{N}}$ in Z such that the following conditions are satisfied.

(i) Each x_r is a $4 - \ell_1^{k_r}$ average and each f_r is a $4 - c_0^{k_r}$ vector with $f_r(x_r) > 1$ and $\operatorname{ran} f_f = \operatorname{ran} x_r$.

(ii) For every $g \in G$ with $\min \operatorname{supp} g \leq \max \operatorname{supp} x_{r-1}$ we have that $|g(x_r)| < \varepsilon$. Thus for every $g \in G$ we have that $|g(x_r)| > \varepsilon$ for at most one r.

Passing to a subsequence we may additionally assume (Lemma II.24) that $(x_r)_{r\in\mathbb{N}}$ is a $(6, \varepsilon)$ R.I.S. We set

$$\phi = \frac{1}{m_{2j}} \sum_{r=1}^{n_{2j}} f_r \text{ and } x = \frac{m_{2j}}{n_{2j}} \sum_{r=1}^{n_{2j}} x_r.$$

We have that $f \in Z$ with $w(f) = m_{2j}$. Proposition II.19 yields that $\|x\| \leq m_{2j} \cdot \frac{2 \cdot 6}{m_{2j}} = 12$. Thus $1 \leq \phi(x) \leq \|x\| \leq 12$ hence we may select θ with $\frac{1}{12} \leq \theta \leq 1$ such that $\phi(\theta x) = 1$. We set $y = \theta x$. Using Proposition II.19 we easily get that (y, ϕ) is a $(12, 2j)$ exact pair.

It only remains to show that $\|y\|_G \leq \frac{5m_{2j}}{n_{2j}}$. Let $g \in G$. Since $|g(x_r)| > \varepsilon$ for at most one r we get that

$$|g(y)| \leq \frac{m_{2j}}{n_{2j}}(|g(x_1)| + \cdots + |g(x_{n_{2j}})|) \leq \frac{m_{2j}}{n_{2j}}(\max_r \|x_r\| + (n_{2j} - 1)\varepsilon)$$

$$< \frac{4m_{2j}}{n_{2j}} + 4\varepsilon < \frac{5m_{2j}}{n_{2j}}.$$

$\square$

Theorem III.15. *If Y_G has uniformly bounded averages and $\mathfrak{X}[G, \sigma]$ is a strictly singular extension of Y_G, then the space $\mathfrak{X}[G, \sigma]_*$ is HI.*

Proof. It follows from Proposition III.10 and III.14. $\square$

Definition III.16. Let G be a ground set. The space Y_G has the *uniform weak Banach–Saks property* if $\ell_1 \not\hookrightarrow Y_G$ and for every $\varepsilon > 0$ there exists $k \in \mathbb{N}$ such that for every normalized weakly null block sequence $(z_n)_{n \in \mathbb{N}}$ in Y_G there exist $i_1 < i_2 < \cdots < i_k$ such that $\|\frac{z_{i_1} + z_{i_2} + \cdots + z_{i_k}}{k}\|_G < \varepsilon$.

Corollary III.17. *If Y_G has the uniform weak Banach–Saks property, then $\mathfrak{X}[G, \sigma]$ is a strictly singular HI extension of Y_G and the space $\mathfrak{X}[G, \sigma]_*$ is HI.*

Notes and Remarks. There exists the unconditional counterpart of the reflexive HI space $\mathfrak{X}[G_0, \sigma]$ (Theorem III.8). This space is a variation of the space presented by W.T. Gowers in [31]. To define this space we modify (v) of Definition III.2 by adding that for each spacial functional f and each E subset (not necessarily interval) of $\mathbb{N}$, $Ef \in D_G$. Then the resulting space denoted as $\mathfrak{X}_u[G_0, \sigma]$ has an unconditional basis and it has a remarkable space of operators. Namely as it is shown in [34] every $T \in \mathcal{L}(\mathfrak{X}_u[G_0, \sigma])$ is of the form $D + S$ with D a diagonal operator and S strictly singular. This property yields that $\mathfrak{X}_u[G_0, \sigma]$ is not isomorphic to any of its proper subspaces and every bounded linear projection is a strictly singular perturbation of a diagonal one. Other variants of the special sequences lead to indecomposable and unconditionally saturated Banach spaces (e.g. [14], [15]). The dual of a HI space X, even if X is reflexive, does not need to be HI. For example there exists a reflexive HI space X such that ℓ_2 is isomorphic to a subspace of X^* (e.g. [10]). More extreme is a recent result of a reflexive HI space X with X^* unconditionally saturated [17]. It is not known if such a divergent could be preserved in the subspaces of X. The following is open. Assume that X is a reflexive HI space. Does there exist a subspace Y of X such that Y^* is also HI? The HI constructions with the use of higher complexity saturation methods which appeared in [6] and [16] yield asymptotic ℓ_1 HI spaces. Variants of them could derive asymptotic ℓ_p HI spaces for $1 < p < +\infty$ (e.g. [22]).

Chapter IV

The Space of the Operators for Hereditarily Indecomposable Banach Spaces

IV.1 Some General Properties of HI Spaces

An important property of the HI spaces concerns the structure of the spaces of their operators. As we will see these spaces have few operators and also remarkable properties of the spaces itself are obtained as consequence of the structure of their operators. We start with the following fundamental results due to Gowers and Maurey [33].

Theorem IV.1. *Every bounded linear operator $T : X \to X$ with X a complex HI space, is of the form $T = \lambda I + S$ with S a strictly singular operator.*

This result is not extendable in real HI spaces. However, as we will see, in the case of the HI extensions $\mathfrak{X}[G, \sigma]$ the above property remains valid. For an arbitrary HI space the following holds.

Theorem IV.2. *Let X be a HI space (real or complex). Then every Fredholm operator is of index zero.*

For complex HI this is a consequence of the above Theorem and Fredholm's theory. For real HI spaces X it uses the complexification of X.

Corollary IV.3. *Let X be a HI space. Then X is not isomorphic to any proper subspace of it. In particular X is not isomorphic to its hyperplanes.*

For the proof of the above results we refer the reader to the Gowers–Maurey paper and also to Maurey's survey in the Handbook of the Geometry of Banach spaces [46]

This rest of the section contains some results obtained from the definition of HI Banach spaces and equivalent reformulations on it. Corollary IV.8 describes the frame on which we shall build the non separable HI space which will be presented in the next chapters. Also, Theorem IV.6 shows that every HI space is a subspace of $\ell^\infty(\mathbb{N})$.

The next proposition summarizes the equivalent reformulations of the definition of HI Banach spaces.

Proposition IV.4. *Let X be a Banach space. The following assertions are equivalent:*

(1) *The space X is HI.*

(2) *For every pair of infinite dimensional closed subspaces Y,Z of X,* $\operatorname{dist}(S_Y, S_Z) = 0$.

(3) *For every pair of infinite dimensional closed subspaces Y,Z of X and $\delta > 0$ there exist $y \in Y$ and $z \in Z$ such that $\|y - z\| \le \delta\|y + z\|$.*

(4) *[V.D. Milman] For every infinite dimensional closed subspace Y of X, $\varepsilon > 0$ and $W \subset B_{X^*}$ such that W ε-norms Y, the space $W_\perp = \{x \in X : f(x) = 0 \text{ for every } f \in W\}$ is a finite dimensional subspace of X.*

Proof. The assertions (1), (2) and (3) are trivially equivalent by the open mapping theorem and the triangle inequality. We first show that (1) implies (4). If Y is an infinite dimensional closed subspace of X, $\varepsilon > 0$ and $W \subset B_{X^*}$ are such that W ε-norms Y then $Y \cap W_\perp = \{0\}$ and $Y \oplus W_\perp$ is a closed subspace of X. The HI property of X yields that $W_\perp$ is necessarily finite dimensional.

It remains to show that (4) implies (1). Suppose that the space X is not HI Then X has two infinite dimensional closed subspaces Y, Z such that $Y \cap Z = \{0\}$ and $Y + Z$ is a closed subspace of X. Then the projection $P : Y + Z \longrightarrow Y$ is continuous; set $\varepsilon = \frac{1}{\|P\|}$. For every $y^* \in \varepsilon B_{Y^*}$ we may select, by the Hahn–Banach theorem, a functional $\widetilde{y}^* \in B_{X^*}$ such that $\widetilde{y}^*$ extends $y^* \circ P$ and $\|\widetilde{y}^*\|_{X^*} = \|y^* \circ P\|_{(Y+Z)^*}$. We set $W = \{\widetilde{y}^* : y^* \in \varepsilon B_{Y^*}\}$. Then W ε-norms Y thus by (4) the space $W_\perp$ should be finite dimensional. This leads to a contradiction since clearly $W_\perp$ contains the infinite dimensional space Z. Therefore X is HI. $\qquad\square$

Proposition IV.5. *Let X be a HI Banach space, and let $T : X \longrightarrow Y$ be a bounded linear operator where Y is any Banach space. Then exactly one of the following two holds:*

(i) *The operator T is strictly singular.*

(ii) *$\operatorname{Ker} T$ is a finite dimensional subspace of X and X can be written as the direct sum of $\operatorname{Ker} T$ and a subspace Z of X such that $T_{|Z}$ is an isomorphism.*

If $Y = X$ and T is one-to-one, then either T is strictly singular or T is an onto isomorphism.

Proof. Assume that $\|T\| = 1$. Suppose that (i) does not hold. Then there exists an infinite dimensional closed subspace W of X and an $\varepsilon > 0$ such that $\|Tw\| \geq \varepsilon$ for every $w \in S_W$.

Suppose also that (ii) does not hold. Then the restriction of T to any subspace of X of finite codimension is not an isomorphism, thus, by Proposition II.3 there exists an infinite dimensional subspace Z of X such that $\|T_{|Z}\| \leq \dfrac{\varepsilon}{2}$.

Then for every $z \in S_Z$ and $w \in S_W$ we have that

$$\|z - w\| \geq \|Tz - Tw\| \geq \|Tw\| - \|Tz\| \geq \varepsilon - \frac{\varepsilon}{2} = \frac{\varepsilon}{2},$$

so we get that $\mathrm{dist}(S_Z, S_W) \geq \dfrac{\varepsilon}{2}$ which contradicts the HI property of X. Hence $\mathrm{Ker}\,T$ is finite dimensional and has a complement Z such that $T_{|Z}$ is an isomorphism.

If $Y = X$ and T is one-to-one and not strictly singular, then T is an isomorphism. As we will see later no HI space is isomorphic to any proper subspace of it. Therefore T is onto. $\qquad\square$

Theorem IV.6. *Every HI Banach space X embeds into ℓ^∞.*

Proof. Let Y be any separable infinite dimensional closed subspace of X and select a countable subset D of the unit ball of X^* such that $D \frac{1}{2}$ norms Y. Since X is III Proposition IV.4 yields that $D_\perp$ is finite dimensional. We enlarge D by a finite set F such that $(D \cup F)_\perp = \{0\}$. Let $D \cup F = \{x_n^* : n \in \mathbb{N}\}$. We define the operator $T : X \longrightarrow \ell^\infty$ by the rule $T(x) = (x_n^*(x))_{n \in \mathbb{N}}$. Observe that T is one-to-one and T restricted to Y is an isomorphism. Proposition IV.5 yields that T is an isomorphism. $\qquad\square$

Theorem IV.7. *Let X be a Banach space and let Z be an infinite dimensional closed subspace of X such that Z is HI and the quotient map $Q : X \longrightarrow X/Z$ is strictly singular. Then the space X is HI.*

Proof. We begin the proof with the next two claims.

Claim 1. *If Z_0 is a finite dimensional subspace of Z, Y is an infinite dimensional closed subspace of X and $\delta > 0$, then there exist $y \in S_Y$ and $z \in S_Z$ with $\|y - z\| < \delta$ such that $\|z_0\| < (1 + \delta)\|z_0 + \lambda z\|$ for every $z_0 \in Z_0$ and every scalar λ.*

Proof of Claim 1. We may assume that $\delta < 1$. Let $\{x_1, x_2, \ldots, x_k\}$ be a $\dfrac{\delta}{4}$ net in S_{Z_0} and pick for each $i = 1, 2, \ldots, k$ an $f_i \in S_{X^*}$ with $f_i(x_i) = 1$. Then $\bigcap\limits_{i=1}^{k} \mathrm{Ker}\,f_i$ is finite codimensional in X, thus the subspace $Y \cap (\bigcap\limits_{i=1}^{k} \mathrm{Ker}\,f_i)$ is infinite dimensional. From the fact that the operator $Q : X \longrightarrow X/Z$ is strictly singular, we may choose a $y \in Y \cap (\bigcap\limits_{i=1}^{k} \operatorname{Ker} f_i)$ with $\|y\| = 1$ and $\|Qy\| < \dfrac{\delta}{16}$. If $z' \in Z$ with $\|y - z'\| < \dfrac{\delta}{16}$

then setting $z = \dfrac{z'}{\|z'\|}$ we get that $z \in S_Z$ and $\|y - z\| < \dfrac{\delta}{8}$.

To finish the proof of the claim it is enough to show that $\|z_0 + \lambda z\| > \dfrac{1}{1+\delta}$ for every $z_0 \in S_{Z_0}$ and every scalar λ. It is also enough to consider only λ with $|\lambda| < 2$. Let $z_0 \in S_{Z_0}$ and $|\lambda| < 2$. We choose $i \in \{1, 2, \ldots, k\}$ such that $\|z_0 - x_i\| < \dfrac{\delta}{4}$. Then

$$
\begin{aligned}
\|z_0 + \lambda z\| &\geq \|x_i + \lambda z\| - \|z_0 - x_i\| > |f_i(x_i + \lambda z)| - \frac{\delta}{4} \\
&\geq 1 - |\lambda| \cdot |f_i(z)| - \frac{\delta}{4} > 1 - |\lambda| \cdot (|f_i(y)| + \|y - z\|) - \frac{\delta}{4} \\
&> 1 - 2(0 + \frac{\delta}{8}) - \frac{\delta}{4} = 1 - \frac{\delta}{2} > \frac{1}{1+\delta}.
\end{aligned}
$$

$\square$

Claim 2. For every infinite dimensional closed subspace Y of X and every $\varepsilon > 0$ there exists an infinite dimensional closed subspace W of Z such that $\operatorname{dist}(w, S_Y) < \varepsilon$ for every $w \in S_W$.

Proof of Claim 2. Let $(\varepsilon_n)_{n \in \mathbb{N}}$ be a sequence of positive reals with $\prod\limits_{n=1}^{\infty} (1 + \varepsilon_n) \leq 2$ and $\sum\limits_{n=1}^{\infty} \varepsilon_n < \dfrac{\varepsilon}{8}$. From Claim 1 we may inductively select a sequence $(y_n)_{n \in \mathbb{N}}$ in S_Y and a sequence $(z_n)_{n \in \mathbb{N}}$ in S_Z such that $\|y_n - z_n\| < \varepsilon_n$ and $\|\sum\limits_{i=1}^{n} a_i z_i\| < (1 + \varepsilon_{n+1}) \|\sum\limits_{i=1}^{n+1} a_i z_i\|$ for every choice of scalars $(a_i)_{i \in \mathbb{N}}$ and all n. Then $(z_n)_{n \in \mathbb{N}}$ is a Schauder basic sequence with basis constant less or equal to 2. We set $W = \overline{\operatorname{span}}\{z_n : n \in \mathbb{N}\}$.

The space W satisfies the conclusion of the claim. Indeed, let $w \in S_W$, $w = \sum\limits_{n=1}^{\infty} a_n z_n$. Then $|a_n| \leq 4$ for all n. The series $\sum\limits_{n=1}^{\infty} a_n y_n$ converges to some $y' \in Y$ and $\|w - y'\| \leq \sum\limits_{n=1}^{\infty} |a_n| \|y_n - z_n\| \leq 4 \sum\limits_{n=1}^{\infty} \varepsilon_n < \dfrac{\varepsilon}{2}$. Setting $y = \dfrac{y'}{\|y'\|}$ we obtain that

$$
\operatorname{dist}(w, S_Y) \leq \|w - y\| \leq \|w - y'\| + \|y' - y\| \leq \frac{\varepsilon}{2} + |\,\|y'\| - 1| < \varepsilon. \qquad \square
$$

We pass now to the proof that the space X is HI. Let Y_1, Y_2 be a pair of infinite dimensional closed subspaces of X^* and we will show that $\operatorname{dist}(S_{Y_1}, S_{Y_2}) = 0$.

Let $\varepsilon > 0$. From Claim 2, there exist two infinite dimensional closed subspaces W_1, W_2 of Z such that for every $w_1 \in S_{W_1}$ we have that $\operatorname{dist}(w_1, S_{Y_1}) < \dfrac{\varepsilon}{3}$ and for every $w_2 \in S_{W_2}$ we have that $\operatorname{dist}(w_2, S_{Y_2}) < \dfrac{\varepsilon}{3}$. Since the space Z is HI we may choose $w_1 \in S_{W_1}$ and $w_2 \in S_{W_2}$ with $\|w_1 - w_2\| < \dfrac{\varepsilon}{3}$. Let $y_1 \in S_{Y_1}$ such that $\|w_1 - y_1\| < \dfrac{\varepsilon}{3}$ and $w_2 \in S_{W_1}$ such that $\|w_2 - y_2\| < \dfrac{\varepsilon}{3}$. We deduce that

$$
\begin{aligned}
\operatorname{dist}(S_{Y_1}, S_{Y_2}) &\leq \|y_1 - y_2\| \\
&\leq \|y_1 - z_1\| + \|z_1 - z_2\| + \|z_2 - y_2\| \\
&< \frac{\varepsilon}{3} + \frac{\varepsilon}{3} + \frac{\varepsilon}{3} = \varepsilon.
\end{aligned}
$$

Thus $\operatorname{dist}(S_{Y_1}, S_{Y_2}) = 0$. Therefore the space X is HI. $\qquad\square$

With the next result we provide some sufficient conditions yielding the HI property for dual Banach spaces.

Corollary IV.8. *Let X be a Banach space and let Z a subspace of X^* such that the following conditions are fulfilled.*

(i) *The space X contains no isomorphic copy of ℓ_1.*

(ii) *The space Z is HI.*

(iii) *The space X^*/Z is isomorphic to $c_0(\Gamma)$ for some set Γ.*

Then the space X^* is HI.

Proof. From Theorem IV.7 it is enough to observe that the quotient map $Q : X^* \longrightarrow X^*/Z$ is strictly singular.

If Γ is finite there is nothing to be proved. Let Γ be infinite and suppose that Q is not strictly singular. Let Y be an infinite dimensional subspace of X^* such that Q restricted to Y is an isomorphism. Since X^*/Z is isomorphic to $c_0(\Gamma)$ for an infinite set Γ we may assume, passing to a subspace, that $Q(Y)$ is isomorphic to c_0. It follows that X^* contains isomorphically c_0, thus, by Bessaga Pelczynski's Theorem, the space X contains isomorphically ℓ_1, a contradiction. $\qquad\square$

Similar arguments also yield the next.

Corollary IV.9. *Let X be a Banach space and Y a closed subspace of X. If the spaces Y and X/Y are HI saturated, then the same holds for the space X.*

A similar result also holds for somewhat reflexive (or reflexive saturated) Banach spaces.

Proposition IV.10. *Let X be a Banach space such that*

(i) *X does not contain isomorphically ℓ_1.*

(ii) *X^{**} is isomorphic to $X \oplus \ell_1(\Gamma)$ for some infinite set Γ.*

Then every bounded linear operator $T : X^* \longrightarrow X^*$ is of the form $T = Q^* + K$ where Q is an operator on X and K is a compact operator on X^*.

Proof. Let $T : X^* \longrightarrow X^*$ be a bounded linear operator. Consider the conjugate operator $T^* : X^{**} \longrightarrow X^{**}$ and the projections

$$P_1 : X \oplus \ell_1(\Gamma) \longrightarrow X \text{ and } P_2 : X \oplus \ell_1(\Gamma) \longrightarrow \ell_1(\Gamma).$$

Let $Q : X \longrightarrow X$ be the operator defined as $Q = P_1 \circ (T^*|X)$ and $S : X \longrightarrow X^{**}$ defined by $S = P_2 \circ (T^*|X)$. Observe that $S(X) \subset \ell_1(\Gamma)$.

The operator S is compact. Indeed, if S was not compact we could construct an operator from a subspace of X onto ℓ_1, thus by the lifting property of ℓ_1, X contains isomorphically ℓ_1, a contradiction. It follows that the operator $S^* : X^{***} \longrightarrow X^*$ is compact so the same holds for the operator $K = S^*|X^*$. An easy computation yields that $T = Q^* + K$. □

IV.2   The Space of Operators $\mathcal{L}(\mathfrak{X}[G, \sigma])$, $\mathcal{L}(\mathfrak{X}[G, \sigma]_*)$

We pass now to discuss the structure of $\mathcal{L}(\mathfrak{X}[G, \sigma])$ when $\mathfrak{X}[G, \sigma]$ is a HI extension of a ground norm. We begin with the following.

Lemma IV.11. *Let Y be a subspace of $\mathfrak{X}[G, \sigma]$ and let $T : Y \to \mathfrak{X}[G, \sigma]$ be a bounded linear operator. Let $(y_l)_{l \in \mathbb{N}}$ be a block sequence of $2 - \ell_1^{n_l}$ averages with increasing lengths in Y such that $(Ty_l)_{l \in \mathbb{N}}$ is also a block sequence and $\lim_l \|y_l\|_G = 0$. Then*

$$\lim_l \text{dist}(Ty_l, \mathbb{R}y_l) = 0.$$

Proof. Assume on the contrary that there exist $\delta > 0$ and $L \in [\mathbb{N}]$ such that $\text{dist}(Ty_l, \mathbb{R}y_l) > \delta$ for all $l \in L$. The Hahn–Banach theorem yields that there exists $\phi_l \in B_{\mathfrak{X}[G,\sigma]_*}$ such that

a) $\phi_l(y_l) = 0$, $\phi_l(Ty_l) > \delta$ and

b) $\text{ran} \, \phi_l \subset \text{ran}(\text{supp} \, y_l \cup \text{supp} \, Ty_l)$.

For simplicity we may assume that $\phi_l \in D_G$ (the precise argument asserts that we can choose $\phi_l \in D_G$ such that $\phi_l(y_l) \geq \delta$ and $|\phi_l(y_l)|$ being as small as we wish).

First we observe that for every $\varepsilon > 0$ and every $j \in \mathbb{N}$ there exist $l_1 < \ldots < l_{n_j} \in L$ such that setting

$$x = \frac{y_{l_1} + \ldots + y_{l_{n_j}}}{n_j} \text{ and } \phi = \frac{\phi_{l_1} + \ldots + \phi_{l_{n_j}}}{m_{2j}}$$

gives $(m_{2j}x, \phi)$ is a $(0, 3C, 2j)$ exact pair such that $\|m_{2j}x\|_G < \varepsilon$ (Lemmas II.22–II.24, Proposition II.25). Then for a given $j \in \mathbb{N}$ we may define $\{(x_k, x_k^*)\}_{k=1}^{n_{2j-1}}$ to be $(0, 18C, 2j - 1)$-dependent sequence with $\|x_k\|_G < 1/m_{2j-1}^2$ and $x_k^*(Tx_k) \geq \delta$ for $k = 1, \ldots, n_{2j-1}$. Proposition III.6 yields that

$$\left\| \frac{1}{n_{2j-1}} \sum_{k=1}^{n_{2j-1}} x_k \right\| \leq \frac{104C}{m_{2j-1}^2} . \tag{IV.1}$$

On the other hand setting $x^* = \frac{1}{m_{2j-1}} \sum_{k=1}^{n_{2j-1}} x_k^*$ we get,

$$\left\| \frac{1}{n_{2j-1}} \sum_{k=1}^{n_{2j-1}} T x_k \right\| \geq x^* \left(\frac{1}{n_{2j-1}} \sum_{k=1}^{n_{2j-1}} x_k \right) \geq \frac{\delta}{m_{2j-1}} . \qquad \text{(IV.2)}$$

For sufficiently large j, (IV.1) and (IV.2) derive a contradiction. $\square$

Theorem IV.12. *Let Y be an infinite dimensional closed subspace of $\mathfrak{X}[G,\sigma]$. Every bounded linear operator $T : Y \to \mathfrak{X}[G,\sigma]$ takes the form $T = \lambda I_Y + S$ with $\lambda \in \mathbb{R}$ and S a strictly singular operator (I_Y denotes the inclusion map from Y to $\mathfrak{X}[G,\sigma]$).*

Proof. Assume that T is not strictly singular. We shall determine a $\lambda \neq 0$ such that $T - \lambda I_Y$ is strictly singular.

Let Y' be an infinite dimensional closed subspace of Y such that $T : Y' \to T(Y')$ is an isomorphism. By standard perturbation arguments and the fact that $\mathfrak{X}[G,\sigma]$ is a strictly singular extension of Y_G, we may assume, passing to a subspace, that Y' is a block subspace of $\mathfrak{X}[G,\sigma]$ spanned by a normalized block sequence $(y'_n)_{n\in\mathbb{N}}$ such that $(Ty'_n)_{n\in\mathbb{N}}$ is also a block sequence and $\sum_{n=1}^{\infty} \|y'_n\|_G < 1$. From Lemma II.22 we may choose a block sequence $(y_n)_{n\in\mathbb{N}}$ of $2 - \ell_1^{n_i}$ averages of increasing lengths in $\text{span}\{y'_n : n \in \mathbb{N}\}$ with $\|y_n\|_G \to 0$. Lemma IV.11 yields that $\lim_n \text{dist}(Ty_n, \mathbb{R}y_n) = 0$. Thus there exists a $\lambda \neq 0$ such that $\lim_n \|Ty_n - \lambda y_n\| = 0$.

Since the restriction of $T - \lambda I_Y$ to any finite codimensional subspace of $\overline{\text{span}}\{y_n : n \in \mathbb{N}\}$ is clearly not an isomorphism and since also Y is a HI space, it follows from Proposition IV.5 that the operator $T - \lambda I_Y$ is strictly singular. $\square$

Theorem IV.13. *Let Y_G have uniformly bounded averages and $\mathfrak{X}[G,\sigma]$ be the strictly singular extension of Y_G. Then every $T \in \mathcal{L}(\mathfrak{X}[G,\sigma]_*)$ is of the form $T = \lambda I + S$ with S strictly singular operator.*

We start with the following lemma.

Lemma IV.14. *Let X be an HI space with a Schauder basis $(e_n)_n$. Assume that $T : X \to X$ is a bounded linear operator not of the form $T = \lambda I + S$ with S strictly singular. Then there exists n_0 and $\delta > 0$ such that for every $z \in X_{n_0} = \overline{\text{span}}\{e_n : n \geq n_0\}$, $\text{dist}(Tz, \mathbb{R}z) \geq \delta\|z\|$.*

Proof. If not, then there exists a normalized block sequence $(z_n)_n$ such that $\text{dist}(Tz_n, \mathbb{R}z_n) \leq \frac{1}{n}$. Choose $\lambda \in \mathbb{R}$ such that $\|Tz_n - \lambda z_n\|_{n\in L} \to 0$ for a subsequence $(z_n)_{n\in L}$. Then for a further subsequence $(z_n)_{n\in M}$, $M \in [L]$, $T - \lambda I|_{\overline{\text{span}}\{z_n : n\in\mathbb{N}\}}$ is a compact operator. The HI property of X easily yields that $T - \lambda I$ is a strictly singular operator, contradicting our assumption. $\square$

In the following lemma we assume that Y_G and $\mathfrak{X}[G,\sigma]$ are as in the statement of the theorem.

Lemma IV.15. *Let* $T : \mathfrak{X}[G,\sigma]_* \to \mathfrak{X}[G,\sigma]_*$, Z *be a block subspace of* $\mathfrak{X}[G,\sigma]_*$ *such that* $Z = <(z_n^*)_n>$ *and for each* $n < m$ $\operatorname{ran} z_n^* \cup \operatorname{ran} Tz_n^* < \operatorname{ran} z_m^* \cup \operatorname{ran} Tz_m^*$. *Assume further that for each* $z^* \in Z^*$, *with* $\|z^*\| = 1$, $\operatorname{dist}(Tz^*, \mathbb{R}z^*) > \delta$. *Then the following holds:*

(i) *For each* $k \in \mathbb{N}$ *there exists a normalized block sequence* $(w_l^*)_{l \in \mathbb{N}}$ *in* Z *satisfying the following properties:*

For every $l_1 < \ldots < l_k$ *setting* $w^* = \sum_{q=1}^{k} w_{l_q}^*$ *we have that* $\|w^*\| \le 2$. *Further there exists an* $2 - \ell_1^k$-*average* $w \in \mathfrak{X}[G,\sigma]$ *such that* $w^*(w) = 0$, $Tw^*(w) > 1$ *and* $\operatorname{ran}(w) \subset \operatorname{ran}(\operatorname{supp} w^* \cup \operatorname{supp} Tw^*)$.

(ii) *For every* $j \in \mathbb{N}$ *there exists* $(0, \frac{18}{\delta}, 2j)$ *exact pair* (x, ϕ) *with* $\phi \in Z$, $x \in \mathfrak{X}[G,\sigma]$ *and* $T\phi(x) > 1$, $\|x\|_G < \frac{18m_{2j}}{\delta n_{2j}}$.

Proof. (i) The proof follows the arguments of Lemma III.12 using the following observation:

Assume that $w_1^*, \ldots, w_k^*$ is a $2 - c_0^k$-vector. Our assumptions for the operator T yield that for $q = 1, \ldots, k$ there exists $w_q \in \mathfrak{X}[G,\sigma]$ such that $w_q^*(w_q) = 0$, $Tw_q^*(w_q) > 1$ and $\|w_q\| < 1\delta$. Now using this and the arguments of Lemma III.12 we obtain the desired sequence.

(ii) It only requires the adaptation described before the proof of Proposition III.14.   □

Proof of Theorem IV.13. On the contrary assume that there exists $T \in \mathcal{L}(\mathfrak{X}[G,\sigma]_*)$ not of the desired form. Assume further that $\|T\| = 1$ and Te_n^* is finitely supported with $\liminf \min \operatorname{supp} Te_n^* = \infty$. (We may assume the later conditions from the fact that the basis $(e_n^*)_n$ of $(X_G)_*$ is weakly null.) In particular for every $(z_n^*)_n$ block sequence in $\mathfrak{X}[G,\sigma]_*$ there exists a subsequence $(z_n^*)_{n \in L}$ such that $(\operatorname{ran} z_n^* \cup \operatorname{ran} Tz_n^*)_n$ is a sequence of successive subsets of $\mathbb{N}$.

Let $\delta > 0$ and $n_0 \in \mathbb{N}$ as in Lemma IV.14.

Then Lemma IV.15 yields that for every $j \in \mathbb{N}$ there exists $(z_k, z_k^*)_{k=1}^{n_{2j-1}}$, a $(0, \frac{18}{\delta}, 2j - 1)$ dependent sequence such that $z_k^*(z_k) = 0$, $Tz_k^*(z_k) > 1$, $\operatorname{ran} z_k \subset \operatorname{ran} z_k^* \cup \operatorname{ran} Tz_k^*$, $(\operatorname{ran} z_k^* \cup \operatorname{ran} Tz_k)_{k=1}^{n_{2j-1}}$ successive subsets of $\mathbb{N}$ and $\|z_k\|_G \le \frac{1}{m_{2j-1}^2}$.

Proposition III.6 yields that

$$\|\frac{1}{n_{2j-1}} \sum_{k=1}^{n_{2j-1}} z_k\| \le \frac{144}{m_{2j-1}^2 \delta}.$$

Finally $\|\frac{1}{m_{2j-1}} \sum_{k=1}^{n_{2j-1}} Tz_k^*\| \le 1$ (since $\|T\| \le 1$) and also

$$1 \ge \|\frac{1}{m_{2j-1}} \sum_{k=1}^{n_{2j-1}} Tz_k^*\| \ge \frac{m_{2j-1}^2 \delta}{144 m_{2j-1}} \frac{1}{n_{2j-1}} \sum_{k=1}^{n_{2j-1}} Tz_k^*(z_k) \ge \frac{m_{2j-1}\delta}{144}.$$

This yields a contradiction for sufficiently large $j \in \mathbb{N}$.   □

Notes and Remarks. Our approach showing that the spaces $\mathfrak{X}[G,\sigma]$ have few operators follows the lines of Gowers and Maurey [34]. In a recent paper ([4]) is shown that there exists a HI space with no reflexive subspace such that every operator T is of the form $\lambda I + W$ with W a weakly compact operator. The central remaining open problem is the existence of a Banach space with very few operators (i.e. $T = \lambda I + K$ with K compact). The recent results [14], [15] show that a Banach space X could admit few operators and also could have rich unconditional structure. It is not known if there exists a subspace Y of a space X with an unconditional basis, such that Y admits few operators. Actually it is unknown if there exists such a Y which is indecomposable.

Chapter V

Examples of Hereditarily Indecomposable Extensions

In the present chapter we shall present HI extensions of Y_G when G is of specific form. Thus we shall show how we can obtain a quasi-reflexive HI space $\mathfrak{X}_{qr}$, a HI space $\mathfrak{X}_p$ which has ℓ_p as a quotient ($1 < p < \infty$) and also a non-separable HI space.

V.1 A Quasi-reflexive HI Space

We start with the quasi-reflexive HI space. We recall that a Banach space X is said to be quasi-reflexive if $\dim(X^{**}/X) < \infty$.

The set G_{qr}

We consider the set

$$G_{qr} = \{\pm I^* : I \text{ is a finite interval of } \mathbb{N}\}.$$

Here we denote by I^* the function $\chi_I \in c_{00}$. It follows readily that G_{qr} is a ground set. Moreover the basis of $Y_{G_{qr}}$ is equivalent to the summing basis of c_0, hence $Y_{G_{qr}}$ is c_0 saturated and $\mathfrak{X}[G_{qr}, \sigma]$ is a HI reflexive extension of $Y_{G_{qr}}$.

Theorem V.1. *The space $\mathfrak{X}[G_{qr}, \sigma]$ is a quasi-reflexive and HI space.*

Proof. Since $\mathfrak{X}[G_{qr}, \sigma]$ is a reflexive extension of $Y_{G_{qr}}$ it follows that $\mathfrak{X}[G_{qr}, \sigma]^* = \overline{\operatorname{span}}(\overline{G_{qr}}^p)$ (Proposition II.26). This yields that $\mathfrak{X}[G_{qr}, \sigma]^* = \overline{\operatorname{span}}(\{e_n^*\}_n \cup \{\mathbb{N}^*\})$ from which we obtain that $\mathfrak{X}[G_{qr}, \sigma]^*/\mathfrak{X}[G_{qr}, \sigma]_* \cong \mathbb{R}$. Since the basis of $\mathfrak{X}[G_{qr}, \sigma]$ is boundedly complete we obtain that $\mathfrak{X}[G_{qr}, \sigma]^{**}/\mathfrak{X}[G_{qr}, \sigma] \cong \mathbb{R}$. $\qquad\square$

Remark. Since every normalized weakly null sequence in $Y_{G_{qr}}$ contains a subsequence equivalent to the c_0 basis, we conclude that $Y_{G_{qr}}$ has the uniform weak Banach–Saks property which yields that the predual of $\mathfrak{X}[G_{qr}, \sigma]$ is also HI.

V.2   The Spaces ℓ_p, $1 < p < \infty$, are Quotients of HI Spaces

Next we define the HI space $\mathfrak{X}_p$ which has ℓ_p ($1 < p < +\infty$) as a quotient. Let $\{M_i\}_{i \in \mathbb{N}}$ be a disjoint partition of $\mathbb{N}$ into infinite sets, $M_i = \{m_1^i < m_2^i < \cdots < m_n^i < \cdots\}$. We consider the partially ordered set $(\mathcal{L}, \prec)$ where $\mathcal{L} = \mathbb{N}$ and $l \prec d$ iff $l < d$ and there exists $i \in \mathbb{N}$ such that $l, d \in M_i$. Clearly $(\mathcal{L}, \prec)$ is a tree. A segment s of $\mathcal{L}$ has the property that it is a segment of M_i for some $i \in \mathbb{N}$. The set G_p is defined as follows.

$$G_p = \left\{ \sum_{i=1}^{n} a_i s_i^* : \sum_{i=1}^{n} |a_i|^q \leq 1 \text{ and each } s_i \text{ is a finite segment of } \mathcal{L} \right\}.$$

Here q is the conjugate of p. Clearly G_p is a ground set and the space Y_{G_p} has the following properties.

Proposition V.2. *Let Z be a closed infinite dimensional subspace of Y_{G_p}. Then either $c_0 \hookrightarrow Z$ or $\ell_p \hookrightarrow Z$. Moreover the space Y_{G_p} has the uniform weak Banach–Saks property.*

Proof. It follows easily from the definition of Y_{G_p} that $Y_{G_p} = (\sum \oplus Y_i)_p$ where $Y_i = \overline{< e_n >}_{n \in M_i}$ and $(e_n)_{n \in M_i}$ is equivalent to the summing basis of c_0. In particular each Y_i is isomorphic to c_0. Clearly for each $n \in \mathbb{N}$, $(\sum_{i=1}^{n} \oplus Y_i)_p$ remains isomorphic to c_0 hence it is c_0 saturated. Let Z be a subspace of Y_{G_p}. Then one of the following two alternatives holds:

(a) There exists $n_0 \in \mathbb{N}$ such that $P_{n_0}|_Z : Z \to \sum_{i=1}^{n_0} \oplus Y_i$ is not strictly singular.

(b) For every $n \in \mathbb{N}$, $P_n|_Z$ is a strictly singular operator.

The preceding remarks immediately yield that in the first case c_0 is isomorphic to a subspace of Z.

If (b) occurs then a "sliding hump" argument derives a normalized sequence $(z_n)_n$ in Z of finite supported vectors and a sequence $(w_n)_{n \in \mathbb{N}}$ in Y_{G_p} such that $\sum_n \|z_n - w_n\| < 1$, and if we denote $I_n = \{i \in \mathbb{N} : \operatorname{supp} w_n \cap M_i \neq \emptyset\}$, then $(I_n)_n$ consists of successive finite subsets of $\mathbb{N}$. The later property yield that $(w_n)_n$ is equivalent to the usual ℓ_p basis and hence, $(z_n)_n$ is equivalent to the ℓ_p basis. This completes the proof of the first part.

Next we show that Y_{G_p} satisfies the uniform weak Banach–Saks property (Definition III.16). First we prove the following:

Claim. *Let $(y_n)_n$ be a normalized weakly null block sequence in Y_{G_p}. Then for every $\varepsilon > 0$ and every $M \in [\mathbb{N}]$ there exists $L \in [M]$ such that the following holds:*
For every s segment of the tree $\mathcal{L}$,

$$\#\{\ell \in L : |s^*(y_\ell)| \geq \varepsilon\} \leq 2\,.$$

Proof of the claim. Assume on the contrary that for some M no such L exists. Then applying the classical Ramsey theorem for triples we obtain an $L \in [M]$ such that for $\ell_1 < \ell_2 < \ell_3$ in L there exists a segment s of $\mathcal{L}$ with $|s^*(y_{\ell_i})| \geq \varepsilon$ for $i = 1, 2, 3$. Let $L = \{\ell_1 < \ell_2 < \ldots\}$. For each $n \in \mathbb{N}$ and $1 < k < n$ choose $s^*_{k,n}$ with $|s^*_{k,n}(\ell_i)| \geq \varepsilon$ for $i = 1, \ldots, n$.

Observe that there exists $n_0 \in \mathbb{N}$ such that for each $n \in \mathbb{N}$ and each $k < n$, $s_{k,n} \subset M_i$ for some $i < n_0$. Indeed, $s_{k,n} \cap \operatorname{supp} y_{\ell_1} \neq \emptyset$; now choose n_0 such that for every $i > n_0$, $\max \operatorname{supp} x_{\ell_1} < \min M_i$. To finish the proof, passing if required to a subsequence, we may assume that for every $n \in \mathbb{N}$, $w^* - \lim_n s^*_{k,n} = s^*_k$ with $\min s_k < \max \operatorname{supp} x_{\ell_1}$.

Clearly s_k is an infinite segment and $s_k \subset M_{i_n}$ for some $i \leq n_0$. Hence there exists $Q \in [\mathbb{N}]$ and $i < n_0$ such that for each $k \in Q$, $|M^*_i(y_{\ell_k})| > \varepsilon$, a contradiction since $(y_\ell)_\ell$ is weakly null. $\qquad\square$

The uniform weak Banach–Saks property of Y_{G_p} is obtained by the above claim in a similar manner as this property is established in ℓ_p. $\qquad\square$

As consequence we obtain the following.

Proposition V.3. (i) *The space $\mathfrak{X}[G_p, \sigma]$ is a HI reflexive extension of Y_{G_p}.*
(ii) *The predual of $\mathfrak{X}[G_p, \sigma]_*$ is also HI.*

The next theorem describes the basic properties of $\mathfrak{X}[G_p, \sigma]$.

Theorem V.4. *There exists a surjective bounded linear operator $Q : \mathfrak{X}[G_p, \sigma] \to \ell_p$. Additionally $Q^*[\ell_q]$ is a complemented subspace of $\mathfrak{X}[G_p, \sigma]^*$. In particular the quotient map Q is described by the rule $Q(e_{m^i_k}) = e_i$ for all $i, k \in \mathbb{N}$. Also $Q^*(e^*_i) = M^*_i$ and $\mathfrak{X}[G_p, \sigma]^* \cong \mathfrak{X}[G_p, \sigma]_* \oplus \ell_q$.*

Proof. The proof follows easily from the next equation.
For all $\{a_i\}^n_{i=1}$ the following holds:

$$(*) \qquad \left\| \sum_{i=1}^n a_i I^*_i \right\| = \left(\sum_{i=1}^n |a_i|^q \right)^{1/q},$$

where $\{I_i\}^n_{i=1}$ are infinite subsegments of $\{M_i\}^n_{i=1}$. This is a consequence of the next lemma.

Lemma V.5. *For all $\{a_i\}^n_{i=1}$, $\varepsilon > 0$, $n_0 \in \mathbb{N}$ there exist $\{J_i\}^n_{i=1}$ finite segments of $(\mathcal{L}, \prec)$ such that $n_0 < J_i \subset M_i$, and setting $x_i = \frac{1}{\#J_i} \sum_{m \in J_i} e_m$, we have that*

$$\left\| \sum_{i=1}^d a_i x_i \right\|_{[G_p, \sigma]} \leq (1 + \varepsilon) \left(\sum_{i=1}^d |a_i|^p \right)^{1/p}.$$

Let us see how we finish the proof of the theorem. First we establish the equality $(*)$. Indeed the definition of G_p yields that

$$\left\| \sum_{i=1}^{d} a_i I_i^* \right\| \leq \left(\sum_{i=1}^{d} |a_i|^q \right)^{1/q}.$$

The inverse inequality is obtained by the lemma and a simple duality argument. As consequence we have that $\overline{\mathrm{span}}\{M_i^* : i \in \mathbb{N}\}$ is isometric to ℓ_q and further

$$\mathfrak{X}[G_p, \sigma]^* / \mathfrak{X}[G_p, \sigma]_* = \ell_q.$$

This yields the proof of the theorem.   □

Proof of Lemma V.5. Assume that $\sum_{i=1}^{d} a_i^p = 1$ and ε, $n_0 \in \mathbb{N}$ are given. Choose $j \in \mathbb{N}$ such that

(i) $\frac{1}{m_j} < \frac{\varepsilon}{2d}$ and

(ii) $\frac{n_{j-1}^{\log_2(m_j - 1)}}{n_j} < \frac{\varepsilon}{2d}$.

Next for $i \leq d$ choose $n_0 \leq \ell_1^i < \ldots < \ell_{n_j}^i$ with

(iii) $J_i = \{\ell_1^i, \ldots, \ell_{n_j}^i\} \subset M_i$

(iv) $\ell_t^1 < \ell_t^2 < \ldots < \ell_t^d < \ell_{t+1}^1 < \ldots$, $t = 1, \ldots, n_j$.

We set $x_i = \frac{1}{n_j} \sum_{k=1}^{n_j} e_{\ell_k^i}$ and we show that

$$\left\| \sum_{i=1}^{d} a_i x_i \right\| \leq 1 + \varepsilon.$$

Since the norming set D of the space $\mathfrak{X}[G_p, \sigma]$ is a subset of the set $W_p = W[G_p, \left(A_{n_j}, \frac{1}{m_j} \right)_j]$ it suffices to show that $f(\sum_{i=1}^{d} a_i x_i) \leq 1 + \varepsilon$ for every $f \in W_p$ of type I. Let $f \in W_p$ of type I. Using similar arguments as in Lemma II.9 we may assume that there exists a tree $T_f = (f_a)_{a \in \mathcal{A}}$ of f such that each f_a is not of type II (Definition II.12). Let $(g_{a_s})_{s=1}^{s_0}$ be the functionals corresponding to the maximal elements of the tree $\mathcal{A}$. We denote by $\preceq$ the ordering of the tree $\mathcal{A}$. Let

$$A = \left\{ s \in \{1, 2, \ldots, s_0\} : \prod_{\gamma \prec a_s} \frac{1}{w(f_\gamma)} \leq \frac{1}{m_j} \right\}$$

$$B = \{1, 2, \ldots, s_0\} \setminus A$$

and set $f_A = f|_{\bigcup_{s \in A} \mathrm{supp}\, g_{a_s}}$, $f_B = f|_{\bigcup_{s \in B} \mathrm{supp}\, g_{a_s}}$.

We have $f_A(x_i) \le \dfrac{1}{m_j}$ for each i thus

$$f_A(\sum_{i=1}^{d} a_i x_i) \le \frac{1}{m_j} \sum_{i=1}^{d} |a_i| \le \frac{1}{m_j} \cdot d < \frac{\varepsilon}{2}. \qquad \text{(V.1)}$$

It remains to estimate the value $f_B(\sum_{i=1}^{d} a_i x_i)$. We observe that

$$\sum_{i=1}^{d} a_i x_i = \sum_{i=1}^{d} a_i \Big(\sum_{t=1}^{n_j} \frac{e_{l_t^i}}{n_j}\Big) = \sum_{t=1}^{n_j} \frac{1}{n_j}\Big(\sum_{i=1}^{d} a_i e_{l_t^i}\Big).$$

Set

$$E_1 = \Big\{t \in \{1,2,\ldots,n\}: \quad \text{the set } \{l_t^1, l_t^2, \ldots, l_t^d\} \text{ is contained in } \operatorname{ran} g_{a_s} \text{ for}$$

$$\text{some } s \in B \text{ or does not intersect any } \operatorname{ran} g_{a_s}, \quad s \in B\Big\}$$

$$E_2 = \{1,2,\ldots,n\} \setminus E_1.$$

For each $s = 1,2,\ldots,s_0$ set $\theta_s = \sum \Big\{\frac{1}{n_j}: \ \{l_t^1, l_t^2, \ldots, l_t^d\} \subset \operatorname{ran} g_{a_s}\Big\}$ and observe that $\sum_{s \in B} \theta_s \le 1$.

We first estimate the quantity $g_{a_s}\big(\sum_{t \in E_1} \frac{1}{n_j}(\sum_{i=1}^{d} a_i e_{l_t^i})\big)$ for $s \in B$. We may assume that g_{a_s} is of the form $g_{a_s} = \sum_i \sum_j c_{i,j} s_{i,j}^*$ where $\{s_{i,j}: i,j\}$ is a family of pairwise disjoint segments with each $s_{i,j} \subset b_i$ and $\sum_i \sum_j |c_{i,j}|^q \le 1$. For each $i' = 1,2,\ldots,d$ we get that $\big(\sum_j c_{i',j} s_{i',j}^*\big)\big(\sum_{t \in E_1} \frac{1}{n_j}(\sum_{i=1}^{d} a_i e_{l_t^i})\big) = a_{i'}\big(\sum_j c_{i',j} s_{i',j}^*\big)\big(\sum_{t \in E_1} \frac{1}{n_j} e_{l_t^{i'}}\big)$ $\le |a_{i'}| \max_j |c_{i',j}| \theta_s$. Thus

$$g_{a_s}\big(\sum_{t \in E_1} \frac{1}{n_j}(\sum_{i=1}^{d} a_i e_{l_t^i})\big) \le \theta_s \sum_{i=1}^{d}(\max_j |c_{i,j}|)|a_i| \le \theta_s\big(\sum_{i=1}^{d} \max_j |c_{i,j}|^q\big)^{\frac{1}{q}}\big(\sum_{i=1}^{d}|a_i|^p\big)^{\frac{1}{p}} \le \theta_s.$$

Therefore

$$f_B\big(\sum_{t \in E_1} \frac{1}{n_j}(\sum_{i=1}^{d} a_i e_{l_t^i})\big) \le \big(\sum_{s \in B} g_{a_s}\big)\big(\sum_{t \in E_1} \frac{1}{n_j}(\sum_{i=1}^{d} a_i e_{l_t^i})\big) \le \sum_{s \in B} \theta_s \le 1. \qquad \text{(V.2)}$$

By the definition of the set B we may prove, as in the proof of Lemma II.9, that the family of functionals $\{g_{a_s}: \ s \in B\}$ has cardinality less or equal $n_{j-1}^{\log_2(m_j-1)}$. By the definition of the set E_2 for $t \in E_2$ the set $\{l_t^1, l_t^2, \ldots, l_t^d\}$

intersects at least one but is not contained in any ran g_{a_s}, $s \in B$. This easily yields that

$$\#\{l_t^1 : t \in E_2\} \le n_{j-1}^{\log_2(m_j-1)}$$

and thus

$$\sum_{t \in E_2} \frac{1}{n_j} < \frac{n_{j-1}^{\log_2(m_j-1)}}{n_j} < \frac{\varepsilon}{2d}.$$

Therefore

$$f_B\Big(\sum_{t \in E_2} \frac{1}{n_j}\Big(\sum_{i=1}^d a_i e_{l_t^i}\Big)\Big) \le \Big(\sum_{t \in E_2} \frac{1}{n_j}\Big)\Big(\sum_{i=1}^d |a_i|\Big) < \frac{\varepsilon}{2d} \cdot d = \frac{\varepsilon}{2}. \qquad (V.3)$$

From (V.1),(V.2) and (V.3), we conclude that

$$
\begin{aligned}
f\Big(\sum_{i=1}^d a_i x_i\Big) &\le f_A\Big(\sum_{i=1}^d a_i x_i\Big) + f_B\Big(\sum_{t \in E_1} \frac{1}{n_j}\Big(\sum_{i=1}^d a_i e_{l_t^i}\Big)\Big) + f_B\Big(\sum_{t \in E_2} \frac{1}{n_j}\Big(\sum_{i=1}^d a_i e_{l_t^i}\Big)\Big) \\
&\le \frac{\varepsilon}{2} + 1 + \frac{\varepsilon}{2} = 1 + \varepsilon.
\end{aligned}
$$

$\square$

Remark. In a similar manner we could show that c_0 is a quotient of a HI space. The corresponding ground set $G_0^{\mathcal{L}}$ is defined as

$$G_0^{\mathcal{L}} = \{\pm s^* : s \text{ is a segment of } (\mathcal{L}, \prec)\}$$

and $\mathfrak{X}[G_0^{\mathcal{L}}, \sigma]$ has as quotient c_0 and also $\mathfrak{X}^*[G_0^{\mathcal{L}}, \sigma] \cong \mathfrak{X}_*[G_0^{\mathcal{L}}, \sigma] \oplus \ell_1$. The well-known lifting property of ℓ_1 does not permit us to have a similar result for ℓ_1.

V.3   A Non Separable HI Space

In this part we shall provide a ground set G_{ns} with the property that $\mathfrak{X}[G_{ns}, \sigma]^*$ is a non separable HI space. This requires more effort than the previous examples.

The ground set G_{ns} and the space $\mathfrak{X}[G_{ns}, \sigma]$

Let $(\mathcal{D}, \prec)$ denote a reorder of $\mathbb{N}$ as a dyadic tree with the property $n \prec m$ implies that $n < m$. We shall denote by s the segments of $(\mathcal{D}, \prec)$. We define

$$G_{ns} = \Big\{ \sum_{i=1}^d \varepsilon_i s_i^* \ : \ (\varepsilon_i)_{i=1}^d \in \{-1, 1\}^d \text{ and } \{s_i\}_{i=1}^d \text{ are pairwise disjoint}$$

$$\text{finite segments of } (\mathcal{D}, \prec) \text{ and } \min\{\min s_i : i = 1, \ldots, d\} \le d \Big\}.$$

The set G_{ns} is a ground set. The following theorem describes the properties of $Y_{G_{ns}}$.

Theorem V.6. (i) *The space $Y_{G_{ns}}$ is c_0 saturated.*
(ii) *The space $Y_{G_{ns}}$ has uniformly bounded averages.*

The proof of this theorem requires some steps described by the next lemmas.

Lemma V.7. *The closure of the set G_{ns} in the topology of pointwise convergence is*

$$\overline{G_{ns}}^p = \Big\{ \sum_{i=1}^{d} \varepsilon_i s_i^* : (\varepsilon)_{i=1}^d \in \{-1,1\}^d, (s_i)_{i=1}^d \text{ are pairwise disjoint}$$

$$\text{segments with } \{\min s_1, \min s_2, \dots, \min s_d\} \in \mathcal{S}, \ d \in \mathbb{N} \Big\}.$$

Proof. Let $g \in \overline{G_{ns}}^p$ and we show that it is of the form described above (the other inclusion is trivial). Let $(g_k)_{k\in\mathbb{N}}$ be a sequence in G_{ns} such that $g_k \xrightarrow{p} g$. Set $d = \min \operatorname{supp} g$. Since $g_k(n) \xrightarrow{k\to\infty} 0$ for $n = 1, \dots, d-1$ and $g_k(d) \xrightarrow{k\to\infty} g(d) \neq 0$ and from the fact that each $g_k(n)$ belongs to $\{-1, 0, 1\}$ we may assume, passing to a subsequence, that $\min \operatorname{supp} g_k = d$ for all $k \in \mathbb{N}$. Let $g_k = \sum_{i=1}^{d} \varepsilon_i^k (s_i^k)^*$ for $k = 1, 2, \dots$.

Observe that the limit of a sequence of finite segments in the pointwise topology is a segment which can be either finite or infinite. We thus may select $L_1 \in [\mathbb{N}]$, $\varepsilon_1 \in \{-1, 1\}$ and a segment s_1 such that $\varepsilon_1 s_1^*$ is the pointwise limit of the sequence $(\varepsilon_1^k (s_1^k)^*)_{k\in L_1}$. After d consecutive applications of the same argument we may select infinite sets of natural numbers $L_1 \supset L_2 \supset \cdots \supset L_d$, $\varepsilon_1, \varepsilon_2, \dots \varepsilon_d \in \{-1, 1\}$ and disjoint segments $s_1, s_2, \dots, s_d$ such that $(\varepsilon_i^k (s_i^k)^*)_{k\in L_i} \xrightarrow{p} \varepsilon_i s_i^*$ for $i = 1, 2, \dots, d$. We deduce that $g = \sum_{i=1}^{d} \varepsilon_i s_i^*$. $\qquad\square$

We remind the reader at this point that $Y_{G_{ns}}$ denotes the completion of the space $(c_{00}, \| \ \|_{G_{ns}})$.

Proposition V.8. *The space $Y_{G_{ns}}$ does not contain isomorphically ℓ^1.*

Proof. It is enough to show that every bounded sequence in $Y_{G_{ns}}$ has a weakly Cauchy subsequence. Let $(x_n)_{n\in\mathbb{N}}$ be a bounded sequence in $Y_{G_{ns}}$. By Rainwater's Theorem [54] is enough to find a subsequence $(x_n)_{n\in M}$ such that $(f(x_n))_{n\in M}$ is convergent for every $f \in \operatorname{Ext}(B_{Y_{G_{ns}}^*})$. Since $B_{Y_{G_{ns}}^*} = \overline{\operatorname{conv}(G_{ns})}^{w^*}$ by Milman's Theorem we get that $\operatorname{Ext}(B_{Y_{G_{ns}}^*}) \subset \overline{G_{ns}}^{w^*}$. Therefore, by the form of elements of $\overline{G_{ns}}^{w^*}$ it is enough to show that there exists $M \in [\mathbb{N}]$ such that $(s^*(x_n))_{n\in M}$ is convergent for every segment s. This is done in the following two lemmas. $\qquad\square$

Lemma V.9. *Let $(x_n)_{n\in\mathbb{N}}$ be a bounded sequence in $Y_{G_{ns}}$ and $\varepsilon > 0$. Then there exists a finite set $\{s_1, s_2, \dots, s_k\}$ of pairwise disjoint segments and an $L \in [\mathbb{N}]$ such* *that*

$$\limsup_{n \in L} |s^*(x_n)| \leq \varepsilon$$

for every segment s with $s \cap (\bigcup_{i=1}^{k} s_i) = \emptyset$.

Proof. Assume the contrary. Then for every finite set $\{t_1, t_2, \ldots, t_m\}$ of pairwise disjoint segments and every $L \in [\mathbb{N}]$ there exists a segment t with $t \cap (\bigcup_{i=1}^{m} t_i) = \emptyset$ such that $\limsup_{n \in L} |t^*(x_n)| > \varepsilon$.

Using this fact we may inductively construct a sequence $(s_j)_{j \in \mathbb{N}}$ of pairwise disjoint segments of $\mathcal{D}$ and a decreasing sequence $(L_j)_{j \in \mathbb{N}}$ of infinite subsets of the natural numbers such that

$$|s_j^*(x_n)| > \varepsilon \qquad \forall j \in \mathbb{N} \quad \forall n \in L_j.$$

We set $r = \sup\{\|x_n\| : n \in \mathbb{N}\}$ and choose $k \in \mathbb{N}$ with $k > \dfrac{r}{\varepsilon}$. Since the segments $s_1, s_2, s_3, \ldots$ are pairwise disjoint we may choose an $i_0 \in \mathbb{N}$ such that $\min s_i \geq k$ for every $i > i_0$.

Let $n \in L_{i_0+k}$. Then $n \in L_{i_0+t}$ for each $t \in \{1, 2, \ldots, k\}$ i.e. $|s_{i_0+t}^*(x_n)| > \varepsilon$ for each $t \in \{1, 2, \ldots, k\}$. Setting $\varepsilon_t = \operatorname{sgn}(s_{i_0+t}^*(x_n))$ we have that

$$f = \sum_{t=1}^{k} \varepsilon_t s_{i_0+t}^* \in \overline{G_{ns}}^{w^*}$$

thus $f \in B_{Y_{G_{ns}}^*}$. It follows that

$$r \geq \|x_n\| \geq f(x_n) = \sum_{t=1}^{k} \varepsilon_t s_{i_0+t}^*(x_n) = \sum_{t=1}^{k} |s_{i_0+t}^*(x_n)| > k\varepsilon,$$

a contradiction of the choice of k. $\qquad\square$

Lemma V.10. *Let $(x_n)_{n \in \mathbb{N}}$ be a bounded sequence in $Y_{G_{ns}}$. There exists a subsequence $(x_n)_{n \in M}$ of $(x_n)_{n \in \mathbb{N}}$ such that for every segment s the sequence $(s^*(x_n))_{n \in M}$ is convergent.*

Proof. From Lemma V.9 we may inductively construct a decreasing sequence $(L_k)_{k \in \mathbb{N}}$ of infinite subsets of the natural numbers and a sequence $(F_k)_{k \in \mathbb{N}}$ of finite sets of pairwise disjoint segments, $F_k = \{s_1^k, s_2^k, \ldots, s_{m_k}^k\}$, such that for each $k \in \mathbb{N}$ and segment s with $s \cap (\bigcup_{i=1}^{m_k} s_i^k) = \emptyset$ we have that

$$\limsup_{n \in L_k} |s^*(x_n)| < \frac{1}{k}.$$

Let F be the countable set consisting of all the finite segments of $\mathcal{D}$ and all subsegments of segments contained in $\bigcup\limits_{k=1}^{\infty} F_k$. We choose a diagonal set L of the decreasing sequence $(L_k)_{k\in\mathbb{N}}$. Since $\{s^* : s \in F\}$ is a countable subset of $B_{Y^*_{G_{ns}}}$ and the sequence $(x_n)_{n\in L}$ is bounded we may choose, by a diagonal argument, an $M \in [L]$ such that the sequence $(s^*(x_n))_{n\in M}$ is convergent for every $s \in F$. It remains to show that the sequence $(s^*(x_n))_{n\in M}$ is convergent for every segment s.

Let s be a segment. We show that $(s^*(x_n))_{n\in M}$ is a Cauchy sequence. Let $\varepsilon > 0$ and choose $k \in \mathbb{N}$ with $\dfrac{1}{k} < \dfrac{\varepsilon}{4}$. We have

$$s = \bigcup_{i=1}^{m_k}(s_i^k \cap s) \cup (s \setminus \bigcup_{i=1}^{m_k} s_i^k)$$

and $s \setminus \bigcup\limits_{i=1}^{m_k} s_i^k$ can be written as the finite union of pairwise disjoint segments with at most one of them being infinite. We assume that one of them is infinite and let

$$s \setminus \bigcup_{i=1}^{m_k} s_i^k = (\bigcup_{j=1}^{n_k} t_j) \cup t$$

where $t_1, t_2, \ldots, t_{n_k}, t$ are pairwise disjoint segments, t is infinite while $t_1, t_2, \ldots, t_{n_k}$ are finite.

It is clear that $t \cap (\bigcup\limits_{i=1}^{m_k} s_i^k) = \emptyset$ thus $\limsup\limits_{n\in L_k} |t^*(x_n)| < \dfrac{1}{k}$ and by the construction of M as a (subset of a) diagonal set we have that $\limsup\limits_{n\in M} |t^*(x_n)| < \dfrac{1}{k}$. Thus we may choose an $m_a \in M$ such that

$$|t^*(x_n)| < \frac{1}{k} \quad \forall n \in M \text{ with } n \geq m_a.$$

Since $s_i^k \cap s \in F$ for each $i = 1, 2, \ldots, m_k$ and $t_j \in F$ for each $j = 1, 2, \ldots, n_k$, the sequences $\left((s_i^k \cap s)^*(x_n)\right)_{n\in M}$ and $\left(t_j^*(x_n)\right)_{n\in M}$ are convergent. Thus we may select an $m_b \in M$ such that

$$\sum_{i=1}^{m_k} |(s_i^k \cap s)^*(x_m) - (s_i^k \cap s)^*(x_n)| + \sum_{j=1}^{n_k} |t_j^*(x_m) - t_j^*(x_n)| < \frac{\varepsilon}{2}$$

for each $m, n \in M$ with $m > n \geq m_b$.

Therefore for each $m, n \in M$ with $m > n \geq \max\{m_a, m_b\}$ we have that

$$|s^*(x_m) - s^*(x_n)| < \frac{\varepsilon}{2} + \frac{2}{k} < \varepsilon.$$

Thus $(s^*(x_n))_{n\in M}$ is a Cauchy, and hence convergent, sequence.   □

Lemma V.11. *Let* $\varepsilon > 0$ *and* $y \in Y_{G_{ns}}$ *such that* $\min \operatorname{supp} y \geq [\frac{\|y\|}{\varepsilon}] + 1$. *If* $s_1, s_2, \ldots, s_t$ *are pairwise disjoint segments such that* $s_i \subset \operatorname{ran} y$ *and* $|s_i^*(y)| > \varepsilon$ *for each* $i = 1, 2, \ldots, t$ *then* $t \leq \frac{\|y\|}{\varepsilon}$.

Proof. Suppose that $t > \frac{\|y\|}{\varepsilon}$ and choose $A \subset \{1, 2, \ldots, t\}$ with $\#A = [\frac{\|y\|}{\varepsilon}] + 1$. Set $\varepsilon_j = \operatorname{sgn}(s_j^*(y))$ for each $j \in A$ and $f = \sum_{j \in A} \varepsilon_j s_j^*$. We have $f \in \overline{G_{ns}}^p \subset B_{Y_{G_{ns}}^*}$. Indeed, the segments s_j, $j \in A$ are pairwise disjoint and $\min s_j \geq \min \operatorname{supp} y \geq [\frac{\|y\|}{\varepsilon}] + 1 = \#A$. Thus

$$\|y\| \geq f(y) = \sum_{j \in A} \varepsilon_j s_j^*(y) = \sum_{j \in A} |s_j^*(y)| > \varepsilon(\#A) = \varepsilon([\frac{\|y\|}{\varepsilon}] + 1) > \|y\|,$$

a contradiction.

Therefore $t \leq \frac{\|y\|}{\varepsilon}$. $\qquad\qquad\qquad\qquad\qquad\qquad\qquad\qquad\qquad\qquad\qquad$ □

The following also holds.

Lemma V.12. *Let* $(x_n)_n$ *be a bounded block sequence such that* $b^*(x_n) \to 0$ *for every* b *branch of* $\mathcal{D}$. *Then for every* $\varepsilon > 0$ *there exists* $L \in [\mathbb{N}]$ *such that for every segment* s *of* $\mathcal{D}$ *the following holds:*

$$\#\{n \in L : |s^*(x_n)| \geq \varepsilon\} \leq 2.$$

Proof. We set

$$\mathcal{A} = \Big\{ \{k_1, k_2, k_3\} \in [\mathbb{N}]^3 : \quad \text{there exists a segment } s$$
$$\text{such that } |s^*(y_{k_l})| > \varepsilon \text{ for } l = 1, 2, 3 \Big\}$$

and

$$\mathcal{B} = [\mathbb{N}]^3 \setminus \mathcal{A}.$$

It follows from Ramsey's theorem that there exists an $L \in [\mathbb{N}]$ such that $[L]^3 \subset \mathcal{A}$ or $[L]^3 \subset \mathcal{B}$. The conclusion of the lemma is exactly that there exists an $L \in [\mathbb{N}]$ such that $[L]^3 \subset \mathcal{B}$. Therefore it is enough to exclude the possibility of existing a $L \in [\mathbb{N}]$ with $[L]^3 \subset \mathcal{A}$.

Suppose that there exists an $L \in [\mathbb{N}]$, $L = \{l_1 < l_2 < l_3 < \cdots\}$ such that $[L]^3 \subset \mathcal{A}$. We set $r = \sup_n \|y_n\|$. We may assume that $\min L > \frac{2r}{\varepsilon}$.

For each $k \geq 3$ we apply the following procedure. We consider the set

$$T_k = \Big\{ \operatorname{ran}(y_{l_2} + y_{l_{k-1}}) \cap s : s \text{ is a segment }, \text{ such that } |s^*(y_{l_1})| > \varepsilon,$$
$$|s^*(y_{l_k})| > \varepsilon, \text{ and there exists a } t \in \{2, \ldots, k-1\} \text{ such that } |s^*(y_{l_t})| > \varepsilon \Big\}.$$

We claim that $\#T_k \leq \frac{r}{\varepsilon}$. Indeed, let $t_1, t_2, \ldots, t_m$ be pairwise different elements of T_k and for each $i = 1, 2, \ldots, m$ let s_i be a segment such that $\mathrm{ran}(y_{l_2} + y_{l_{k-1}}) \cap s_i = t_i$ and $|s_i(y_{l_k})| > \varepsilon$. The segments $r_i = (\mathrm{ran}\, y_{l_k}) \cap s_i$ $i = 1, 2, \ldots, m$ are pairwise disjoint. We also have that $|r_i^*(y_{l_k})| > \varepsilon$ and $r_i \subset \mathrm{ran}\, y_{l_k}$ for each $i = 1, 2, \ldots, m$. From the fact that $\|y_{l_k}\| \geq |r_i^*(y_{l_k})| > \varepsilon$ we get that

$$\min \mathrm{supp}\, y_{l_k} \geq l_k > l_1 > \frac{2r}{\varepsilon} \geq \frac{2\|y_{l_k}\|}{\varepsilon} > [\frac{\|y_{l_k}\|}{\varepsilon}] + 1.$$

We easily get that $m \leq \frac{\|y_{l_k}\|}{\varepsilon}$, thus $m \leq \frac{r}{\varepsilon}$. We conclude that $\#T_k \leq \frac{r}{\varepsilon}$. We set $m_0 = [\frac{r}{\varepsilon}]$. Then

$$T_k = \{s_1^k, s_2^k, \ldots, s_{t_k}^k\}$$

for some $t_k \leq m_0$.
For $k \geq 3$ and $i = 1, 2, \ldots, t_k$ we set

$$A_i^k = \Big\{ j \in \{2, \ldots, k-1\} : |(s_i^k)^*(y_{l_j})| > \varepsilon \Big\}$$

while for $t_k < i \leq m_0$ we set $A_i^k = \emptyset$. Hence

$$\{2, \ldots, k-1\} = \bigcup_{i=1}^{m_0} A_i^k. \tag{V.4}$$

We consider the following partition of $[\mathbb{N} \setminus \{1\}]^2$. For $j = 1, 2, \ldots, m_0$ we set

$$B_j = \Big\{ \{p, q\} \in [\mathbb{N} \setminus \{1\}]^2 : p < q \text{ and } p \in A_j^q \Big\}.$$

Observe that (V.4) yields that $[\mathbb{N} \setminus \{1\}]^2 = \bigcup_{j=1}^{m_0} B_j$; thus by Ramsey's theorem there exist an $M \in [\mathbb{N} \setminus \{1\}]$ and a $j_0 \in \{1, 2, \ldots, m_0\}$ such that $[M]^2 \subset B_{j_0}$.

Let $M = \{m_1 < m_2 < m_3 < \cdots\}$. Then for each $t \in \mathbb{N}$ and $i \leq t$ we have that $\{m_i, m_{t+1}\} \in B_{j_0}$, thus $m_i \in A_{j_0}^{m_{t+1}}$. So there exists a segment s_t such that

$$|s_t^*(y_{m_i})| > \varepsilon \qquad \forall i = 1, 2, \ldots, t.$$

The sequence $(s_t^*)_{t \in \mathbb{N}}$ has a w^* convergent subsequence; its w^* limit is of the form s^* for some infinite segment s. We deduce that $|s^*(y_{m_i})| > \varepsilon$ for all i which contradicts to the assumption that the sequence $(y_n)_{n \in \mathbb{N}}$ is weakly null. $\square$

The next result uses the above lemmas.

Lemma V.13. *For every Z block subspace of $Y_{G_{ns}}$ and every $\varepsilon > 0$ there exists $z \in Z$ with $\|z\| = 1$ and $|s^*(z)| < \varepsilon$ for every s segment of $(\mathcal{D}, \prec)$.*

Proof. Assume that the conclusion fails. Then there exists a Z block subspace of $Y_{G_{ns}}$ and $\varepsilon_0 > 0$ such that for every $z \in Z$ there exists a segment s of $\mathcal{D}$ with $s^*(z) \geq \varepsilon_0 \|z\|$. Choose $n_0 > 4/e_0$ and let $(y_n)_{n \in \mathbb{N}}$ be a normalized weakly null block sequence in Z. Lemma V.12 yields that we may assume that for every segment s of $\mathcal{D}$

$$\#\{n \in \mathbb{N} : s^*(x_n) \geq \frac{\varepsilon_0}{n}\} \leq 2. \tag{V.5}$$

Select $y_m, \ldots, y_{m+n_0}$ with $n_0 < y_m$. Further for every $m \leq k \leq m+n_0$ choose a segment s_k of $\mathcal{D}$ with $\operatorname{supp} s_k \subset \operatorname{ran}(y_k)$ and $s_k^*(y_k) \geq \varepsilon_0$. Since $\min\{\min \operatorname{supp} s_k^*\} > n_0$ the family $\{\min \operatorname{supp} s_k^*\}_{k=1}^{n_0}$ is Schreier admissible, hence $\sum_{k=1}^{n_0} s_k^* \in G_{ns}$. Therefore

$$\|\sum_{k=1}^{n_0} y_{m+k}\| \geq (\sum_{k=1}^{n_0} s_k^*) \left(\sum_{k=1}^{n_0} y_{m+k}\right) \geq n_0 \varepsilon_0 > 4. \tag{V.6}$$

Also from (V.5) for a segment s of $\mathcal{D}$

$$|s^*(\sum_{k=1}^{n_0} y_{m+n})| \leq 3\varepsilon_0. \tag{V.7}$$

(V.6) and (V.7) derive a contradiction completing the proof.   □

Proof of Theorem V.6. (i) Let Z be a block subspace of $Y_{G_{ns}}$ and let $\varepsilon > 0$. Using Lemma V.13 we may inductively select a normalized block sequence $(y_n)_{n \in \mathbb{N}}$ in Y such that, setting $d_n = \max \operatorname{supp} y_n$ for each n and $d_0 = 1$, $|x^*(y_n)| < \frac{\varepsilon}{2^n d_{n-1}}$ for every σ_F special functional x^*.

We claim that $(y_n)_{n \in \mathbb{N}}$ is $1 + \varepsilon$ isomorphic to the standard basis of c_0. Indeed, let $(\beta_n)_{n=1}^N$ be a sequence of scalars. We shall show that $\max_{1 \leq n \leq N} |\beta_n| \leq$ $\|\sum_{n=1}^N \beta_n y_n\|_{G_{ns}} \leq (1+\varepsilon) \max_{1 \leq n \leq N} |\beta_n|$. We may assume that $\max_{1 \leq n \leq N} |\beta_n| = 1$. The left inequality follows directly from the bimonotonicity of the Schauder basis $(e_n)_{n \in \mathbb{N}}$ of $Y_{G_{ns}}$. To see the right inequality we consider a $g \in G_{ns}$, $g = \sum_{i=1}^d a_i s_i^*$, where $(s_i)_{i=1}^d$ are finite segments with $\min \operatorname{supp} s_i \geq d$. Let n_0 be the minimum integer n such that $d \leq d_n$. Since $\min \operatorname{supp} g \geq d > d_{n_0 - 1}$ we get that $g(y_n) = 0$ for $n < n_0$. Therefore

$$g(\sum_{n=1}^N \beta_n y_n) \leq |g(y_{n_0})| + \sum_{n=n_0+1}^N |g(y_n)| \leq 1 + \sum_{n=n_0+1}^N \sum_{i=1}^d |s_i^*(y_n)|$$

$$< 1 + \sum_{n=n_0+1}^N d\frac{\varepsilon}{2^n d_{n-1}} < 1 + \sum_{n=n_0+1}^N \frac{\varepsilon}{2^n} < 1 + \varepsilon.$$

(ii) We pass now to show that $Y_{G_{ns}}$ has uniformly bounded averages. Let $0 < \varepsilon < 1$ and $n \in \mathbb{N}$. We set $k_0 = k(n, \varepsilon) > 3n/\varepsilon$ and we claim that k_0 satisfies the requirements of Definition III.13.

Indeed, let $(y_l)_l$ be a normalized weakly null block sequence in $Y_{G_{ns}}$. It follows from Lemma V.12 that we can assume the following:

For every segment s of $\mathcal{D}$,

$$\#\{l : |s^*(y_l)| > \frac{\varepsilon}{k_0^2}\} \leq 2.$$

Consider $g \in G_{ns}$, $g = \sum_{j=1}^d s_j^*$ with $\min g < n$. This yields that $d < n$. Consider also $x = \frac{1}{k_0} \sum_{i=1}^{k_0} y_{l_i}$, $l_1 < \ldots < l_{k_0}$. Then

$$|g(x)| \leq \sum_{j=1}^d |s_j^*(x)| \leq \sum_{i=1}^d \frac{3}{k_0} = \frac{3d}{k_0} < \frac{3n}{k_0} < \varepsilon,$$

and this completes the proof of (ii) and the entire proof of the theorem. $\qquad\square$

Proposition V.14. *Let $\mathcal{B}$ denote the set of all branches of the binary tree $\mathcal{D}$. Then we have the following:*

(i) $\mathfrak{X}[G_{ns}, \sigma]^* = \overline{\mathrm{span}}(\{e_n^* : n \in \mathbb{N}\} \cup \{b^* : b \in \mathcal{B}\})$.

(ii) $\mathfrak{X}[G_{ns}, \sigma]^* / \mathfrak{X}[G_{ns}, \sigma]_* - \overline{\mathrm{span}}\{b^* \mid \mathfrak{X}[G_{ns}, \sigma]_* : b \in \mathcal{B}\} = c_0(\mathcal{B})$.

(iii) *The space $\mathfrak{X}[G_{ns}, \sigma]^{**}$ is isomorphic to $\mathfrak{X}[G_{ns}, \sigma] \oplus \ell_1(\mathcal{B})$.*

Proof. (i) It is easy to see that the set $\overline{G_{ns}}^{w^*}$ is equal to the set

$$\overline{G_{ns}}^p = \{\sum_{i=1}^d \varepsilon_i s_i^* : \varepsilon_i \in \{-1, 1\}, \ i = 1, 2, \ldots, d, \ (s_i)_{i=1}^d \text{ are pairwise}$$

disjoint segments and $\{\min s_1, \min s_2, \ldots, \min s_d\} \in \mathcal{S}_1, d \in \mathbb{N}\}$.

Therefore, from Proposition II.26 we get that $\mathfrak{X}[G_{ns}, \sigma]^* = \overline{\mathrm{span}}(\{e_n^* : n \in \mathbb{N}\} \cup \{b^* : b \in \mathcal{B}\})$.

(ii) By Remark II.29 the basis $(e_n)_{n \in \mathbb{N}}$ is boundedly complete and the space $\mathfrak{X}[G_{ns}, \sigma]_* = \overline{\mathrm{span}}\{e_n^* : n \in \mathbb{N}\}$ is the predual of $\mathfrak{X}[G_{ns}, \sigma]$. It follows from (i) that $\mathfrak{X}[G_{ns}, \sigma]^* / \mathfrak{X}[G_{ns}, \sigma]_* = \overline{\mathrm{span}}\{b^* + \mathfrak{X}[G_{ns}, \sigma]_* : b \in \mathcal{B}\}$ so it remains to show that $\overline{\mathrm{span}}\{b^* + (\mathfrak{X}[G_{ns}, \sigma])_* : b \in \mathcal{B}\} = c_0(\mathcal{B})$. It is enough to show that if $b_1, b_2, \ldots, b_n$ are pairwise different elements of $\mathcal{B}$ and $\varepsilon_1, \varepsilon_2, \ldots, \varepsilon_n \in \{-1, 1\}$ then

$$\|\varepsilon_1(b_1^* + \mathfrak{X}[G_{ns}, \sigma]_*) + \varepsilon_2(b_2^* + \mathfrak{X}[G_{ns}, \sigma]_*) + \cdots + \varepsilon_n(b_n^* + \mathfrak{X}[G_{ns}, \sigma]_*)\| = 1.$$

For $k = 1, 2, \ldots$ we denote by E_k the interval $E_k = \{n \in \mathbb{N} : n \geq 2^k\}$. Since $b_1, b_2, \ldots, b_n$ are pairwise different we may select a $k_0 \geq n$ such that $b_1 \cap E_k, b_2 \cap E_k, \ldots, b_n \cap E_k$ are incomparable segments for all $k \geq k_0$. Thus

$$
\begin{aligned}
\| \varepsilon_1 & (b_1^* + \mathfrak{X}[G_{ns}, \sigma]_*) + \cdots + \varepsilon_n(b_n^* + \mathfrak{X}[G_{ns}, \sigma]_*) \| \\
&= \operatorname{dist}(\varepsilon_1 b_1^* + \cdots + \varepsilon_n b_n^*, \ \mathfrak{X}[G_{ns}, \sigma]_*) \\
&= \operatorname{dist}(\varepsilon_1 b_1^* + \cdots + \varepsilon_n b_n^*, \ \overline{\operatorname{span}}\{e_j^* : \ j \in \mathbb{N}\}) \\
&= \operatorname{dist}(\varepsilon_1 b_1^* + \cdots + \varepsilon_n b_n^*, \ \operatorname{span}\{e_j^* : \ j \in \mathbb{N}\}) \\
&= \lim_k \operatorname{dist}(\varepsilon_1 b_1^* + \cdots + \varepsilon_n b_n^*, \ \operatorname{span}\{e_j^* : \ j \leq 2^k - 1\}) \\
&= \lim_k \| \varepsilon_1 (b_1 \cap E_k)^* + \cdots + \varepsilon_n (b_n \cap E_k)^* \| \\
&= 1.
\end{aligned}
$$

The last equality holds since for every $k \geq k_0$ the functional $g_k = \sum_{i=1}^{n} \varepsilon_i (b_i \cap E_k)^*$ belongs to G_{ns}, thus $\|g_k\| \leq 1$, while for $n \in b_1 \cap E_k$ we have that $|g_k(e_n)| = 1$.

(iii) As is well known, for every Banach space Z the space Z^{***} is isomorphic to $Z^* \oplus (Z^{**}/Z)^*$. For $Z = \mathfrak{X}[G_{ns}, \sigma]_*$ we get that the space $\mathfrak{X}[G_{ns}, \sigma]^{**}$ is isomorphic to the space
$$
\mathfrak{X}[G_{ns}, \sigma] \oplus (\mathfrak{X}[G_{ns}, \sigma]^*/\mathfrak{X}[G_{ns}, \sigma]_*)^* = \mathfrak{X}[G_{ns}, \sigma] \oplus (c_0(\mathcal{B}))^* = \mathfrak{X}[G_{ns}, \sigma] \oplus \ell_1(\mathcal{B}).
$$
$\qquad\square$

Theorem V.15. *The space $\mathfrak{X}[G_{ns}, \sigma]^*$ is a non separable HI Banach space.*

Proof. The space $\mathfrak{X}[G_{ns}, \sigma]$, being HI, contains no isomorphic copy of ℓ_1, the space $\mathfrak{X}[G_{ns}, \sigma]_* = \overline{\operatorname{span}}\{e_n^* : n \in \mathbb{N}\}$ is HI and the quotient space $\mathfrak{X}[G_{ns}, \sigma]^*/\mathfrak{X}[G_{ns}, \sigma]_*$ is isometric to $c_0(\mathcal{B})$. From Corollary IV.8 we deduce that $\mathfrak{X}[G_{ns}, \sigma]^*$ is HI while it is clear that $\mathfrak{X}[G_{ns}, \sigma]^*$ is non separable. $\qquad\square$

Remark. The three examples presented in this chapter are from [16] where they have been constructed with the use of higher complexity saturation methods and have the additional property that they are asymptotic ℓ_1 spaces. In particular for the space corresponding to $\mathfrak{X}[G_{ns}, \sigma]^*$, it is shown that every bounded linear operator is of the form $\lambda I + W$ with W weakly compact. Since the space $\mathfrak{X}[G_{ns}, \sigma]^*$ is the dual of a separable space we conclude that the weakly non-compact operator W has separable range. We do not include this property here as it requires much more efforts.

Chapter VI

The Space $\mathfrak{X}_{\omega_1}$

In this part we present a reflexive Banach space $\mathfrak{X}_{\omega_1}$ with a transfinite basis $(e_\alpha)_{\alpha<\omega_1}$ not containing any unconditional basic sequence. As it is well known every non separable reflexive Banach space is decomposable. In particular it admits a resolution of the identity. Hence any non separable reflexive space not containing an unconditional basic sequence is not HI. On the other side Gowers dichotomy yields that such a space is HI saturated. Constructing the space $\mathfrak{X}_{\omega_1}$ we shall follow the general approach we have developed in the previous parts, namely $\mathfrak{X}_{\omega_1}$ will be $\mathfrak{X}_{\omega_1}[D_{G_0}]$, where $G_0 = \{\pm c_\alpha : \alpha < \omega_1\}$, D_{G_0} is a tree complete set and it will be closed for all $(\mathcal{A}_{n_{2j+1}}, \frac{1}{m_{2j+1}})$ operations acting on n_{2j+1}-special sequences. The main difficulty that we have to overcome is that defining the n_{2j+1}-special sequences we are not able to use a one-to-one coding σ, a crucial property for the ω constructions. To solve this problem we employ Todorcevic's ρ-function. Thus changing one of the fundamental ingredients of the ω-construction (i.e. the injection of the coding σ) we arrive to a non separable reflexive space where we control the structure of all separable subspaces of it.

The norming set K_{ω_1}

The space $\mathfrak{X}_{\omega_1}$ will be defined as the completion of $(c_{00}(\omega_1), \|\cdot\|_*)$ under the norm $\|\cdot\|_*$ induced by a set of functionals $K_{\omega_1} \subseteq c_{00}(\omega_1)$.

The set K_{ω_1} is the minimal subset of $c_{00}(\omega_1)$ satisfying that:

(1) It contains $(e_\gamma^*)_{\gamma<\omega_1}$, is symmetric (i.e., $\phi \in K$ implies $-\phi \in K$) and is closed under the restriction on intervals of ω_1.

(2) For every $\{\phi_i : i = 1,\ldots,n_{2j}\} \subseteq K_{\omega_1}$ with $\operatorname{supp}\phi_1 < \cdots < \operatorname{supp}\phi_{n_{2j}}$, it holds that $\phi = (1/m_{2j})\sum_{i=1}^{n_{2j}} \phi_i \in K_{\omega_1}$. We say that ϕ is a result of a $(\mathcal{A}_{n_{2j}}, \frac{1}{m_{2j}})$-operation.

(3) For every special sequence $(\phi_1,\ldots,\phi_{n_{2j+1}})$ (see later for the definition), the

functional $\phi = (1/m_{2j+1}) \sum_{i=1}^{n_{2j+1}} \phi_i$ is in K_{ω_1}. We call such a ϕ *special functional* and say that ϕ is a result of a $(\mathcal{A}_{n_{2j+1}}, \frac{1}{m_{2j+1}})$-operation.

(4) It is rationally convex.

Remark VI.1. From the definition of the norming set K_{ω_1} it follows easily that $(e_\alpha)_{\alpha<\omega_1}$ is a bimonotone basis of $\mathfrak{X}_{\omega_1}$. Also, it is not difficult to see using (2) from the definition of K_{ω_1} that the basis $(e_\alpha)_{\alpha<\omega_1}$ is boundedly complete. Indeed, for $x \in c_{00}(\omega_1)$ and $E_1 < \cdots < E_{n_{2j}}$ intervals of ω_1, property (2) of the norming set K_{ω_1} yields that $\|x\| \geq (1/m_{2j}) \sum_{i=1}^{n_{2j}} \|E_i x\|$. Also, from the choice of the sequence $(m_j)_j$, $(n_j)_j$, it follows that n_{2j}/m_{2j} increases to infinity. From these observations it follows that the basis $(e_\alpha)_{\alpha<\omega_1}$ is boundedly complete. To prove that the space $\mathfrak{X}_{\omega_1}$ is reflexive we need to show that the basis is shrinking.

The definition of the special sequences, as in the spaces $\mathfrak{X}[G,\sigma]$, depends crucially on a certain coding σ_ϱ. The essential difference is that now σ_ϱ is not an injection, a crucial property on which the proofs in the case of $\mathfrak{X}[G,\sigma]$ rely (the injectivity of σ is required for the proof of Proposition III.3). Our proofs on the other hand will rely on a "tree-like property" of our coding which we now describe. First we notice that each $2j+1$-special sequence $\Phi = (\phi_1, \phi_2, \ldots, \phi_{n_{2j+1}})$ is of the form $\operatorname{supp} \phi_1 < \cdots < \operatorname{supp} \phi_{n_{2j+1}}$ with each ϕ_i of type I. The *tree-like property* is the following: For any pair of $2j+1$-special sequences $\Phi = (\phi_1, \phi_2, \ldots, \phi_{n_{2j+1}})$, $\Psi = (\psi_1, \psi_2, \ldots, \psi_{n_{2j+1}})$ there exist $1 \leq \kappa_{\Phi,\Psi} \leq \lambda_{\Phi,\Psi} \leq n_{2j+1}$ such that

(i) If $1 \leq k < \kappa_{\Phi,\Psi}$ then $\phi_k = \psi_k$ and if $\kappa_{\Phi,\Psi} < k < \lambda_{\Phi,\Psi}$, then $w(\phi_k) = w(\psi_k)$.

(ii) $(\cup_{\kappa_{\Phi,\Psi}<k<\lambda_{\Phi,\Psi}} \operatorname{supp} \phi_k) \cap (\cup_{\kappa_{\Phi,\Psi}<k<\lambda_{\Phi,\Psi}} \operatorname{supp} \psi_k) = \emptyset$.

(iii) $\{w(\phi_k) : \lambda_{\Phi,\Psi} < k \leq n_{2j+1}\} \cap \{w(\psi_k) : \lambda_{\Phi,\Psi} < k \leq n_{2j+1}\} = \emptyset$.

Comparing the above tree-like property with the corresponding property in $\mathfrak{X}[G,\sigma]$ (Proposition III.3), we notice that the new ingredient is the number $\kappa_{\Phi,\Psi}$. Its occurrence is a byproduct of the fact that the coding σ_ϱ is not one-to-one. Property (ii) will however give a sufficient control of our special functionals. The coding σ_ϱ is based on the ϱ-functions introduced by S. Todorcevic.

ϱ-functions

A function $\varrho : [\omega_1]^2 \to \omega$ such that:

1. $\varrho(\alpha, \gamma) \leq \max\{\varrho(\alpha, \beta), \varrho(\beta, \gamma)\}$ for all $\alpha < \beta < \gamma < \omega_1$,

2. $\varrho(\alpha, \beta) \leq \max\{\varrho(\alpha, \gamma), \varrho(\beta, \gamma)\}$ for all $\alpha < \beta < \gamma < \omega_1$,

3. $\{\alpha < \beta : \varrho(\alpha, \beta) \leq n\}$ is finite for all $\beta < \omega_1$ and $n \in \mathbb{N}$,

is called a ϱ-function. Later we shall give yet another construction of a ϱ-function with a certain universality property.

Let $\varrho : [\omega_1]^2 \to \omega$ be a ϱ-function fixed from now on, and all definitions and facts that follow should be relative to this choice of ϱ.

Definition VI.2. Recall that given a finite set $F \subseteq \omega_1$, we let $p_F = p_\varrho(F) = \max_{\alpha,\beta \in F} \varrho(\alpha, \beta)$.

For a finite set $F \subseteq \omega_1$ and $p \in \mathbb{N}$, let

$$\overline{F}^p = \{\alpha \leq \max F : \text{there is } \beta \in F \text{ s.t. } \alpha \leq \beta \text{ and } \varrho(\alpha, \beta) \leq p\}.$$

Notice that by condition 3., $\overline{F}^p$ is a finite set of countable ordinals. We say that F is *p-closed* iff $\overline{F}^p = F$, and that F is ϱ-closed iff it is p_F-closed.

Remark VI.3. 1. Note that $\overline{\cdot}^p$ is a monotone and idempotent operator and so, in particular, every $\overline{F}^p$ is a p-closed set: It is clear that if $F \subseteq G$, then $\overline{F}^p \subseteq \overline{G}^p$. Let us show now that $\overline{\overline{F}^p}^p = \overline{F}^p$. Let $\alpha \in \overline{\overline{F}^p}^p$. This implies that $\varrho(\alpha, \alpha_0) \leq p$, for some $\alpha_0 \in \overline{F}^p$, $\alpha \leq \alpha_0$. Choose $\alpha_1 \geq \alpha_0$, $\alpha_1 \in F$ such that $\varrho(\alpha_0, \alpha_1) \leq p$. Then, $\varrho(\alpha, \alpha_1) \leq \max\{\varrho(\alpha, \alpha_0), \varrho(\alpha_0, \alpha_1)\} \leq p$.

2. Suppose that $F \subseteq \omega_1$ is finite and suppose that $p \geq p_F$. Then $p_{\overline{F}^p} \leq p$: Fix F finite, and $p \geq p_F$. Suppose that $\alpha < \beta$ are both in $\overline{F}^p$. Let $\alpha' \geq \alpha$, $\beta' \geq \beta$ such that $\alpha, \beta' \in F$ and $\varrho(\alpha, \alpha'), \varrho(\beta, \beta') \leq p$. Then one of the following cases occurs:

(a) If $\alpha \leq \alpha' \leq \beta \leq \beta'$, then

$$\varrho(\alpha, \beta) \leq \max\{\varrho(\alpha, \alpha'), \varrho(\alpha', \beta)\} \leq \max\{\varrho(\alpha, \alpha'), \varrho(\alpha', \beta'), \varrho(\beta, \beta')\} < p.$$

(b) If $\alpha \leq \beta \leq \alpha' \leq \beta'$, then

$$\varrho(\alpha, \beta) \leq \max\{\varrho(\alpha, \alpha'), \varrho(\beta, \alpha')\} \leq \max\{\varrho(\alpha, \alpha'), \varrho(\beta, \beta'), \varrho(\alpha', \beta')\} \leq p.$$

(c) If $\alpha \leq \beta \leq \beta' < \alpha'$, use a similar proof to case (a).

Proposition VI.4. *Let* $F, G \subseteq \omega_1$ *be two finite sets and* $p \geq p_F, p_G$. *Then:*

1. *For every ordinal* $\alpha \leq \omega_1$, $\overline{F \cap \alpha}^p = \overline{F}^p \cap \alpha$ *and* $\overline{F \cap \alpha}^p$ *is an initial part of* $\overline{F}^p$. *Therefore, if* F *is p-closed, so is* $F \cap \alpha$.

2. *For every* $\alpha \in F \cap G$, *we have that* $\overline{F \cap (\alpha + 1)}^p = \overline{G \cap (\alpha + 1)}^p$. *Hence, if* F *and* G *are in addition p-closed, then* $F \cap (\alpha + 1) = G \cap (\alpha + 1)$.

3. $\overline{F \cap G}^p = \overline{F}^p \cap \overline{G}^p$. *Therefore, if* F *and* G *are p-closed then* $F \cap G$ *is also p-closed and it is an initial part of both* F *and* G.

Proof. 1. Since $F \cap \alpha \subseteq F, \alpha$, it follows that $\overline{F \cap \alpha}^p \subseteq \overline{F}^p \cap \alpha$. Now let $\beta \in \overline{F}^p \cap \alpha$. Then there is some $\gamma \in F$, $\gamma \geq be$ such that $\varrho(\beta, \gamma) \leq p$. If $\gamma < \alpha$, then we are done. If not, let $\delta = \max F \cap \alpha \in F$ and since $\beta \leq \delta < \gamma$ we have that

$$\varrho(\beta, \delta) \leq \max\{\varrho(\beta, \gamma), \varrho(\delta, \gamma)\} \leq \max\{p, p_F\} = p, \tag{VI.1}$$

the last equality using our assumption that $p \geq p_F$. (VI.1) shows that $\beta \in \overline{F \cap \alpha}^p$. Suppose now that F is p-closed. Then we have just shown that $\overline{F \cap \alpha}^p = \overline{F}^p \cap \alpha = F \cap \alpha$, and we are done.

2. Fix $\alpha \in F \cap G$. Let $\beta \in \overline{F \cap (\alpha + 1)}^p = \overline{F}^p \cap (\alpha+1)$. Let $\gamma \in F \cap (\alpha+1)$, $\gamma \geq \beta$ be such that $\varrho(\beta, \gamma) \leq p$. Then $\varrho(\beta, \alpha) \leq \max\{\varrho(\beta, \gamma), \varrho(\gamma, \alpha)\} \leq \max\{p, p_F\} = p$. Since G is p-closed, and $\alpha \in G$, we can conclude that $\beta \in \overline{G \cap (\alpha + 1)}^p$. This shows that $\overline{F \cap (\alpha + 1)}^p \subseteq \overline{G \cap (\alpha + 1)}^p$. The other inclusion follows by symmetry. The last part of 2. follows easily.

3. Let $\alpha = \max F \cap G$. Then by 2., $\overline{F \cap G}^p = \overline{F \cap G \cap (\alpha + 1)}^p = \overline{F \cap (\alpha + 1)}^p = \overline{F}^p \cap (\alpha + 1)$ and $\overline{F \cap G}^p = \overline{G}^p \cap (\alpha + 1)$. Combining the above equalities we get $\overline{F \cap G}^p = \overline{F}^p \cap \overline{G}^p \cap (\alpha + 1) = \overline{F}^p \cap \overline{G}^p$, the last equality because $\overline{F}^p \cap \overline{G}^p \subseteq F \cap G \subseteq \max(F \cap G) + 1 = \alpha + 1$. $\square$

The σ_ϱ-coding and the special sequences

We denote by $\mathbb{Q}_s(\omega_1)$ the set of finite sequences $(\phi_1, w_1, p_1, \phi_2, w_2, p_2, \ldots, \phi_d, w_d, p_d)$ such that

1. for all $i \leq d$, $\phi_i \in c_{00}(\omega_1)$ and $\phi_1 < \phi_2 < \cdots < \phi_d$,

2. $(w_i)_{i=1}^d$, $(p_i)_{i=1}^d \in \mathbb{N}^d$ are strictly increasing, and

3. $p_i \geq p_{(\cup_{k=1}^i \operatorname{supp} \phi_k)}$ for every $i \leq d$.

Let $\mathbb{Q}_s$ be the set of finite sequences $(\phi_1, w_1, p_1, \phi_2, w_2, p_2, \ldots, \phi_d, w_d, p_d)$ satisfying 1., and 2. above and in addition for every $i \leq d$, $\phi_i \in c_{00}(\mathbb{N})$. Notice that $\mathbb{Q}_s$ is a countable set. Fix a one-to-one function $\sigma : \mathbb{Q}_s \to \{2j : j \text{ odd}\}$ such that

$$\sigma(\phi_1, w_1, p_1, \phi_2, w_2, p_2, \ldots, \phi_d, w_d, p_d) > \max\{p_d^2, \tfrac{1}{\varepsilon^2}, \max \operatorname{supp} \phi_d\},$$

where $\varepsilon = \min\{|\phi_k(e_\alpha)| : \alpha \in \operatorname{supp} \phi_k, \ k = 1, \ldots, d\}$. Given a finite subset F of ω_1, we denote by $\pi_F : \{1, 2, \ldots, \#F\} \to F$ the natural order preserving map. Given

$$\Phi = (\phi_1, w_1, p_1, \phi_2, w_2, p_2, \ldots, \phi_d, w_d, p_d) \in \mathbb{Q}_s(\omega_1)$$

we denote $G_\Phi = \overline{\cup_{i=1}^d \operatorname{supp} \phi_i}^{p_d}$ and then we consider the family

$$\pi_{G_\Phi}(\Phi) = (\pi_G(\phi_1), w_1, p_1, \pi_G(\phi_2), w_2, p_2, \ldots, \pi_G(\phi_d), w_d, p_d) \in \mathbb{Q}_s,$$

where

$$\pi_G(\phi_k)(n) = \begin{cases} \phi_k(\pi_{G_\Phi}(n)) & \text{if } n \in G_\Phi \\ 0 & \text{otherwise.} \end{cases}$$

Finally, $\sigma_\varrho : \mathbb{Q}_s(\omega_1) \to \{2j : j \text{ odd}\}$ is defined as $\sigma_\varrho(\Phi) = \sigma(\pi_G(\Phi))$.

A sequence $\Phi = (\phi_1, \ldots, \phi_{n_{2j+1}})$ of functionals of K_{ω_1} is said to be a $2j + 1$-*special sequence* if:

(1) $\operatorname{supp} \phi_1 < \operatorname{supp} \phi_2 < \cdots < \operatorname{supp} \phi_{n_{2j+1}}$, each ϕ_k is of type I, $w(\phi_k) = m_{2j_k}$ and $w(\phi_1) = m_{2j_1}$ with j_1 even and satisfying $m_{2j_1} > n_{2j+1}^2$.

(2) There exists a strictly increasing sequence $(p_1^\Phi, \ldots, p_{n_{2j+1}-1}^\Phi)$ of natural numbers such that for all $1 \leq i \leq n_{2j+1} - 1$ we have that $w(\phi_{i+1}) = m_{\sigma_\varrho(\Phi_i)}$ where

$$\Phi_i = (\phi_1, w(\phi_1), p_1^\Phi, \phi_2, w(\phi_2), p_2^\Phi, \ldots, \phi_i, w(\phi_i), p_i^\Phi).$$

Notice that for a given $2j + 1$-special sequence Φ, both sequences $(p_i^\Phi)_i$ and $(w(\phi_i))_i$ above are uniquely determined, and that $w(\phi_1) = m_{2j_1}$ with j_1 even, while for every $i > 1$, $w(\phi_i) = m_{2j_i}$ is such that j_i is odd.

Lemma VI.5 (Tree-like interference of a pair of special sequences). *Let* $\Phi = (\phi_1, \ldots, \phi_{n_{2j+1}})$ *and* $\Psi = (\psi_1, \ldots, \psi_{n_{2j+1}})$ *be two* $2j + 1$-*special sequences. Then there are two numbers* $0 \leq \kappa_{\Phi,\Psi} \leq \lambda_{\Phi,\Psi} \leq n_{2j+1}$ *such that the following conditions hold:*

TP.1 *For all* $i \leq \lambda_{\Phi,\Psi}$, $w(\phi_i) = w(\psi_i)$ *and* $p_i^\Phi = p_i^\Psi$.

TP.2 *For all* $i < \kappa_{\Phi,\Psi}$, *it holds that* $\phi_i = \psi_i$,

TP.3 *For all* $\kappa_{\Phi,\Psi} < i < \lambda_{\Phi,\Psi}$, *it holds that*

$$\operatorname{supp} \phi_i \cap \overline{\operatorname{supp} \psi_1 \cup \cdots \cup \operatorname{supp} \psi_{\lambda_{\Phi,\Psi}-1}}^{p_{\lambda_{\Phi,\Psi}-1}} = \emptyset$$

$$\operatorname{supp} \psi_i \cap \overline{\operatorname{supp} \phi_1 \cup \cdots \cup \operatorname{supp} \phi_{\lambda_{\Phi,\Psi}-1}}^{p_{\lambda_{\Phi,\Psi}-1}} = \emptyset.$$

TP.4 $\{w(\psi_i) : \lambda_{\Phi,\Psi} < i \leq n_{2j+1}\} \cap \{w(\psi_i) : i \leq n_{2j+1}\} = \emptyset$ *and* $\{w(\psi_i) : \lambda_{\Phi,\Psi} < i \leq n_{2j+1}\} \cap \{w(\phi_i) : i \leq n_{2j+1}\} = \emptyset.$

Proof. Let $\lambda_{\Phi,\Psi}$ be the maximum of all $i \leq n_{2j+1}$ such that $w(\phi_i) = w(\psi_i)$ if defined. If not, we set $\lambda_{\Phi,\Psi} = \kappa_{\Phi,\Psi} = 0$. Notice that for every $i > 1$, $w(\phi_i) \neq w(\psi_i)$, and that for every $1 < i \neq j$, $w(\phi_i) \neq w(\psi_j)$ since they are coding sequences of different length. But $w(\phi_1) = m_{j_1}$ and $w(\psi_1) = m_{j_1'}$ are such that j_1, j_1' are even, so $w(\phi_1) \neq w(\psi_j)$ and $w(\psi_1) \neq w(\phi_j)$ for every $j > 1$. Suppose now that $\lambda_{\Phi,\Psi} > 0$. Define $\kappa_{\Phi,\Psi}$ by

$$\kappa_{\Phi,\Psi} = \min\{i < \lambda_{\Phi,\Psi} : \phi_i \neq \psi_i\},$$

if defined and $\kappa_{\Phi,\Psi} = 0$ if not. For this last case it is trivial to check our requirements. So assume that $\kappa_{\Phi,\Psi} > 0$. (TP.2) and (TP.4) follow easily from the properties of the coding σ_ϱ. Let us show now (TP.3). Let

$$G = \overline{\bigcup_{i=1}^{\lambda_{\Phi,\Psi}-1} \operatorname{supp} \phi_i}^{p_{\lambda_{\Phi,\Psi}-1}} \quad \text{and} \quad G' = \overline{\bigcup_{i=1}^{\lambda_{\Phi,\Psi}-1} \operatorname{supp} \psi_i}^{p_{\lambda_{\Phi,\Psi}-1}},$$

and let $\pi_G : G \to \{1, \ldots, \#G\}$ and $\pi_G' : G' \to \{1, \ldots, \#G\}$ be the unique order-preserving bijections.

Claim. 1. $\#G = \#G'$.

2. $\pi_G|(G \cap G') = \pi_{G'}|(G \cap G')$ *and* $(G \cap G')\phi_{\kappa_{\Phi,\Psi}} = (G \cap G')\psi_{\kappa_{\Phi,\Psi}}$.

3. $\max(G \cap G') < \min\{\max \operatorname{supp} \phi_{\kappa_{\Phi,\Psi}}, \max \operatorname{supp} \psi_{\kappa_{\Phi,\Psi}}\}$

Proof. 1: Notice that

$$\#G = \max \operatorname{supp} \pi_G(\phi_{\lambda_{\Phi,\Psi}-1}) \text{ and } \#G' = \max \operatorname{supp} \pi_{G'}(\psi_{\lambda_{\Phi,\Psi}-1}). \qquad \text{(VI.2)}$$

Since $\sigma_\varrho((\phi_i, w(\phi_i), p_i)_{i=1}^{\lambda_{\Phi,\Psi}-1}) = \sigma_\varrho((\psi_i, w(\psi_i), p_i)_{i=1}^{\lambda_{\Phi,\Psi}-1})$, we have $\pi_G(\phi_{\lambda_{\Phi,\Psi}-1}) = \pi_{G'}(\psi_{\lambda_{\Phi,\Psi}-1})$ and hence $\#G = \#G'$, as desired.

2: It follows from the properties of ϱ that $\pi_G|(G \cap G') = \pi_{G'}|(G \cap G')$. Fix now $\alpha \in G \cap G'$. Since $\pi_G(\alpha) = \pi_{G'}(\alpha)$ we have that

$$\phi_{\kappa_{\Phi,\Psi}}(e_\alpha) = \psi_{\kappa_{\Phi,\Psi}}(e_{\pi_G(\pi_{G'}^{-1}\alpha)}) = \psi_{\kappa_{\Phi,\Psi}}(e_\alpha), \qquad \text{(VI.3)}$$

as desired.

3: W.l.o.g we assume that $\max G \cap G' \geq \max \operatorname{supp} \phi_{\kappa_{\Phi,\Psi}}$. Property 2. yields that

$$\phi_{\kappa_{\Phi,\Psi}} = (G \cap G')\phi_{\kappa_{\Phi,\Psi}} = (G \cap G')\psi_{\kappa_{\Phi,\Psi}}, \qquad \text{(VI.4)}$$

and since $\# \operatorname{supp} \phi_{\kappa_{\Phi,\Psi}} = \# \operatorname{supp} \psi_{\kappa_{\Phi,\Psi}}$ we obtain that $\phi_{\kappa_{\Phi,\Psi}} = \psi_{\kappa_{\Phi,\Psi}}$, a contradiction. $\qquad \square$

To complete the proof choose $\kappa_{\Phi,\Psi} < i < \lambda_{\Phi,\Psi}$. Then previous Claim yields that $\operatorname{supp} \phi_i \subseteq G \setminus (G \cap G')$ and hence $\operatorname{supp} \phi_i \cap G' = \emptyset$. $\qquad \square$

The space $\mathfrak{X}_{\omega_1}$ has no unconditional basic sequence

Definition VI.6. Let $j \in \mathbb{N}$. A sequence $(x_1, \phi_1, \ldots, x_{n_{2j+1}}, \phi_{n_{2j+1}})$ is said to be a $(1, j)$-*dependent sequence* if:

DS.1 $\operatorname{supp} x_1 \cup \operatorname{supp} \phi_1 < \cdots < \operatorname{supp} x_{n_{2j+1}} \cup \operatorname{supp} \phi_{n_{2j+1}}$.

DS.2 The sequence $\Phi = (\phi_1, \ldots, \phi_{n_{2j+1}})$ is a $2j + 1$-special sequence.

DS.3 (x_i, ϕ_i) is a $(6, 2j_i)$-exact pair for $1 \leq i \leq n_{2j+1}$, with $\# \operatorname{supp} x_i \leq m_{2j+1}/n_{2j+1}^2$ for every $1 \leq i \leq n_{2j+1}$.

DS.4 For every $\Psi = (\psi_1, \ldots, \psi_{n_{2j+1}})$ $(2j + 1)$-special sequence we have that

$$\bigcup_{\kappa_{\Phi,\Psi} < i < \lambda_{\Phi,\Psi}} \operatorname{supp} x_i \cap \bigcup_{\kappa_{\Phi,\Psi} < i < \lambda_{\Phi,\Psi}} \operatorname{supp} \psi_i = \emptyset. \qquad \text{(VI.5)}$$

Proposition VI.7. *For every block sequence $(y_n)_n$ of $\mathfrak{X}_{\omega_1}$ and every $j \in \mathbb{N}$ there exists a $(1, j)$-dependent sequence $(x_1, \phi_1, \ldots, x_{n_{2j+1}}, \phi_{n_{2j+1}})$ such that $x_i \in \langle y_n \rangle_n$ for every $i = 1, \ldots, n_{2j+1}$.*

Proof. Let $(y_n)_n$ and j be given. We inductively produce $\{(x_i, \phi_i)\}_{i=1}^{n_{2j+1}}$ as follows. For $i = 1$ we choose a $(6, 2j_1)$-exact pair (x_1, ϕ_1) such that $m_{2j_1} > m_{2j+1}^2$, j_1 even (see the definition of special sequences) and $x_1 \in \langle y_n \rangle_n$. Assume that $\{(x_l, \phi_l)\}_{l=1}^{i-1}$ has been chosen such that there exists $(p_l)_{l=1}^{i-2}$ satisfying

(a) $\operatorname{supp} x_1 \cup \operatorname{supp} \phi_1 < \cdots < \operatorname{supp} x_{i-1} \cup \operatorname{supp} \phi_{i-1}$, each $x_l \in \langle y_n \rangle_n$ and (x_l, ϕ_l) being a $(6, 2j_l)$-exact pair.

(b) For $1 < l \leq i - 1$, $w(\phi_l) = \sigma_\varrho(\phi_1, w(\phi_1), p_1, \ldots, \phi_{l-1}, w(\phi_{l-1}), p_{l-1})$

(c) For $1 \leq l < i - 1$, $p_l \geq \max\{p_{l-1}, p_{F_l}\}$, where $F_l = \bigcup_{k=1}^l \operatorname{supp} \phi_k \cup \operatorname{supp} x_k$.

To define (x_i, ϕ_i) we choose $p_{i-1} \geq \max\{p_{i-2}, p_{F_{i-1}}, n_{2j+1}^2 \cdot \#\operatorname{supp} x_i\}$ and we set

$$2j_i = \sigma_\varrho(\phi_1, w(\phi_1), p_1, \ldots, \phi_{i-1}, w(\phi_{i-1}), p_{i-1}).$$

Choose a $(6, 2j_i)$-exact pair (x_i, ϕ_i) such that $x_i \in \langle y_n \rangle_n$ and $\operatorname{supp} x_{i-1} \cup \operatorname{supp} \phi_{i-1} < \operatorname{supp} x_i \cup \operatorname{supp} \phi_i$. This completes the inductive construction. (DS.1)–(DS.3) easily holds, while (DS.4) follows from (c) and (TP.3) of Lemma VI.5. $\qquad \square$

Remark VI.8. Suppose that $(y_n)_n$ and $(z_n)_n$ are block sequences such that $\sup_n \max \operatorname{supp} y_n = \sup_n \max \operatorname{supp} z_n$. Then for every $j \in \mathbb{N}$ there is a $(1, j)$-dependent sequence $(x_1, \phi_1, \ldots, x_{n_{2j+1}}, \phi_{n_{2j+1}})$ with the property that $x_{2i-1} \in \langle y_n \rangle_n$ and $x_{2i} \in \langle z_n \rangle_n$ for every $i = 1, \ldots, n_{2j+1}/2$.

Lemma VI.9. *Fix a $(1, j)$-dependent sequence $(x_1, \phi_1, \ldots, x_{n_{2j+1}}, \phi_{n_{2j+1}})$, and a sequence $(\lambda_i)_{i=1}^{n_{2j+1}}$ of scalars such that $\max_i |\lambda_i| \leq 1$. Suppose that for every $\psi \in K_{\omega_1}$ with $w(\psi) = m_{2j+1}$, and every interval of integers $E \subseteq [1, n_{2j+1}]$ it holds that*

$$|\psi(\sum_{i \in E} \lambda_i x_i)| \leq 12(1 + \frac{\#E}{n_{2j+1}^2}). \qquad (\text{VI.6})$$

Then,

$$\|\frac{1}{n_{2j+1}} \sum_{i=1}^{n_{2j+1}} \lambda_i x_i\| \leq \frac{1}{m_{2j+1}^2}. \qquad (\text{VI.7})$$

Proposition VI.10. *If $(x_1, \phi_1, \ldots, x_{n_{2j+1}}, \phi_{n_{2j+1}})$ is a $(1, j)$-dependent sequence, then*

$$\|\frac{1}{n_{2j+1}} \sum_{i=1}^{n_{2j+1}} x_i\| \geq \frac{1}{m_{2j+1}} \quad \text{and} \quad \|\frac{1}{n_{2j+1}} \sum_{i=1}^{n_{2j+1}} (-1)^{i+1} x_i\| \leq \frac{1}{m_{2j+1}^2}. \qquad (\text{VI.8})$$

Proof. The first estimation is clear since the functional $\psi = (1/m_{2j+1}) \sum_{i=1}^{n_{2j+1}} \phi_i \in K_{\omega_1}$ and $\psi((1/n_{2j+1}) \sum_{i=1}^{n_{2j+1}} x_i) = 1/m_{2j+1}$. For the second, we use Lemma VI.9 applied to the sequence of scalars $((-1)^{i+1})_i$, and the desired estimation will follow from (VI.7). Fix $\psi \in K_{\omega_1}$ with $w(\psi) = m_{2j+1}$, and an interval $E \subseteq [1, n_{2j+1}]$. Set $\Psi = (\psi_1, \ldots, \psi_{n_{2j+1}})$ and $x = \sum_{i \in E} (-1)^{i+1} x_i$, where $\psi = (1/m_{2j+1}) \sum_{i \in E} \psi_i$. Notice that

$$|\psi(x)| = |\frac{1}{m_{2j+1}} \sum_{i=1}^{\kappa_{\Phi,\Psi}-1} \phi_i(x) + \frac{1}{m_{2j+1}} \sum_{i=\kappa_{\Phi,\Psi}}^{n_{2j+1}} \psi_i(x)| \leq \qquad (\text{VI.9})$$

$$\leq \frac{1}{m_{2j+1}} + |\frac{1}{m_{2j+1}} \sum_{i=\kappa_{\Phi,\Psi}}^{n_{2j+1}} \psi_i(x)|.$$

We shall show that the following hold:

(a) $|\psi_{\kappa_{\Phi,\Psi}}(\sum_{i\in E}(-1)^{i+1}x_i)| \leq 1 + 12(\#E-1)/n_{2j+1}^2$,

(b) $|\psi_{\lambda_{\Phi,\Psi}}|(\sum_{i\in E}(-1)^{i+1}x_i) \leq 1 + 12(\#E-1)/n_{2j+1}^2$, and

(c) $|(\sum_{l>\kappa_{\Phi,\Psi}, l\neq\lambda_{\Phi,\Psi}} \psi_l)(x_i)| \leq 12/n_{2j+1}$ for every $1 \leq i \leq n_{2j+1}$.

Let us show first (a). Let $2j_i$ be such that $w(\phi_i) = m_{2j_i}$. Notice that for $i \neq \kappa_{\Phi,\Psi}$ we have that

$$|\psi_{\kappa_{\Phi,\Psi}}(x_i)| \leq \begin{cases} \dfrac{12}{w(\psi_{\kappa_{\Phi,\Psi}})} & \text{if } i > \kappa_{\Phi,\Psi} \\ \dfrac{6}{m_{2j_i}^2} & \text{if } i < \kappa_{\Phi,\Psi}. \end{cases} \qquad (\text{VI.10})$$

By the properties of the sequences $(m_l)_l$, $(n_l)_l$ and the fact that $n_{2j+1}^2 < w(\psi_{\kappa_{\Phi,\Psi}})$, m_{2j_i}, (VI.10) yields that $|\psi_{\kappa_{\Phi,\Psi}}(x_i)| \leq \frac{12}{n_{2j+1}^2}$ for $i \neq \kappa_{\Phi,\Psi}$. Hence

$$|\psi_{\kappa_{\Phi,\Psi}}(\sum_{i\in E} x_i)| \leq |\psi_{\kappa_{\Phi,\Psi}}(x_{\kappa_{\Phi,\Psi}})| + |\psi_{\kappa_{\Phi,\Psi}}(\sum_{i\in E, i\neq\kappa_{\Phi,\Psi}} x_i)| \leq 1 + \frac{12(\#E-1)}{n_{2j+1}^2}.$$
$$(\text{VI.11})$$

(b) has a proof similar to the one of (a). We check now (c). Fix $l > \kappa_{\Phi,\Psi}$, $l \neq \lambda_{\Phi,\Psi}$. Suppose that $l > \lambda_{\Phi,\Psi}$. Since $w(\psi_l) \neq w(\phi_i)$ for all $i \leq n_{2j+1}$, we obtain that $|\psi_l(x_i)| \leq \frac{12}{n_{2j+1}^2}$. Now suppose that $\kappa_{\Phi,\Psi} < l < \lambda_{\Phi,\Psi}$. By (DS.4) we have that $\psi_l(x_i) = 0$ for every $\kappa_{\Phi,\Psi} < i < \lambda_{\Phi,\Psi}$. And for $i \notin (\kappa_{\Phi,\Psi}, \lambda_{\Phi,\Psi})$, using the fact that $w(\psi_l) \neq w(\phi_i)$, we can conclude that $|\psi_l(x_i)| \leq 12/n_{2j+1}^2$. Hence, $(\sum_{l>\kappa_{\Phi,\Psi}, l\neq\lambda_{\Phi,\Psi}} \psi_l)(x_i) \leq 12/n_{2j+1}$ for every $1 \leq i \leq n_{2j+1}$, as desired.

Combining (a), (b) and (c) we obtain that

$$|\frac{1}{m_{2j+1}} \sum_{i=\kappa_{\Phi,\Psi}}^{n_{2j+1}} \psi_i(x)| \leq 1 + \frac{\#E}{n_{2j+1}^2}. \qquad (\text{VI.12})$$

From (VI.9) and (VI.12) we conclude that $|\psi(x)| \leq 12(1 + \#E/n_{2j+1}^2)$, as desired. $\qquad \square$

Proposition VI.11. *Let* $(y_n)_n$ *be a block sequence of vectors of* $\mathfrak{X}_{\omega_1}$. *Then the closed linear span of* $(y_n)_n$ *is hereditarily indecomposable.*

Proof. Fix a block sequence $(y_n)_n$ of $\mathfrak{X}_{\omega_1}$, two block subsequences $(z_n)_n$ and $(w_n)_n$ of $(y_n)_n$ and $\varepsilon > 0$. Let j be large enough such that $m_{2j+1}\varepsilon > 1$. By Proposition VI.7 we can choose a $(1,j)$-dependent sequence $(x_1, \phi_1, \ldots, x_{n_{2j+1}}, \phi_{n_{2j+1}})$ such that $x_{2i-1} \in \langle z_n\rangle_n$, and $x_{2i} \in \langle w_n\rangle_n$. Set $z = (1/n_{2j+1})\sum_{i=1, i \text{ odd}}^{n_{2j+1}} x_i$ and $w = (1/n_{2j+1})\sum_{i=1, i \text{ even}}^{n_{2j+1}} x_i$. Notice that $z \in \langle z_n\rangle_n$ and $w \in \langle w_n\rangle_n$. By Proposition VI.9, we know that $\|z+w\| \geq 1/m_{2j+1}$ and $\|z-w\| \leq 1/m_{2j+1}^2$. Hence $\|z-w\| \leq \varepsilon\|z+w\|$. $\qquad \square$

Corollary VI.12. (a) *The distance between the unit spheres of every two normalized block sequences* (x_n) *and* (y_n) *in* $\mathfrak{X}_{\omega_1}$ *such that* $\sup_n \max \operatorname{supp} x_n = \sup_n \max \operatorname{supp} y_n$ *is* 0.

(b) *There is no unconditional basic sequence in* $\mathfrak{X}_{\omega_1}$.

(c) *Every infinite dimensional closed subspace of* $\mathfrak{X}_{\omega_1}$ *contains a hereditarily indecomposable subspace.*

(d) *The distance between the unit spheres of two non separable subspaces of* $\mathfrak{X}_{\omega_1}$ *is equal to* 0.

Proof. (b) follows from Proposition VI.11 and 4. of Proposition A.3.

(c) This result follows from the previous corollary and Gowers' dichotomy. Moreover, every subspace of $\mathfrak{X}_{\omega_1}$ isomorphic to the closed linear span of a block sequence with respect to the basis $(e_\alpha)_{\alpha<\omega_1}$ is hereditarily indecomposable.

(d) Fix two non separable closed subspaces X and Y of $\mathfrak{X}_{\omega_1}$. Now we can find a sequence $(z_n)_n$ of normalized vectors such that for every n (a) $z_{2n-1} \in X$, $z_{2n} \in Y$ and (b) $\operatorname{supp} z_n < \operatorname{supp} z_{n+1}$. Notice that the supports $\operatorname{supp} z_n$ are not necessarily finite. Now approximate $(z_n)_n$ by a normalized block sequence $(w_n)_n$ as closed as needed and we are done. $\qquad \square$

Chapter VII

The Finite Representability of J_{T_0} and the Diagonal Space $\mathcal{D}(\mathfrak{X}_\gamma)$

Since the space $\mathfrak{X}_{\omega_1}$ has a transfinite basis $(e_\alpha)_{\alpha<\omega_1}$ it also admits a transfinite family of naturally defined projections. Thus we could not expect that the operators on $\mathfrak{X}_{\omega_1}$ or on the spaces $\mathfrak{X}_\gamma$, $\gamma < \omega_1$, have small spaces of non-strictly singular operators. In this part we shall discuss the structure of the step diagonal operators which are the non strictly singular diagonal operators and we shall show that for all infinite $\gamma < \omega_1$, $\mathcal{D}(\mathfrak{X}_\gamma)$ is isomorphic to $J_{T_0}^*(\Lambda_\gamma)$, where A_γ is a closed subset of γ. This result, quite unexpectedly, is a consequence of the finite interval representability of J_{T_0} in the transfinite block subspaces of $\mathfrak{X}_{\omega_1}$.

James-like spaces

Definition VII.1. Let X be a reflexive space with a 1-subsymmetric basis $(x_n)_n$, and let A be a set of ordinals. $J_X(A)$ is the completion of $(c_{00}(A), \|\cdot\|_{J_X(A)})$, where for $x \in c_{00}(A)$,

$$\|x\|_{J_X(A)} = \sup\{\| \textstyle\sum_{n=1}^{l} \left(\sum_{i\in I_n} x(i)\right) x_n\|_X : I_1 < \cdots < I_n \text{ intervals of } A\}.$$

The natural Hamel basis $(v_\alpha)_{\alpha\in A}$ of $c_{00}(A)$ is a bimonotone 1-subsymmetric transfinite basis of $J_X(A)$. Also, for every interval I of A the functional $I^* : J_X(A) \to \mathbb{R}$, $I^*(x) = \sum_{\alpha\in A} x(\alpha)$ belongs to $J_X^*(A)$ and $\|I^*\| = 1$.

Remark VII.2. Spaces of the above form for $A = \mathbb{N}$, have been introduced in [19]. The transfinite analogue of James quasi-reflexive space was defined by G.A. Edgar in [23]. As we shall see next ℓ_1 does no embed into $J_X(A)$ hence the basis $(v_\alpha)_{\alpha\in A}$ is not unconditional.

Proposition VII.3. *Let $(y_n)_n$ be a semi-normalized block sequence in $J_X(A)$ with $\sum_{\alpha \in A} y_n(\alpha) = 0$ for every n. Then $(y_n)_n$ is equivalent to the basis $(x_n)_n$ of X.*

Proof. Let $0 < c < C$ be such that $c \leq \|y_n\| \leq C$ for all n. It is easy to see that

$$c \| \sum_n a_n x_n \|_X \leq \| \sum_n a_n y_n \|_{J_X(A)} \leq \sup_{i_1 \leq i_2 \leq \cdots \leq i_l} \| \sum_{q=1}^{l-1} (|a_{i_q}|$$

$$+ |a_{i_{q+1}}|) x_q \|_X \leq (2CK) \| \sum_n a_n x_n \|_X, \qquad \text{(VII.1)}$$

where K is the unconditional constant of $(x_n)_n$. The first inequality holds for any block sequence and the second uses our assumptions. $\qquad \square$

Corollary VII.4. *The space ℓ_1 does not embed into $J_X(A)$.*

Proof. Assume the contrary, then from Proposition A.3 we could find a semi-normalized block sequence $(y_n)_n$ equivalent to the ℓ_1-basis. Therefore, passing if necessary to a further block sequence, we may assume that for all $n \in \mathbb{N}$, $\sum_{\alpha \in A} y_n(\alpha) = 0$. Hence Proposition VII.3 yields that $(y_n)_n$ is equivalent to $(x_n)_n$, a contradiction. $\qquad \square$

Remark VII.5. Suppose that A and B are two sets of ordinals with the same order type. Then the unique order-preserving mapping $f : A \to B$ defines naturally an isometry between $\tilde{f} : J_{T_0}(A) \to J_{T_0}(B)$ by $\tilde{f}(\sum_{\alpha \in H} r_\alpha v_\alpha) = \sum_{\alpha \in H} r_\alpha v_{f(\alpha)}$.

Proposition VII.6. *For every ordinal γ the space $J_X^*(\gamma)$ is generated in norm by $\{[0, \alpha)^*\}_{\alpha < \gamma+1}$.*

Proof. We proceed by induction. It is clear that the successor ordinal case follows immediately from the inductive assumption. So we assume that γ is limit ordinal and for all $\lambda < \gamma$ the conclusion holds. Assume on the contrary that $Y = \overline{\langle [0, \alpha)^* \rangle_{\alpha < \gamma+1}}^{\|\cdot\|} \subsetneq J_X^*(\gamma)$, then there exists $x^* \in J_X^*(\gamma)$ with $\|x^*\| = 1$ and $\varepsilon > 0$ such that $d(x^*, Y) > \varepsilon$. Observe also that the inductive assumption yields that for all $\alpha < \gamma$ if x_α^* denotes the functional defined by

$$x_\alpha^*(v_\beta) = \begin{cases} 0 & \text{if } \beta < \alpha \\ x^*(v_\beta) & \text{if } \beta \geq \alpha \end{cases}$$

satisfies that $\|x_\alpha^*\| \leq 1$ and $d(x_\alpha^*, Y) > \varepsilon$. In particular for all $\alpha < \gamma$, $d(x_\alpha^*, \langle [\alpha, \gamma)^* \rangle) > \varepsilon$ and from the Hahn–Banach and Goldstime theorems there exists a finitely supported $\tilde{y}_\alpha \in J_X(\gamma)$, $\|\tilde{y}_\alpha\| \leq 1$, $\alpha \leq \min \operatorname{supp} \tilde{y}_\alpha$, $x^*(\tilde{y}_\alpha) > \varepsilon$ and $|\sum_{\beta < \gamma} \tilde{y}_\alpha(\beta)| \leq \varepsilon/4$. Assuming further that α is a successor ordinal we consider the vector $y_\alpha = \tilde{y}_\alpha - (\sum_{\beta \geq \alpha} \tilde{y}_\alpha(\beta)) v_{\alpha^-}$. Observe that $\alpha^- \leq \min \operatorname{supp} y_\alpha$, $x^*(y_\alpha) > \varepsilon - \varepsilon/4 > \varepsilon/2$ and $\sum_{\beta < \gamma} y_\alpha(\beta) = 0$. Hence we may inductively choose a block sequence $(z_n)_n$ such that $\varepsilon/2 \leq \|z_n\| \leq 1$, $\sum_{\alpha < \gamma} z_n(\alpha) = 0$ and $x^*(z_n) > \varepsilon/2$. Observe that $(z_n)_n$ is unconditional (Proposition VII.3) therefore equivalent to ℓ_1-basis which yields a contradiction. $\qquad \square$

Corollary VII.7. *For every set of ordinals A we have that $\dim J_X^*(A) = \#A$.* $\qquad \square$

Finite interval representability of J_{T_0} and the space of diagonal operators

Definition VII.8. Let X and Y be Banach spaces and let $(x_\alpha)_{\alpha<\gamma}$ and $(y_n)_n$ be a transfinite basis for X and a Schauder basis of Y respectively. We say that Y is *finitely interval representable* in X if there exists a constant $C > 0$ such that for every integer n and intervals $I_1 \leq I_2 \leq \cdots \leq I_n$ successive, not necessarily distinct, intervals of γ there exists $z_i \in \langle (x_\alpha)_{\alpha \in I_i} \rangle$ $(i = 1, \ldots, n)$ with $\text{supp } z_1 <$ $\text{supp } z_2 < \cdots < \text{supp } z_n$ and such that the natural order preserving isomorphism $H : \langle (y_i)_{i=1}^n \rangle \to \langle (z_i)_{i=1}^n \rangle$ satisfies $\|H\| \cdot \|H^{-1}\| \leq C$.

Theorem VII.9. *Let $(y_\alpha)_{\alpha<\gamma}$ be a normalized transfinite block sequence in $\mathfrak{X}_{\omega_1}$, and Y its closed linear span. Then J_{T_0} is finitely interval representable in the space Y, where T_0 is the mixed Tsirelson space $T[(\mathcal{A}_{n_{2j}}, \frac{1}{m_{2j}})_j]$.*

We present the proof in Appendix B. Throughout all this section C will denote the finitely block representability constant of J_{T_0} in $\mathfrak{X}_{\omega_1}$. We will see in Appendix B that $C < 121$.

Remark VII.10. 1. Let us observe that since, as we will show, the basis of J_{T_0} is not unconditional and it is finitely block representable in any block subsequence of the basis $(e_\alpha)_{\alpha<\omega_1}$, $\mathfrak{X}_{\omega_1}$ cannot have any unconditional basic sequence. In other words the finite interval representability of J_{T_0} in the block subsequences of $\mathfrak{X}_{\omega_1}$ must make use of the conditional structure of $\mathfrak{X}_{\omega_1}$. Indeed we get more. Suppose that $\mathfrak{X}$ has a transfinite basis, and suppose that a Banach space Y with a conditional basis $(y_n)_n$ is finite block representable in every block sequence of $\mathfrak{X}$. Then $\mathfrak{X}$ does not contain unconditional basic sequences and from Gowers dichotomy, $\mathfrak{X}$ is HI saturated.

2. The James like space J_{T_0} has the following alternative description. It is the mixed Tsirelson extension $T[G, (\mathcal{A}_{n_{2j}}, \frac{1}{m_{2j}})_j]$, where $G = \{I^* : I \subseteq \mathbb{N} \text{ interval}\}$. We recall that the norming set of this space is the minimal subset K_0 of $c_{00}(\mathbb{N})$ which is symmetric, contains G, and is closed under the $(\mathcal{A}_{n_{2j}}, \frac{1}{m_{2j}})_j$-operations.

Proposition VII.11. *Let $x_1 < \cdots < x_n$ be finitely supported, $\phi \in K_{\omega_1}$ and set $r_i = \phi(x_i)$ for each $i = 1, \ldots, n$. Then $\|\sum_{i=1}^n r_i v_i\|_{J_{T_0}} \leq \|x_1 + \cdots + x_n\|$.*

Proof. Fix a functional f of K_0 with support contained in $\{1, \ldots, n\}$, and a tree-analysis $(f_t)_{t \in \mathcal{T}}$ of f. We show by induction over the tree $\mathcal{T}$ that for every $t \in \mathcal{T}$ there is some $\phi_t \in K_{\omega_1}$ such that $f_t(\sum_{i=1}^n r_i v_i) = \phi_t(x_1 + \cdots + x_n)$. In particular $f_0(\sum_i r_i v_i) = \phi_0(x_1 + \cdots + x_n)$, and hence the desired result holds. If $t \in \mathcal{T}$ is a terminal node, then $f_t = \pm I^*$, $I \subseteq \{1, \ldots, n\}$ interval. We set $\phi_t = \pm\phi|[\min \text{supp } x_{\min I}, \max \text{supp } x_{\max I}]$. It is clear that $\phi_t \in K_{\omega_1}$, and

$$\phi_t(x_1 + \cdots + x_n) = \pm\sum_{i \in I} \phi x_i = \pm\sum_{i \in I} r_i = f_t(\sum_i r_i v_i). \tag{VII.2}$$

If $t \in \mathcal{T}$ is not a terminal node, then $f_t = (1/m_{2j})\sum_{i=1}^d f_{s_i}$, where $S_t = \{s_1, \ldots, s_d\}$ is ordered by $f_{s_1} < \cdots < f_{s_d}$. Then $\phi_t = (1/m_{2j})\sum_{i=1}^d \phi_{s_i}$ clearly satisfies our inductive requirements. $\qquad\square$

The next result shows that J_{T_0} is minimal in a precise sense.

Corollary VII.12. *Suppose that X is a Banach space with a normalized Schauder basis $(x_n)_n$ which dominates the summing basis of c_0 and finitely block represented in $\mathfrak{X}_{\omega_1}$. Then $(x_n)_n$ also dominates the basis $(v_n)_n$ of J_{T_0}.*

Proof. Fix scalars $(a_i)_{i=1}^n$. Choose a normalized block sequence $(w_i)_i^n$ of $\mathfrak{X}_{\omega_1}$, C-equivalent to $(x_i)_{i=1}^n$. Fix $f \in K_0$ with supp $f \subseteq \{1,\ldots,n\}$ and a tree-analysis $(f_t)_{t \in \mathcal{T}}$ of it. We are going to find $\phi_t \in K_{\omega_1}$ such that $|f_t(\sum_{i=1}^n a_i v_i)| \leq C|\phi_t(\sum a_i w_i)|$, for each $t \in \mathcal{T}$. This will show that

$$\| \textstyle\sum_{i=1}^n a_i v_i \|_{J_{T_0}} \leq C \| \textstyle\sum_{i=1}^n a_i w_i \|_{\mathfrak{X}_{\omega_1}} \leq C^2 \| \textstyle\sum_{i=1}^n a_i w_i \|_X, \qquad \text{(VII.3)}$$

as desired. If $t \in \mathcal{T}$ is a terminal node, then $f_t = \pm I^*$, $I \subseteq [1,n]$ an interval. Since $(x_n)_n$ dominates the summing basis of c_0, we can find $\phi_t \in K_{\omega_1}$ such that

$$\phi_t(\textstyle\sum_{i=1}^n a_i w_i) = \| \textstyle\sum_{i=1}^n a_i w_i \|_{\mathfrak{X}_{\omega_1}} \geq \frac{1}{C} \| \sum_{i=1}^n a_i x_i \|_X \geq \qquad \text{(VII.4)}$$

$$\geq \frac{1}{C} | \sum_{i \in I} a_i | = | f_t(\sum_{i=1}^n a_i v_i)|.$$

If t is not terminal node, then we use the appropriate $(\mathcal{A}_{n_{2j}}, \frac{1}{m_{2j}})$-operation. $\square$

Definition VII.13. Let $(x_\alpha)_{\alpha<\gamma}$ be a normalized transfinite block sequence, X its closed linear span. We denote by $\mathcal{D}(X)$ the space of all bounded diagonal operators $D : X \to X$ satisfying the property that for all $\alpha < \gamma$ limit there exists some $\lambda_\alpha \in \mathbb{R}$ such that $D(x_\beta) = \lambda_\alpha x_\beta$ for every $\beta \in [\alpha, \alpha+\omega)$. We also denote by $\widetilde{\mathcal{D}}(X)$ the space of all diagonal operators (not necessarily bounded) satisfying the above condition acting on $\langle x_\alpha \rangle_{\alpha<\gamma}$.

Notice the following (linear) decomposition of $\langle x_\alpha \rangle_{\alpha<\gamma}$,

$$\langle x_\alpha \rangle_{\alpha<\gamma} = \bigoplus_{\alpha \in \Lambda(\gamma)} \langle x_\beta \rangle_{\beta \in [\alpha,\alpha+\omega)}. \qquad \text{(VII.5)}$$

The *canonical decomposition* of $y \in \langle x_\alpha \rangle_{\alpha<\gamma}$ in X is $y = y_1 + \cdots + y_n$ given by (VII.5).

Remark VII.14. $\mathcal{D}(X)$ is a closed subalgebra of $\mathcal{L}(X)$.

For an ordinal μ we denote by $\Lambda(\mu)^{(0)}$ the set of limit ordinals $\alpha < \mu$ such that $\alpha = \beta \dotplus \omega$ for a limit ordinal β. We denote this β by α^-. For technical reasons, we assume that 0 is a limit ordinal.

Remark VII.15. Notice that for γ limit, $\Lambda(\gamma + 1)^{(0)}$ is order isomorphic to $\Lambda(\gamma)$ considering the predecessor map.

Definition VII.16. Let $D \in \widetilde{\mathcal{D}}(X)$. We define the map $\xi_D : \Lambda(\gamma + 1)^{(0)} \to \mathbb{R}$ via $D(x_{\alpha^-}) = \xi_D(\alpha)x_{\alpha^-}$. Namely, $\xi_D(\alpha)$ is the eigenvalue of D associated to the eigenvectors $(x_\beta)_{\beta \in [\alpha^-, \alpha)}$.

We consider the linear map $\Xi : \widetilde{D}(X) \to c_{00}(\Lambda(\gamma+1)^{(0)})^{\#}$ defined by

$$\Xi(D)(v_\alpha) = \xi_D(\alpha), \tag{VII.6}$$

where $c_{00}(\Lambda(\gamma+1)^{(0)})^{\#}$ denotes the algebraic conjugate of $c_{00}(\Lambda(\gamma+1)^{(0)})$. The main goal here is to show that Ξ defines an isomorphism between $\mathcal{D}(X)$ and $J_{T_0}^*(\Lambda(\gamma+1)^{(0)})$. For $D \in \widetilde{\mathcal{D}}(X)$, let us denote

$$\|D\| = \sup\{\|Dx\|_{\mathfrak{X}_{\omega_1}} : x \in \langle x_\alpha \rangle_{\alpha<\gamma}, \|x\|_{\mathfrak{X}_{\omega_1}} \leq 1\} \leq \infty,$$

and for $f \in c_{00}(\Lambda(\gamma+1)^{(0)})^{\#}$,

$$\|f\| = \sup\{f(x) : x \in c_{00}(\lambda(\gamma+1)^{(0)}), \|x\|_{J_{T_0}} \leq 1\} \leq \infty.$$

Proposition VII.17. $\|D\| \leq \|\Xi(D)\| \leq C\|D\|$ *for every* $D \in \widetilde{\mathcal{D}}(X)$.

Proof. Fix $D \in \widetilde{\mathcal{D}}(X)$, and $\varepsilon > 0$. Let $y \in \langle x_\alpha \rangle_{\alpha<\gamma}$ with $\|y\| \leq 1$ be such that $|\|D\| - \|Dy\|| < \varepsilon$. Let $y = y_1 + \cdots + y_n$ be the canonical decomposition of y in X, and $\alpha_1, \ldots, \alpha_n$ be such that $y_i \in \langle x_\beta \rangle_{\beta \in [\alpha_i^-, \alpha_i)}$ for every $1 \leq i \leq n$. Let $\phi \in K$ be such that $\|Dy\| = \phi(Dy)$, and set $r_i = \phi y_i$ for $i = 1, \ldots, n$. By Proposition VII.11, $\|\sum_{i=1}^n r_i v_i\|_{J_{T_0}} \leq \|x\|$, and since $(v_\alpha)_\alpha$ is 1-subsymmetric we have that $\|\sum_{i=1}^n r_i v_{\alpha_i}\|_{J_{T_0}} \leq \|y\| \leq 1$. Hence

$$\|\Xi(D)\| \geq \|\Xi(D)(\sum_{i=1}^n r_i v_{\alpha_i})\|_{J_{T_0}} = \|\sum_{i=1}^n \xi_D(\alpha_i) r_i v_{\alpha_i}\|_{J_{T_0}} \tag{VII.7}$$

$$\geq \sum_{i=1}^n \xi_D(\alpha_i) = \phi(Dy) \geq \|D\| - \varepsilon.$$

This shows that $\|D\| \leq \|\Xi(D)\|$. Fix $v = \sum_{i=1}^n a_i v_{\alpha_i} \in J_{T_0}$ with $\|v\|_{J_{T_0}} \leq 1$, and choose a finite normalized block sequence $(w_i)_{i=1}^n$ C-equivalent to $(v_{\alpha_i})_{i=1}^n$ with $w_i \in \langle x_\beta \rangle_{\beta \in [\alpha_i^-, \alpha_i)}$ for every $i = 1, \ldots, n$ (indeed we may assume that the natural isomorphism $F : \langle w_i \rangle_{i=1}^n \to \langle v_i \rangle_{i=1}^n$ satisfies that $\|F\| \leq 1$, $\|F^{-1}\| \leq C$; see Corollary B.15). Then,

$$\|\Xi(D)(v)\|_{J_{T_0}} = \|\sum_{i=1}^n \xi_D(\alpha_i) a_i v_{\alpha_i}\|_{J_{T_0}} \leq \|\sum_{i=1}^n \xi_D(\alpha_i) a_i w_i\|_{\mathfrak{X}_{\omega_1}}$$

$$= \|D(\sum_{i=1}^n a_i w_i)\|_{\mathfrak{X}_{\omega_1}} \leq \|D\| \|\sum_{i=1}^n a_i w_i\|_{\mathfrak{X}_{\omega_1}} \leq C\|D\|. \tag{VII.8}$$

$\square$

Theorem VII.18. *The spaces* $\mathcal{D}(X)$ *and* $J_{T_0}^*(\Lambda(\gamma+1)^{(0)})$ *are isomorphic.*

Proof. By Proposition VII.17, $\Xi|\mathcal{D}(X) : \mathcal{D}(X) \to J_{T_0}^*(\Lambda(\gamma+1)^{(0)})$ is an isomorphism. To see that it is also onto consider $f \in J_{T_0}^*(\Lambda(\gamma+1)^{(0)})$ and define $D_f \in \widetilde{\mathcal{D}}(X)$ as follows. For $\beta \in [\alpha^-, \alpha)$ set $D_f(x_\beta) = f(v_\alpha)x_\beta$. It is easy to check that $\Xi(D_f) = f$. This completes the proof. $\square$

Corollary VII.19. *Let X and Y be the closed linear span of two transfinite block sequences of the same length γ. Then the natural mapping $\psi_\gamma : \mathcal{D}(X) \to \mathcal{D}(Y)$ defined by $\psi_\gamma(D) = D_{\xi_D}$ is an isomorphism.* $\qquad\qquad\square$

Our intention now is to compare $\mathcal{D}(X)$ and $\mathcal{D}(\mathfrak{X}_{\omega_1})$.

Definition VII.20. 1. Given a closed $A \subseteq \Lambda(\omega_1 + 1)$, let $\widetilde{\mathcal{D}}_A(\mathfrak{X}_{\omega_1})$ be the subalgebra of $\widetilde{\mathcal{D}}(\mathfrak{X}_{\omega_1})$ consisting on all $D \in \mathcal{D}(\mathfrak{X}_{\omega_1})$ satisfying that for every $\alpha \in A^{(0)}$, there is some λ_α such that $D|\mathfrak{X}_{[\alpha^-,\alpha)} = \lambda_\alpha i_{\mathfrak{X}_{[\alpha^-,\alpha)},\mathfrak{X}_{\omega_1}}$ and $D|\mathfrak{X}_{[\max A,\omega_1)} = 0$. Let $\mathcal{D}_A(\mathfrak{X}_{\omega_1})$ be the subalgebra of bounded operators of $\widetilde{\mathcal{D}}_A(\mathfrak{X}_{\omega_1})$.
2. Given a transfinite block sequence $(x_\alpha)_{\alpha<\gamma}$, let $\Gamma_X \subseteq \Lambda(\omega_1 + 1)$ be defined as follows. Let

$$\Gamma' = \{\sup_{n\to\infty} \max\operatorname{supp} x_{\alpha_n} : (\alpha_n)_n \uparrow, \alpha_n < \gamma\}, \qquad\qquad \text{(VII.9)}$$

and let $\Gamma_X = \Gamma' \cup \{0, \sup \Gamma'\}$. Another interpretation of Γ_X is to consider the map $f_X : \Lambda(\gamma + 1) \to \omega_1$ defined by $f_X(\alpha) = \sup_{\beta<\alpha} \max\operatorname{supp} x_\beta$ and Γ_X is nothing else but the image $f(\Lambda(\gamma+1))$, and hence $\Gamma_X \setminus \max\{\Gamma_X\}$ and $\Lambda(\gamma+1)^{(0)}$ are order isomorphic.
3. Given $D \in \mathcal{D}(X)$, let $E(D) \in \widetilde{\mathcal{D}}_{\Gamma_X}(\mathfrak{X}_{\omega_1})$ be the unique extension of D. Notice that $D|X \in \mathcal{D}(X)$ for every $D \in \mathcal{D}_{\Gamma_X}(\mathfrak{X}_{\omega_1})$.

Proposition VII.21. $\|ED\| \leq C\|D\|$ *for every $D \in \mathcal{D}(X)$. Moreover the restriction $\mathcal{D}_{\Gamma_X}(\mathfrak{X}_{\omega_1}) \to \mathcal{D}(X)$, $D \mapsto D|X$ is an isomorphism onto with inverse $E : \mathcal{D}(X) \to \mathcal{D}_{\Gamma_X}(\mathfrak{X}_{\omega_1})$.*

Proof. We show that $\|E(D)\| \leq C\|D\|$ for every $D \in \mathcal{D}(X)$. Fix a finitely supported $y \in \mathfrak{X}_{\omega_1}$ such that $\|y\| \leq 1$ and $\|E(D)\| = \|E(D)(y)\|$. Since $\mathcal{I} = \{[\alpha^-_{\Gamma_X}, \alpha) : \alpha \in \Gamma^{(0)}_X\} \cup \{[\max \Gamma_X, \omega_1)\}$ is a partition of ω_1, y has a unique decomposition $y = y_1 + \cdots + y_n$ for $I_1 < \cdots < I_n$ in $\mathcal{I}$ and $y_i \in \langle e_\alpha\rangle_{\alpha\in I_i}$. Notice that $E(D)|\mathfrak{X}_{[\max \Gamma_X,\omega_1)} = 0$, so we may assume that $I_n \neq [\max \Gamma_X, \omega_1)$. By definition of $E(D)$ we have that $E(D)(y) = \sum_{i=1}^n \xi_D(\beta_i) y_i$ where $\beta_i = f_X^{-1}(\alpha_i)$ for every $i = 1, \ldots, n$. Choose $\phi \in K_{\omega_1}$ such that $\|E(D)(y)\| = \phi(E(D)(y))$. By Proposition VII.17,

$$\|E(D)\| = \phi(\textstyle\sum_{i=1}^n \xi_D(\beta_i)y_i) = \sum_{i=1}^n \xi_D(\beta_i)\phi(y_i) = \Xi(D)(\textstyle\sum_{i=1}^n \phi(y_i)v_{\beta_i})$$

$$\text{(VII.10)}$$

$$\leq \|\Xi(D)\|_{J_{T_0}^*(\Lambda(\gamma+1)^{(0)})} \|\textstyle\sum_{i=1}^n \phi(y_i)v_i\|_{J_{T_0}} \leq C\|D\|. \qquad\qquad \text{(VII.11)}$$

$$\square$$

Chapter VIII

The Spaces of Operators $\mathcal{L}(\mathfrak{X}_\gamma)$, $\mathcal{L}(X, \mathfrak{X}_{\omega_1})$

This is a continuation of the previous part, and the main result is that for every subspace X of $\mathfrak{X}_{\omega_1}$ there exists a closed subset Γ_X of ω_1 such that $\mathcal{L}(\mathfrak{X}_{\omega_1})$ is isomorphic to $J^*_{T_0}(\Gamma_X) \oplus \mathcal{S}(X, \mathfrak{X}_{\omega_1})$. This representation yields some further properties for the space $\mathfrak{X}_{\omega_1}$, for example for I, J disjoint intervals of ω_1, the correspon ding spaces $\mathfrak{X}_I$, $\mathfrak{X}_J$ are totally incomparable, which yields that $\mathfrak{X}_{\omega_1}$ is arbitrarily distortable.

The spaces $\mathcal{L}(\mathfrak{X}_\gamma)$

Definition VIII.1. A sequence $(x_1, \phi_1, \ldots, x_{n_{2j+1}}, \phi_{n_{2j+1}})$ is called a $(0, j)$-*dependent sequence* if the following conditions are fulfilled:

DS0.1 $\Phi = (\phi_1, \ldots, \phi_{n_{2j+1}})$ is a $2j + 1$-special sequence and $\phi_i(x_{i'}) = 0$ for every $1 \leq i, i' \leq n_{2j+1}$.

DS0.2 There exists $\{\psi_1, \ldots, \psi_{n_{2j+1}}\}$ such that $w(\psi_i) = w(\phi_i)$, $\#\operatorname{supp} x_i \leq \leq w(\phi_{i+1})/n_{2j+1}^2$ and (x_i, ψ_i) is a $(6, 2j_i)$-exact pair for every $1 \leq i \leq n_{2j+1}$.

DS0.3 If $H = (h_1, \ldots, h_{n_{2j+1}})$ is an arbitrary $2j + 1$-special sequence, then

$$\left(\bigcup_{\kappa_{\Phi, H} < i < \lambda_{\Phi, H}} \operatorname{supp} x_i\right) \cap \left(\bigcup_{\kappa_{\Phi, H} < i < \lambda_{\Phi, H}} \operatorname{supp} h_i\right) = \emptyset. \qquad \text{(VIII.1)}$$

Proposition VIII.2. *For every $(0, j)$-dependent sequence $(x_1, \phi_1, \ldots, x_{n_{2j+1}}, \phi_{n_{2j+1}})$ we have that*

$$\left\|\frac{1}{n_{2j+1}}(x_1 + \cdots + x_{n_{2j+1}})\right\| \leq \frac{1}{m_{2j+1}^2}.$$

Proof. The proof is rather similar to the proof of Proposition VI.9. One first shows that

$$|\psi(1/n_{2j+1} \sum_{i \in E} x_i)| \leq 12(1 + \#E/n_{2j+1}^2)$$

for every special functional ψ with $w(\psi) = m_{2j+1}$, and then the result follows from the basic inequality, since, by condition (DS0.2), $(x_i)_{i=1}^{n_{2j+1}}$ is a $(12, 1/n_{2j+1}^2)$-RIS.　　　　　$\square$

Proposition VIII.3. *Suppose that $(y_k)_k$ is a (C, ε)-RIS, and suppose that $T : \langle y_k \rangle_k \rightarrow \mathfrak{X}_{\omega_1}$ is a linear function (not necessarily bounded) such that $\lim_{n \to \infty} d(Ty_n, \mathbb{R}y_n) \neq 0$. Then for every $\varepsilon > 0$ there is some $z \in \langle y_k \rangle_k$ such that $\|z\| < \varepsilon \|Tz\|$.*

Proof. We may assume that there is some $\delta > 0$ such that $\inf_n d(Ty_n, \mathbb{R}y_n) > \delta > 0$, and also that $(Ty_n)_n$ is a block sequence (hint: consider the following limit ordinal $\gamma_0 = \min\{\gamma < \omega_1 : \exists A \in [\mathbb{N}]^\infty \inf_{n \in A} d(P_\gamma Ty_n, \mathbb{R}y_n) > 0\}$, and pass, if necessary, to a subsequence of $(y_n)_n$ and replace T by $P_{\gamma_0}T$).

Claim. *There exist an infinite set $A \subseteq \mathbb{N}$ and a block sequence $(f_n)_{n \in A}$ of functionals in K such that:*
(a) For every $n \in A$, $f_n Ty_n \geq \delta$, $f_n y_n = 0$, $\operatorname{ran} f_n \subseteq \operatorname{ran} Ty_n$ and $\operatorname{supp} f_n \cap \operatorname{supp} y_m = \emptyset$ for every $m \neq n$.
(b) Either for every $n \in A$, $\max \operatorname{supp} y_n \geq \max \operatorname{supp} f_n$ or for every $n \in A$, $\max \operatorname{supp} y_n \leq \max \operatorname{supp} f_n$.

Proof of the claim. By the Hahn–Banach theorem, for each $n \in \mathbb{N}$ we can find a functional f_n of norm 1 such that $f_n(Ty_n) \geq \delta$ and $f_n(y_n) = 0$. Since the w^*-closure of K is $B_{\mathfrak{X}_{\omega_1}^*}$ (notice that K by definition is closed under rational convex combinations) and K is closed under intervals, we may assume that $f_n \in K$ and $\operatorname{ran} f_n \subseteq \operatorname{ran} Ty_n$. Let $\alpha = \max_n \operatorname{supp} y_n$ and $\beta = \max_n \operatorname{supp} f_n$. If $\alpha \neq \beta$, it is rather easy to achieve the desired result. If $\alpha = \beta$, then we can pass to a subsequence A and distort f_n such that for every $n \in A$, $\max \operatorname{supp} f_n \geq \max \operatorname{supp} y_n$.　　　　　$\square$

So, we may assume that $(f_n)_{n \geq 1}$ satisfies the requirements of the previous claim. Fix now j such that $m_{2j+1} > 12/(\varepsilon\delta)$.

Claim. *There is a $(0, j)$-dependent sequence $(z_1, \phi_1, \ldots, z_{n_{2j+1}}, \phi_{n_{2j+1}})$ such that for every $k \leq n_{2j+1}$, $z_k \in X$, $\operatorname{ran} \phi_k \subseteq \operatorname{ran} Tz_k$ and $\phi_k Tz_k > \delta$.*

Proof of the claim. Choose j_1 even such that $m_{2j_1} > n_{2j+1}^2$, and choose $F_1 \subseteq \mathbb{N}$ of size n_{2j_1} such that $(y_k)_{k \in F_1}$ is a $(3, 1/n_{2j_1}^2)$-RIS (going to a subsequence of $(y_k)_k$ if necessary). Set

$$\phi_1 = \frac{1}{m_{2j_1}} \sum_{i \in F_1} f_i \in K_{\omega_1} \text{ and } z_1 = \frac{m_{2j_1}}{n_{2j_1}} \sum_{k \in F_1} y_k.$$

Note that $\phi_1 T z_1 = (1/n_{2j_1}) \sum_{k \in F_1} f_k T y_k > \delta$ and by (a) from the above claim, we have that $\phi_1 z_1 = (1/n_{2j_1}) \sum_{k \in F_1} \sum_{l \in F_1} f_k(y_l) = 0$. Pick $p_1 \geq \max\{p_\varrho(\operatorname{supp} z_1 \cup \operatorname{supp} T z_1 \cup \operatorname{supp} \phi_1), \# \operatorname{supp} z_1 \cdot n_{2j+1}^2\}$ and set $2j_2 = \sigma_\varrho(\Phi_1, m_{2j_1}, p_1)$. Now choose $F_2 > F_1$ finite of length n_{2j_2} such that $(x_k)_{k \in F_2}$ is a $(3, 1/n_{2j_2}^2)$-RIS. Set $\phi_2 = (1/m_{2j_2}) \sum_{k \in F_2} f_k \in K_{\omega_1}$ and $z_2 = (m_{2j_2}/n_{2j_2}) \sum_{k \in F_2} y_k$. Notice that $\phi_2 > \phi_1$, $\phi_2 T z_2 > \delta$ and $\phi_2 z_2 = 0$. Pick $p_2 \geq \max\{p_1, p_\varrho(\operatorname{supp} z_1 \cup \operatorname{supp} z_2 \cup \operatorname{supp} T z_1 \cup \operatorname{supp} T z_2 \cup \operatorname{supp} \Phi_1 \cup \operatorname{supp} \Phi_2), \# \operatorname{supp} z_2 \cdot n_{2j+1}^2\}\}$ and set $2j_3 = \sigma_\varrho(\phi_1, m_{2j_1}, p_1, \phi_2, m_{2j_2}, p_2)$, and so on. Let us check that $(z_1, \phi_1, \ldots, z_{n_{2j+1}}, \phi_{n_{2j+1}})$ is a $(0, j)$-dependent sequence: Conditions (DS0.1) and (DS0.2) are rather easy to check from the definition of this sequence. Let us check (DS0.3). There are two cases: (a) Suppose that $\max \operatorname{supp} z_k \leq \max \operatorname{supp} \phi_k$ for every $1 \leq k \leq n_{2j+1}$. Then $\operatorname{supp} z_k \subseteq \operatorname{supp} \overline{\phi_{\lambda_{\Phi,H}-1}}^{p_{\lambda_{\Phi,H}-1}}$ for every $\kappa_{\Phi,H} < k < \lambda_{\Phi,H}$. Then part 2 of (TP.3) gives the desired result. (b) Suppose that $\max \operatorname{supp} \phi_k \leq \max \operatorname{supp} z_k$ for every $1 \leq k \leq n_{2j+1}$, then $\operatorname{supp} \phi_k \subseteq \operatorname{supp} \overline{z_{\lambda_{\Phi,H}-1}}^{p_{\lambda_{\Phi,H}-1}}$ for every $\kappa_{\Phi,H} < k < \lambda_{\Phi,H}$, and we are done by part 1 of (TP.3). $\qquad\square$

Fix a $(0, j)$-dependent sequence $(z_1, \phi_1, \ldots, z_n, \phi_{n_{2j+1}})$ as in the claim, and set

$$z = \tfrac{1}{n_{2j+1}} \sum_{k=1}^{n_{2j+1}} (-1)^{k+1} z_k \quad \text{and} \quad \phi = \tfrac{1}{m_{2j+1}} \sum_{k=1}^{n_{2j+1}} \phi_k.$$

Then $\phi T z = 1/n_{2j+1} \sum_{k=1}^{n_{2j+1}} (-1)^{k+1} \phi T z_k \geq \delta/m_{2j+1}$ and $\|z\| \leq 12/m_{2j+1}^2$. So, $\|T(z)\| \geq \delta/m_{2j+1} > \delta m_{2j+1} \|z\|/12 > \varepsilon \|z\|$ as desired. $\qquad\square$

Corollary VIII.4. *Let* $(y_k)_k$ *be a* (C, ε)-RIS, Y *its closed linear span and* $T : Y \to \mathfrak{X}_{\omega_1}$ *be a bounded operator. Then* $\lim_{n \to \infty} d(T y_k, \mathbb{R} y_k) = 0$.

Proof. If not, by previous Proposition VIII.3, we can find a vector $z \in \langle y_k \rangle_k$ such that $\|z\| < (1/\|T\|) \|T z\|$ which is impossible if T is bounded. $\qquad\square$

Lemma VIII.5. *Let* $(x_n)_n$ *be a* (C, ε)-RIS, X *its closed span and* $T : X \to \mathfrak{X}_{\omega_1}$ *be a bounded operator. Then* $\lambda_T : \mathbb{N} \to \mathbb{R}$ *defined by* $d(T x_n, \mathbb{R} x_n) = \|T x_n - \lambda_T(n) x_n\|$ *is a convergent sequence.*

Proof. Fix any two strictly increasing sequences $(\alpha_n)_n$ and $(\beta_n)_n$ with $\sup_n \alpha_n = \sup_n \beta_n$, and suppose that $\lambda_T(\alpha_n) \to_n \lambda_1$, $\lambda_T(\beta_n) \to_n \lambda_2$. By going to a subsequences, we can assume that $x_{\alpha_n} < x_{\beta_n}$ for every n. Since the closed linear span of $\{x_{\alpha_n}\}_n \cup \{x_{\beta_n}\}_n$ is an HI space, we can find for every ε two normalized vectors $w_1 \in \langle x_{\alpha_n} \rangle_n$ and $w_2 \in \langle x_{\beta_n} \rangle_n$ such that $\|T w_1 - \lambda_1 w_1\| \leq \varepsilon/3$, $\|T w_2 - \lambda_2 w_2\| \leq \varepsilon/3$ and $\|w_1 - w_2\| \leq \varepsilon/3\|T\|$. Then we have that

$$\|\lambda_1 w_1 - \lambda_2 w_2\| \leq \|T w_1 - \lambda_1 w_1\| + \|T w_1 - T w_2\| + \|T w_2 - \lambda_2 w_2\| \leq \varepsilon, \quad \text{(VIII.2)}$$

and hence,

$$\varepsilon \geq \|\lambda_1 w_1 - \lambda_2 w_2\| \geq |\lambda_1 - \lambda_2| \|w_1\| - |\lambda_2| \|w_1 - w_2\| \geq |\lambda_1 - \lambda_2| - |\lambda_2| \varepsilon. \quad \text{(VIII.3)}$$

So, $|\lambda_1 - \lambda_2| \leq \varepsilon(1 + |\lambda_2|)$ for every ε. This implies that $\lambda_1 = \lambda_2$. $\qquad\square$

Definition VIII.6. Recall that for a set A of ordinals, $A^{(0)}$ is the set of isolated points of A. Fix a transfinite block sequence $(x_\alpha)_{\alpha<\gamma}$, let X be the closed linear span of it and let $T : X \to \mathfrak{X}_{\omega_1}$ be a bounded operator. We define the *step function* ξ_T of T, $\xi_T : \Lambda(\gamma+1)^{(0)} \to \mathbb{R}$ as follows: Let γ be a successor limit ordinal less than γ. Let $\xi_T(\gamma) = \xi \in \mathbb{R}$ be such that $\lim_{n\to\infty} \|Ty_n - \xi y_n\| = 0$ for every $(3,\varepsilon)$-RIS $(y_n)_n$ satisfying that $\sup_n \max \operatorname{supp} y_n = \gamma$. Lemma VIII.5 shows that ξ exists and is unique, and that ξ_T can be extended to a continuous $\widetilde{\xi}_T : \Lambda(\gamma+1) \to \mathbb{R}$.

Given a mapping $\xi : \Lambda(\gamma+1)^{(0)} \to \mathbb{R}$ we define the diagonal, not necessarily bounded, operator $D_\xi : X \to X$ in the natural way by $D_\xi(x_\alpha) = \xi(\alpha+\omega)x_\alpha$. Given a bounded $T : X \to \mathfrak{X}_{\omega_1}$ we define the *diagonal step operator* $D_T : \langle x_\alpha \rangle_{\alpha<\gamma} \to \mathfrak{X}_{\omega_1}$ of T as $D_T = D_{\xi_T}$.

Remark VIII.7. ξ_T has only countable many values, since it can be extended to a continuous mapping $\widetilde{\xi}_T$ defined in $\Lambda(\gamma+1)$.

Proposition VIII.8. *The sequence* $(\|(T-D_T)(y_n)\|)_{\alpha<\gamma}$ *belongs to $c_0(\mathbb{N})$ for every RIS $(y_n)_n$ in X.*

Proof. This is just a consequence of the definition of D_T. $\square$

Proposition VIII.9. *A bounded operator $T : X \to \mathfrak{X}_{\omega_1}$ is strictly singular iff $\xi_T = 0$.*

Proof. Suppose that T is not strictly singular. Then there is a block sequence $(y_n)_n$ such that T is an isomorphism restricted to the closed linear span Y of $(y_n)_n$. Going to a block subsequence if necessary we assume that $(y_n)_n$ is a RIS. Since $T|Y$ is an isomorphism, $\lim_{n\to\infty} \|Ty_n\| > 0$. This implies that $\xi_T|\Lambda(\alpha+1)^{(0)} \neq 0$, since otherwise $\widetilde{\xi}_T(\alpha) = 0$, contradicting the above inequality.

Suppose now that $\xi_T \neq 0$. Choose some successor limit γ such that $\xi_T(\gamma) \neq 0$. Then we can find a block sequence $(y_n)_n \subseteq X_\gamma$ such that T is close enough to $\xi_T(\gamma)i_{Y,\mathfrak{X}_{\omega_1}}$, where Y is the closed linear span of $(y_n)_n$. Hence, T is not strictly singular. $\square$

Proposition VIII.10. *Let $(x_\alpha)_{\alpha<\gamma}$ be a transfinite block sequence, X its closed linear span of $(x_\alpha)_{\alpha<\gamma}$ and a bounded operator $T : X \to \mathfrak{X}_{\omega_1}$. Then $\|D_T\| \leq C\|T\|$ and hence $D_T \in \mathcal{D}(X)$.*

Proof. Fix a normalized $y \in \langle x_\alpha \rangle_{\alpha<\gamma}$, let $y = y_1 + \cdots + y_n$ be its decomposition in X, $y_i \in \langle x_\beta \rangle_{\beta \in [\alpha_i^-, \alpha_i)}$ for $i = 1, \ldots, n$. Choose $\phi \in K_{\omega_1}$ such that $\phi(D(y)) = \|D(y)\|$. Then,

$$\|D(y)\| = \sum_{i=1}^n \xi_T(\alpha_i)\phi(y_i) = \left(\sum_{i=1}^n \xi_T(\alpha_i)v_i^*\right)\left(\sum_{i=1}^n \phi(y_i)v_i\right) \leq \left\| \sum_{i=1}^n \xi_T(\alpha_i)v_i \right\|, \tag{VIII.4}$$

the last inequality because $\|\sum_{i=1}^n \phi(y_i)v_{\alpha_i}\|_{J_{T_0}} \leq \|y\|_{\mathfrak{X}_{\omega_1}} \leq 1$. We finish with the next claim. $\square$

Claim. $\|\sum_{i=1}^n \xi_T(\alpha_i)v_i^*\|_{J_{T_0}^*} \leq C\|T\|$.

Proof of the claim. Fix $\varepsilon > 0$. By the finite block representability of J_{T_0} in $\mathfrak{X}_{\omega_1}$ and Proposition VIII.8 we can produce inductively $w_1, \ldots, w_n$ such that

(1) $w_i \in \langle x_\beta \rangle_{\beta \in [\alpha_i^-, \alpha_i)}$,

(2) the natural isomorphism $F : \langle w_i \rangle_{i=1}^n \to \langle v_i \rangle_{i=1}^n$ is such that $\|F\| \leq 1$ and $\|F^{-1}\| \leq C$, and

(3) $\sum_{i=1}^n \|\xi_T(\alpha_i) w_i - T w_i\| < \varepsilon$.

Choose $x = \sum_{i=1}^n r_i v_i \in J_{T_0}$ of norm 1 such that $\|\sum_{i=1}^n \xi_T(\alpha_i) v_i^*\|_{J_{T_0}^*} = \sum_{i=1}^n \xi_T(\alpha_i) r_i$. Then $\|\sum_{i=1}^n r_i w_i\|_{\mathfrak{X}_{\omega_1}} \leq C$ and hence

$$\|D_T(\sum_{i=1}^n r_i w_i)\| \geq \|\sum_{i=1}^n r_i \xi_T(\alpha_i) v_i\|_{J_{T_0}} \qquad \text{(VIII.5)}$$

$$\geq \sum_{i=1}^n \xi_T(\alpha_i) r_i = \|\sum_{i=1}^n \xi_T(\alpha_i) v_i^*\|_{J_{T_0}^*}.$$

Hence $\|\sum_{i=1}^n \xi_T(\alpha_i) v_i^*\|_{J_{T_0}^*} \leq \|T(\sum_{i=1}^n r_i w_i)\| + \|(T - D_T)(\sum_{i=1}^n r_i w_i)\| \leq C\|T\| + \varepsilon$. $\qquad \square$

Theorem VIII.11. *Let* $(x_\alpha)_{\alpha < \gamma}$ *be a normalized block sequence of* $\mathfrak{X}_{\omega_1}$, *X its closed linear span. Then, for every bounded operator* $T : X \to \mathfrak{X}_{\omega_1}$, $D_T : X \to \mathfrak{X}_{\omega_1}$ *is bounded and* $T - D_T$ *is strictly singular.*

Proof. It follows from Proposition VIII.9 and Proposition VIII.10. $\qquad \square$

Corollary VIII.12. *Any bounded operator from the closed linear span* X *of a transfinite block sequence into the space* $\mathfrak{X}_{\omega_1}$ *is the sum of the restriction of a unique diagonal operator* $D \in \mathcal{D}_X(\mathfrak{X}_{\omega_1})$ *and an strictly singular operator.*

Proof. This follows from previous theorem and Proposition VII.21. $\qquad \square$

Corollary VIII.13. (a) *For* $T : X \to \mathfrak{X}_{\omega_1}$ *bounded TFAE:* (i) T *is strictly singular,* (ii) $\xi_T = 0$, *and* (iii) $D_T = 0$.

(b) *The transformation* $T \mapsto D_T$ *is a projection in the operator algebra* $\mathcal{L}(X)$ *of norm* $\leq C$. $\qquad \square$

Proposition VIII.14. *Let* $X \hookrightarrow \mathfrak{X}_{\omega_1}$, $I \subseteq \omega_1$ *an interval such that* $P_I | X$ *is not strictly singular. Then for every* $\varepsilon > 0$ *there exist a normalized sequence* $(x_n)_n$ *in* X *and a normalized block sequence* $(z_n)_n$ *in* $\mathfrak{X}_I$ *such that* $\sum_n \|y_z - z_n\| < \varepsilon$.

Proof. Set $I = [\alpha, \beta]$ and suppose that $P_I | X$ is not strictly singular. Let

$$\gamma_0 = \{\gamma \in (\alpha, \beta] : P_\gamma | X \text{ is not strictly singular}\}.$$

We can find for every $\varepsilon > 0$, $(y_n)_n \subseteq X$ and a block sequence $(w_n)_n \subseteq \mathfrak{X}_{\gamma_0}$ such that P_{γ_0} is an isomorphism when restricted to the closed linear span of $(y_n)_n$,

$\sup_n \max \operatorname{supp} w_n = \gamma_0$ and $\sum_n \|w_n - P_{\gamma_0} y_n\| \leq \varepsilon/2$. Consider $U : \overline{\langle w_n \rangle_n} \to \mathfrak{X}_{[\gamma_0, \omega_1)}$ defined by $Uw_n = P_{[\gamma_0, \omega_1)} y_n$. Notice that U is bounded. Since $\xi_U = 0$, U is strictly singular. Hence we can find a block sequence $(z_n)_n$ of $(w_n)_n$ such that for all n, $\|Uz_n\| \leq \varepsilon/2^{n+1}$ and hence the corresponding block sequence $(x_n)_n$ of $(y_n)_n$ satisfies that $\sum_n \|z_n - x_n\| \leq \varepsilon$. Finally, notice that for n_0 large enough, $(z_n)_{n \geq n_0} \subseteq \mathfrak{X}_I$. $\qquad\square$

Corollary VIII.15. *$\mathfrak{X}_{\omega_1}$ is arbitrarily distortable.*

Proof. For $j \in \mathbb{N}$, and $x \in \mathfrak{X}_{\omega_1}$, let $\|x\|_{2j} = \sup\{\phi(x) : w(\phi) = m_{2j}\}$. Let $X \hookrightarrow \mathfrak{X}_{\omega_1}$. Since for every $\varepsilon > 0$ we can find a subspace of X generated by a Schauder basis $(y_n)_n$ and a normalized block sequence $(z_n)_n$ of $\mathfrak{X}_{\omega_1}$ such that $\sum_n \|y_n - z_n\| \leq \varepsilon$, without loss of generality we can assume that X is generated by a block sequence $(z_n)_n$. Now, we can find an $(6, j)$-exact pair (x, ϕ), with $x \in \langle z_n \rangle_n$ and hence $1 \leq \|x\|_{2j} \leq \|x\| \leq 6$. And for any other $j' > j$, a $(6, 2j')$-exact pair (x', ϕ') with $x' \in \langle z_n \rangle_n$ and hence $1 \leq \|x'\| \leq 6$ and $\|x'\|_{2j} \leq 12/m_{2j}$. So,

$$\frac{\|x/\|x\|\|_{2j}}{\|x'/\|x'\|\|_{2j}} \geq \frac{1/6}{12/m_{2j+1}} = \frac{m_{2j+1}}{72}. \qquad\qquad \text{(VIII.6)}$$
$\qquad\square$

Definition VIII.16. Two Banach spaces X and Y are called *totally incomparable* if and only if no infinite dimensional closed $X_1 \hookrightarrow X$ is isomorphic to $Y_1 \hookrightarrow Y$.

Corollary VIII.17. *$\mathfrak{X}_I$ and $\mathfrak{X}_J$ are totally incomparable for disjoint infinite intervals I and J.*

Proof. Suppose not, and let $X \hookrightarrow \mathfrak{X}_I$, and $Y \hookrightarrow \mathfrak{X}_J$ such that $T : X \to Y$ is an isomorphism onto. By previous Proposition VIII.14, we can assume that X is generated by a block sequence. But since $\xi_T = 0$, T cannot be an isomorphism, a contradiction. $\qquad\square$

Another consequence of the representability of J_{T_0} on each transfinite block sequence is that we can identify the space $\mathcal{D}(X)$ of diagonal step operators on X and hence $\mathcal{L}(X)/\mathcal{S}(X)$ for every closed span X of a transfinite block sequence.

Corollary VIII.18. *$\mathcal{L}(X)/\mathcal{S}(X) \cong \mathcal{L}(X, \mathfrak{X}_{\omega_1})/\mathcal{S}(X, \mathfrak{X}_{\omega_1}) \cong J^*_{T_0}(\Gamma^{(0)}_X)$ for every $X \hookrightarrow \mathfrak{X}_{\omega_1}$ generated by a transfinite block sequence.*

Proof. This follows from Lemma VII.18, since $\Lambda(\gamma + 1)^{(0)}$ and $\Gamma^{(0)}_X$ are order-isomorphic. $\qquad\square$

Remark VIII.19. Note that $\mathcal{L}(X)/\mathcal{S}(X) \cong J^*_{T_0}(\Gamma_X)$ if Γ_X is infinite. To see this, fix a transfinite block sequence $(x_\alpha)_{\alpha < \gamma}$ generating X such that $\gamma \geq \omega^2$. Then $\Gamma_X \setminus \{\max \Gamma_X\}$ and $\Lambda(\gamma + 1)^{(0)} \setminus \{\omega\}$ are order-isomorphic.

Theorem VIII.20. *Every projection P of $\mathfrak{X}_{\omega_1}$ is of the form $P = P_{I_1} + \cdots + P_{I_n} + S$, where I_i are intervals of ordinals, $I_i < I_{i+1}$ and S is strictly singular.*

Proof. Suppose that $P : \mathfrak{X}_{\omega_1} \to \mathfrak{X}_{\omega_1}$ is a projection, $P = D_P + S$. Since $P^2 = P$, we obtain that $D_P^2 - D_P$ is also strictly singular and therefore $(\xi_P(\alpha)^2 - \xi_P(\alpha))i_{\mathfrak{X}_{[\alpha^-, \alpha)}, \mathfrak{X}_{\omega_1}}$ is strictly singular for every successor limit α. This implies that $\xi_P : \Lambda(\omega_1 + 1)^{(0)} \to \{0, 1\}$. And since ξ_P has the continuous extension property, there is no strictly increasing sequence $\{\alpha_n\}_n \subseteq \Lambda(\omega_1+1)^{(0)}$ such that $\xi_P(\alpha_{2n}) = 1$ and $\xi_P(\alpha_{2n+1}) = 0$ for every n. □

Corollary VIII.21. *For every $n \in \mathbb{N}$ there is some $m \in \mathbb{N}$ such that for every projection P of $\mathfrak{X}_{\omega_1}$ with $\|P\| \leq n$, P can be written as $P = P_{I_1} + \cdots + P_{I_k} + S$ such that $k \leq m$ and $I_1 << I_2 << \cdots << I_k$, where $A << B$ denotes that the interval $(\sup A, \inf B)$ is infinite.*

Proof. Fix n, and let $P : \mathfrak{X}_{\omega_1} \to \mathfrak{X}_{\omega_1}$ be a projection such that $\|P\| \leq n$. Let j be the first integer such that $m_{2j} > 2nC$. We claim that $m = n_{2j}$ works. For suppose that $P = P_{I_1} + \cdots + P_{I_k} + S$ with $I_1 << \cdots << I_k$ and $k > n_{2j}$. Fix $\varepsilon > 0$. Find a normalized block sequence $(x_1, y_1, \ldots, x_{n_{2j}/2}, y_{n_{2j}/2})$ such that

(a) $x_i \in \mathfrak{X}_{I_i}$, $y_i \in \mathfrak{X}_{(\sup I_i, \min I_{i+1})}$ for $1 \leq i \leq n_{2j}/2 - 1$, and $y_{n_{2j}/2} > x_{n_{2j}/2}$,

(b) $(x_1, y_1, \ldots, x_{n_{2j}/2}, y_{n_{2j}/2})$ is C-equivalent to $(v_i)_{i=1}^{n_{2j}}$, and

(c) $\|S|F\|$ $\leq$ ε where F $=$ $\langle (x_1,, y_1, \ldots, x_{n_{2j}/2}, y_{n_{2j}/2}) \rangle$.

Set $x = x_1 - y_1 + \cdots + x_{n_{2j}/2} - y_{n_{2j}/2}$. Then,

$$\|x\| < C\| \textstyle\sum_{i=1}^{n_{2j}} (-1)^{i+1} v_i \|_{J_{T_0}} \leq C\| \textstyle\sum_{i=1}^{n_{2j}} l_i \|_{T_0} = C n_{2j}/m_{2j}, \qquad (\text{VIII.7})$$

and

$$\|P(x)\| \geq \| \textstyle\sum_{i=1}^{n_{2j}/2} x_i \| - \varepsilon \geq \| \textstyle\sum_{i=1}^{n_{2j}/2} v_i \|_{J_{T_0}} - \varepsilon = n_{2j}/2 - \varepsilon. \qquad (\text{VIII.8})$$

(VIII.7) and (VIII.8) imply that $\|P\| \geq (m_{2j}/2 - \varepsilon m_{2j}/n_{2j})/C$. Hence, $\|P\| > n$, a contradiction. □

Asymptotically equivalent subspaces and $\mathcal{L}(X, \mathfrak{X}_{\omega_1})$

Our aim here is to extend the results about operators on subspaces generated by a transfinite block sequence to arbitrary subspaces.

Definition VIII.22. Let X be a subspace of $\mathfrak{X}_{\omega_1}$. A subset Γ of $\omega_1 + 1$ is said to be a *critical set* of X if the following hold:

(CS1) Γ is closed of limit ordinals, and $0 \in \Gamma$.

(CS2) For all $\gamma \in \Gamma$, $\gamma < \Omega$, $P_{(\gamma, \gamma^+)}|X$ is not strictly singular and for all $\alpha \in (\gamma, \gamma^+)$, $P_{(\gamma, \alpha)}|X$ is strictly singular, where γ^+ is the successor of γ in Γ and $\Omega = \max \Gamma$.

(CS3) $P_{[\Omega, \omega_1)}|X$ is strictly singular (we use $P_\emptyset = 0$).

Notice that from the above definition, it follows easily that if Γ is a critical set of X, then

$$\max \Gamma = \min\{\gamma \leq \omega_1 : P_{[\gamma,\omega_1)}|X \text{ is strictly singular}\}.$$

Proposition VIII.23. *For every* $X \hookrightarrow \mathfrak{X}_{\omega_1}$ *a critical set* Γ *is uniquely defined, denoted by* Γ_X.

Proof. Fix $X \hookrightarrow \mathfrak{X}_{\omega_1}$. We show first that a critical set X exists. We proceed by induction defining an increasing sequence $(\gamma_\alpha)_{\alpha<\omega_1}$ as follows: We set $\gamma_0 = 0$. Suppose defined $(\gamma_\beta)_{\beta<\alpha}$ satisfying conditions (CS1) and (CS2). If α is limit, then we set $\gamma_\alpha = \sup_{\beta<\alpha} \gamma_\beta$. Suppose now that α is successor. If $P_{[\gamma_{\alpha-},\omega_1)}|X$ is strictly singular, then we set $\gamma_\alpha = \gamma_{\alpha-}$. If not, let

$$\gamma_\alpha = \min\{\gamma \in (\gamma_{\alpha-}, \omega_1) : P_{[\gamma_{\alpha-},\gamma)}|X \text{ is not strictly singular}\}.$$

Let us observe that if X is separable, then the sequence $(\gamma_\alpha)_{\alpha<\omega_1}$ is eventually constant and we set $\Gamma_X = \{\gamma_\alpha\}_{\alpha<\omega_1}$. If X is non separable, then the sequence $(\gamma_\alpha)_{\alpha<\omega_1}$ is strictly increasing and $\Gamma_X = \{\gamma_\alpha\}_{\alpha<\omega_1} \cup \{\omega_1\}$.

Next we prove the uniqueness of Γ_X. Suppose the opposite, and fix $\Gamma \neq \Gamma'$ two different critical sets. Set $\gamma = \max(\Gamma \cap \Gamma')$. First notice that $\max \Gamma = \max \Gamma'$. So, either $\gamma_\Gamma^+ < \gamma_{\Gamma'}^+$ or $\gamma_{\Gamma'}^+ < \gamma_\Gamma^+$. This leads to a contradiction using the fact that both Γ and Γ' satisfy (CS2). $\square$

Remark VIII.24. 1. The critical set Γ_X provides information concerning the structure of the space X. For example the space X is HI if and only if $\Gamma_X = \{0, \Omega_X\}$. Also, two subspaces $X, Y \hookrightarrow \mathfrak{X}_{\omega_1}$ are totally incomparable if and only if $\Gamma_X \cap \Gamma_Y = \{0\}$.
2. For a transfinite block sequence $(x_\alpha)_{\alpha<\gamma}$ its critical set is nothing else but the set introduced from Definition VII.20 (2).

Proposition VIII.25. *For every* $Y \hookrightarrow X$, *the corresponding critical set* Γ_Y *is a subset of* Γ_X.

Proof. It follows by an easy inductive argument. $\square$

Proposition VIII.26. *For every separable* $X \hookrightarrow \mathfrak{X}_{\omega_1}$ *and for every* $\varepsilon > 0$ *there exist an ordinal* $\gamma < \omega_1$, *a normalized sequence* $(y_\alpha)_{\alpha<\gamma}$ *in* X *and a normalized transfinite block sequence* $(z_\alpha)_{\alpha<\gamma}$ *such that* (a) $\sum_{\alpha<\gamma} \|z_\alpha - x_\alpha\| < \varepsilon$ *and* (b) $\Gamma_X = \Gamma_Z$ *where* Z *is the closed linear span of* $(z_\alpha)_{\alpha<\gamma}$.

Proof. Use Proposition VIII.14, and a standard gliding hump argument. $\square$

Definition VIII.27. Let $X, Y \hookrightarrow \mathfrak{X}_{\omega_1}$.

  (i) We say that X is *asymptotically finer* to Y, $X \leq_a Y$, if and only if $\Gamma_X \subseteq \Gamma_Y$.

  (ii) We say that X is *asymptotically equivalent* to Y, $X \equiv_a Y$, if and only if $\Gamma_X = \Gamma_Y$.

It follows easily from the above definition that the relation $\leq_a$ is a quasi-ordering in the class of the subspaces of $\mathfrak{X}_{\omega_1}$ which from Proposition VIII.25 extends the natural inclusion. Notice also that $\equiv_a$ is an equivalence relation.

We now give two alternative formulation of these notions.

Proposition VIII.28. *For $X, Y \hookrightarrow \mathfrak{X}_{\omega_1}$ the following are equivalent:* (1) $X \leq_a Y$, (2) *if $P_I | X$ is not strictly singular, then $P_I | Y$ is not strictly singular, for every interval $I \subseteq \omega_1$, and* (3) $d(S_{X'}, S_Y) = 0$ *for every $X' \hookrightarrow X$.*

Proof. Let us observe that for a closed infinite interval I, $P_I | X$ is not strictly singular iff there is some $\gamma^+_{\Gamma_X} \in \Gamma_X$ with $\min \Gamma_X < \gamma^+_{\Gamma_X} \leq \max I$. The inverse direction follows immediately from the definition of the critical sets. So assume now that $P_I | X$ is not strictly singular. Set $\gamma_0 = \max\{\gamma \in \Gamma_X : \gamma \leq \min I\}$. Observe that $\gamma_0 \leq \min I < \Omega_X$, hence $\min \Gamma_X < \gamma^+_{\Gamma_X} \leq \max I$ by minimality of $\gamma^+_{\Gamma_X}$ (Property (CS2)). It is easy to see that the above observation implies easily the equivalence (1) $\Leftrightarrow$ (2). (1) $\Rightarrow$ (3): Suppose that $X' \hookrightarrow X$. Then by Proposition VIII.25 and our assumption, $\Gamma_{X'} \subseteq \Gamma_Y$. By Proposition VIII.26, we can find two block sequences $(z_n)_n$ and $(w_n)_n$ in $\mathfrak{X}_{0^+_{\Gamma_{X'}}}$ such that

(a) $\sup_n \max \operatorname{supp} z_n = \sup_n \max \operatorname{supp} w_n = 0^+_{\Gamma_{X'}}$, and

(b) $d(S_Z, S'_X) = d(S_W, S_Y) = 0$ where Z and W are the closed linear span of $(z_n)_n$ and $(w_n)_n$ respectively.

By Corollary VI.12, $d(Z, W) = 0$ and we are done. (3) $\rightarrow$ (2): Since for every $X' \hookrightarrow X$, $d(S_{X'}, S_Y) = 0$, we obtain that for every $\varepsilon > 0$, and every $X' \hookrightarrow X$ there exists two basic sequences $(z_n)_n$ and $(w_n)_n$ such that $z_n \in S_{X'}$ and $w_n \in S_Y$ for all n and $\sum_n \|z_n - w_n\| < \varepsilon$. Assume now that $P_I | X$ is not strictly singular. Choose $X' \hookrightarrow X$ such that $P_I | X'$ is isomorphism. Let $(z_n)_n \subseteq X'$ and $(w_n)_n \subseteq Y$ as above. Then $P_I | W$ is isomorphism and hence $P_I | Y$ is not strictly singular. $\quad\square$

Proposition VIII.29. *For $X, Y \hookrightarrow \mathfrak{X}_{\omega_1}$ the following are equivalent:* (1) $X \equiv_a Y$, (2) $P_I | X$ *is not strictly singular if and only if $P_I | Y$ is not strictly singular, for every interval $I \subseteq \omega_1$, and* (3) $d(S_{X'}, S_Y) = d(S_{Y'}, S_X) = 0$ *for every $X' \hookrightarrow X$, $Y' \hookrightarrow Y$.* $\quad\square$

Corollary VIII.30.   1. *For every $X \hookrightarrow \mathfrak{X}_{\omega_1}$ and every $A \subseteq \Gamma_X$ there is $X_A \hookrightarrow X$ such that $\Gamma_{X_A} = A$.*

   2. *For every non separable $X, Y \hookrightarrow \mathfrak{X}_{\omega_1}$ there are non separable $X_1 \hookrightarrow X$, $Y_1 \hookrightarrow Y$ such that $X_1 \equiv_a Y_1$.* $\quad\square$

We also need the following well known result.

Lemma VIII.31. *For every $Z \hookrightarrow X \hookrightarrow \mathfrak{X}_{\omega_1}$ with Z separable there exist $W \hookrightarrow Z$ and $\gamma < \omega_1$ such that $P_\gamma | X$ is a projection onto W.* $\quad\square$

Remark VIII.32. Notice that for W and X as in the lemma, Γ_W is an initial part of Γ_X.

Proposition VIII.33. *Let X be a subspace of $\mathfrak{X}_{\omega_1}$ and $T : X \to \mathfrak{X}_{\omega_1}$ a bounded operator. Then there exists a unique $D_T \in \mathcal{D}_{\Gamma_X}(\mathfrak{X}_{\omega_1})$ such that* (a) $\|D_T\| \leq 2C^2\|T\|$ *and* (b) $T - D_T|X$ *is strictly singular.*

Proof. Fix $X \hookrightarrow \mathfrak{X}_{\omega_1}$ and a bounded operator $T : X \to \mathfrak{X}_{\omega_1}$. First suppose that X is separable. Then we can find a transfinite basic sequence $(y_\alpha)_{\alpha<\gamma} \subseteq X$ and a transfinite block sequence $(z_\alpha)_{\alpha<\gamma}$ of $\mathfrak{X}_{\omega_1}$ such that $\sum_{\alpha<\gamma} \|y_\alpha - z_\alpha\| < 1$ and $X \equiv_a Z$, where Z denotes the closed linear span of $(z_\alpha)_{\alpha<\gamma}$. Consider now $T' :$ $Z \overset{U}{\to} Y \overset{T|Y}{\to} \mathfrak{X}_{\omega_1}$ where Y is the closed linear span of $(y_\alpha)_{\alpha<\gamma}$ and $U : Y \to Z$ is the isomorphism defined by $U(\sum_{\alpha<\gamma} a_\alpha z_\alpha) = \sum_{\alpha<\gamma} a_\alpha y_\alpha$. Notice that $\|U\| \leq 2$. Then there is a unique $D \in \mathcal{D}(Y)$ such that $T' - D$ is strictly singular, or equivalently there is unique $D_{T'} \in \mathcal{D}_{\Gamma_Z}(\mathfrak{X}_{\omega_1})$ such that $T' - D_{T'}|Z$ is strictly singular. Notice that $\|D_{T'}\| \leq C\|D\| \leq C^2\|T'\| \leq C^2\|U\|\|T\| \leq 2C^2\|T\|$. Let us show that $T - D_{T'}$ is strictly singular. Let $X' \hookrightarrow X$ and $\varepsilon > 0$.

Choose $Z' \hookrightarrow Z$ such that $(\Gamma_{Z'} \setminus \{0\}) \cap (\Gamma_{X'} \setminus \{0\}) \neq \emptyset$, $\|U|Z' - i_{Z',\mathfrak{X}_{\omega_1}}\| \leq \varepsilon/(4\|T\|)$ and $\|(T'-D_{T'})|Z'\| \leq \varepsilon/4$. Pick $z' \in Z'$ and $x' \in X'$ such that $\|z'-x'\| \leq \varepsilon/(2(\|D_{T'}\| + \|T\|))$. Then

$$
\begin{aligned}
\|(T - D_{T'})x'\| &\leq \|(T - D_{T'})x' - (T' - D_{T'})z'\| + \|(T' - D_{T'})z'\| \\
&\leq \|T\|\|x' - Uz'\| + \|D_{T'}\|\|x' - z'\| + \frac{\varepsilon}{4} \\
&\leq (\|T\| + \|D_{T'}\|)\|x' - z'\| + \frac{\varepsilon}{2} \leq \varepsilon.
\end{aligned}
\tag{VIII.9}
$$

Now suppose that X is non separable. By Lemma VIII.31, we can find a sequence $(X_\gamma)_{\gamma<\omega_1}$ of separable complemented subspaces of X such that Γ_{X_γ} is an initial part of Γ_X for every $\gamma < \omega_1$. Now the result for X easy follows from the result for the corresponding $T_\gamma = T|X_\gamma$ and the fact that $D_T \in \mathcal{D}_{\Gamma_X}(\mathfrak{X}_{\omega_1})$ and $D_{T_\gamma} \in \mathcal{D}_{\Gamma_{X_\gamma}}(\mathfrak{X}_{\omega_1})$ are unique. The uniqueness of $D_T \in \mathcal{D}_{\Gamma_X}(\mathfrak{X}_{\omega_1})$ is clear from the analogous result for transfinite block sequences. $\qquad\square$

Theorem VIII.34. $\mathcal{L}(X, \mathfrak{X}_{\omega_1}) \cong \mathcal{D}_{\Gamma_X}(\mathfrak{X}_{\omega_1}) \oplus \mathcal{S}(X, \mathfrak{X}_{\omega_1}) \cong J_{T_0}^*(\Gamma_X^{(0)}) \oplus \mathcal{S}(X, \mathfrak{X}_{\omega_1})$ *for every $X \hookrightarrow \mathfrak{X}_{\omega_1}$. If in addition Γ_X is infinite, then $\mathcal{L}(X, \mathfrak{X}_{\omega_1}) \cong J_{T_0}^*(\Gamma_X) \oplus \mathcal{S}(X, \mathfrak{X}_{\omega_1})$.*

Proof. Let $H : \mathcal{D}_{\Gamma_X} \to \mathcal{L}(X, \mathfrak{X}_{\omega_1})$ be defined by $D \mapsto D|X$. Assume first that X is separable. It is clear that $\|D|X\| \leq \|D\|$. For an appropriate $\varepsilon' > 0$, we can find normalized $(y_\alpha)_{\alpha<\gamma}$ and a normalized block sequence $(z_\alpha)_\alpha$ such that $\Gamma_X = \Gamma_Z$ and $\sum_\alpha \|z_\alpha - y_\alpha\| \leq \varepsilon'$ where Z the closed linear span of $(z_\alpha)_{\alpha<\gamma}$. Since by Proposition VII.21 $\|D|Z\| \geq \|D\|/C$, for C the finite block representability constant of J_{T_0} in $\mathfrak{X}_{\omega_1}$, we get that

$$
\|D\|/C \leq \|D|Z\| \leq (1+\varepsilon)\|D|Y\| \leq (1+\varepsilon)\|D|X\| = (1+\varepsilon)\|H(D)\|.
\tag{VIII.10}
$$

Hence, H defines an isomorphism. To show that H is an isomorphism when X is non separable we use a family $(X_\alpha)_{\alpha<\omega_1}$ of separable complemented subspaces of X defined as in previous proof. Proposition VIII.33 shows that $\mathcal{L}(X, \mathfrak{X}_{\omega_1}) \cong \mathcal{D}_{\Gamma_X}(\mathfrak{X}_{\omega_1}) \oplus \mathcal{S}(X, \mathfrak{X}_{\omega_1})$.

For the latter isomorphism see Remark VIII.19. $\square$

Notes and Remarks. Chapters VI, VII, VIII and the following appendices originate from [13]. It follows easily from the results of Chapter VIII that every $T \in \mathcal{L}(\mathfrak{X}_{\omega_1})$ is of the form $\lambda I + S$ with S an operator with separable range. The first example of a non separable Banach space with few operators, in the above sense, was provided by S. Shelah ([57]) in the late 1970s assuming $V = L$. Ten years later S. Shelah and J. Steprans, using Todorcevic' "square-bracket" coloring [60], were able to provide a ZFC example with few operators [58], and in 2001 H.M. Wark using interpolation methods and the previous space, gave a reflexive space with few operators [64].

None of the above examples controls the separable subspaces as it happens in $\mathfrak{X}_{\omega_1}$. There are variants of the ρ function which impose additional properties in $\mathfrak{X}_{\omega_1}$. Of particular interest is the universal ρ function, introduced in [13] which permits the local structure of $\mathfrak{X}_{\omega_1}$ to be repeated almost everywhere. Using the σ_ρ coding with the universal ρ function one can obtain an HI $\mathfrak{X}$ with a Schauder basis $(e_n)_n$ which belongs to the asymptotic structure of itself in the sense of [47]. A consequence of the representability of $D(\mathfrak{X}_{\omega^2})$ as J_T^* yields that J_T^* with the pointwise multiplication is a Banach algebra. Notice that a similar result for the classical James space [36] is given in [2].

The unconditional counterpart of $\mathfrak{X}_{\omega_1}$ is also presented in [13]. This is a reflexive space with an unconditional basis not isomorphic to any of its proper subspaces. In particular for every infinite $A \subset \omega_1$ denoted by $\mathfrak{X}_A = < \{e_a : a \in A\} >$ the shift operator is not continuous.

As we have mentioned in the introduction it remains open if there exist reflexive Banach spaces not containing an unconditional basic sequence and with dimension greater than ω_1. It is clear that the problem of defining such a space is reduced to defining an efficient coding function σ. S. Todorcevic has extended the notion of a ρ function for cardinals higher than ω_1. However it is not yet clear how we can pass from these ρ functions to an efficient σ_ρ coding.

To make more transparent the meaning of Theorem VIII.34, we state the following:

Theorem. *There exists a separable reflexive Banach space* $\mathfrak{X}$ *admitting an infinite dimensional Schauder decomposition* $\mathfrak{X} = \sum_n \oplus \mathfrak{X}_n$ *such that, denoting by* $\mathcal{D}(\mathfrak{X})$ *the class of bounded operators* $D : \mathfrak{X} \to \mathfrak{X}$ *with the property* $D|_{\mathfrak{X}_n} = \lambda_n I_{\mathfrak{X}_n}$ *for all* n, *the following hold:*

(i) $\mathcal{L}(\mathfrak{X}) \cong \mathcal{D}(\mathfrak{X}) \oplus \mathcal{S}(\mathfrak{X}) \cong J_{T_0}^* \oplus \mathcal{S}(\mathfrak{X})$.

(ii) *For every subspace X of* $\mathfrak{X}$ *there exists* $A \subseteq \mathbb{N}$ *which is either an initial finite interval or is equal to* $\mathbb{N}$ *such that* $\mathcal{L}(X, \mathfrak{X}) \cong J_{T_0}^*(A) \oplus \mathcal{S}(X, \mathfrak{X})$.

For example, the space $\mathfrak{X} = \mathfrak{X}_{\omega^2}$ has all these properties. It is worth pointing out that $D(\mathfrak{X})$ is a natural class of operators which behaves similar to the class of operators of the form $\lambda I + K$ with K compact diagonal. For example if $x_n \in \mathfrak{X}_n$ with $\|x_n\| = 1$ and $X = \overline{\langle (x_n)_n \rangle}$ then for every $D \in \mathcal{D}(\mathfrak{X})$ we have that $D|_X = \lambda I + K$. Let us mention that there exists a reflexive HI space X with a Schauder basis $(e_n)_n$ admitting strictly singular diagonal operators, which are not compact [7]. Moreover the space of strictly singular diagonal operators in this example is non separable. The problem of the existence of a separable reflexive Banach space such that the diagonal operators are compact perturbations of the multiples of the identity seems to be also open.

Appendix A

Transfinite Schauder Basic Sequences

The first appendix concerns the presentation of some results related to transfinite (Schauder) bases. We recall one of the equivalent formulations of their definition.

Definition A.1. Let X be a Banach space, and γ be an ordinal number.
1. A total family $(x_\alpha)_{\alpha<\gamma}$ of elements of X (i.e., a family such that $X = \overline{\langle x_\alpha \rangle_{\alpha<\gamma}}$) is said to be a *transfinite basis* if there exists a constant $C \geq 1$ such that for every interval I of γ the naturally defined map on the linear span of $(x_\alpha)_{\alpha<\gamma}$

$$\textstyle\sum_{\alpha<\gamma} \lambda_\alpha x_\alpha \mapsto \sum_{\alpha \in I} \lambda_\alpha x_\alpha$$

is extended to a bounded projection $P_I : X \to X_I = \overline{\langle x_\alpha \rangle_{\alpha \in I}}$ of norm at most C.
2. A transfinite basis $(x_\alpha)_{\alpha<\gamma}$ of X is said to be *bimonotone* if for each interval I of γ, the corresponding projection P_I has norm 1.
3. A transfinite basis $(x_\alpha)_{\alpha<\gamma}$ of X is said to be *unconditional* if there exists a constant $C \geq 1$ such that for all subset A of γ, the corresponding P_A has norm at most C.
4. A transfinite basis $(x_\alpha)_{\alpha<\gamma}$ of X is said to be *1-subsymmetric* if for every $n \in \mathbb{N}$, every $\alpha_1 < \alpha_2 < \cdots < \alpha_n < \gamma$ and every $(\lambda_i)_{i=1}^n \in \mathbb{R}^n$, $\|\sum_{i=1}^n \lambda_i x_i\| = \|\sum_{i=1}^n \lambda_i x_{\alpha_i}\|$.

Remark A.2. 1. As in the case of the usual Schauder basis (i.e., $\gamma = \omega$) the above definition is equivalent to the fact that each $x \in X$ admits a unique representation as $\sum_{\alpha<\gamma} \lambda_\alpha x_\alpha$, where the convergence of these series is recursively defined.
2. The definition of $\sum_{\alpha<\gamma} \lambda_\alpha x_\alpha$ easily yields that for each convergent series $\sum_{\alpha<\gamma} \lambda_\alpha x_\alpha$ with $(x_\alpha)_{\alpha<\gamma}$ a bounded family, the sequence of coefficients $(\lambda_\alpha)_{\alpha<\gamma}$ belongs to $c_0(\gamma)$. Furthermore, for every $\varepsilon > 0$ there exists a finite subset F of γ such that $\| \sum_{\alpha \notin F} \lambda_\alpha x_\alpha \| < \varepsilon$.

3. For every transfinite basis $(x_\alpha)_{\alpha<\gamma}$ the dual basis $(x_\alpha^*)_{\alpha<\gamma}$ is also well defined. As for the usual Schauder bases, $(x_\alpha^*)_{\alpha<\gamma}$ is a w^*-total subset of X^* and each x^* in X^* has a unique representation of the form $\sum_{\alpha<\gamma} x^*(x_\alpha)x_\alpha^*$ where the series is w^*-convergent. For a detailed study of transfinite Schauder bases we refer the reader to [59].

4. If $(x_\alpha)_{\alpha<\gamma}$ is a transfinite basis for the space $(X, \|\cdot\|)$, then there exists an equivalent norm $\|\|\cdot\|\|$ on X such that $(x_\alpha)_{\alpha<\gamma}$ is a bimonotone basis for the space $(X, \|\|\cdot\|\|)$. This norm is defined by $\|\|x\|\| = \sup\{\|P_I(x)\| : I \text{ interval of } \gamma\}$.

In the sequel, for every ordinal γ we shall denote by $c_{00}(\gamma)$ the vector space of all sequences $(\lambda_\alpha)_{\alpha\in\gamma}$ of real numbers such that the set $\{\alpha < \gamma : \lambda_\alpha \neq 0\}$ is finite. We also denote by $(e_\alpha)_{\alpha<\gamma}$ the natural Hamel basis of $c_{00}(\gamma)$. It is an easy observation that every space X with a transfinite basis $(x_\alpha)_{\alpha<\gamma}$ is isometric to the completion of $c_{00}(\gamma)$ endowed with an appropriate norm. Moreover if K is a subset of $c_{00}(\gamma)$ with the properties (a) $\{e_\alpha^*\}_{\alpha<\gamma} \subseteq K$ and (b) for every $\phi \in K$, $\|\phi\|_\infty \leq 1$ and for every interval I of γ, the restriction $\phi_I = \phi \cdot \chi_I$ of ϕ to I is also a member of K, then the norm defined on $c_{00}(\gamma)$ by

$$\|x\|_K = \sup\{|\phi(x)| = \langle \phi, x \rangle : \phi \in K\}$$

satisfies that the completion of $(c_{00}(\gamma), \|\cdot\|_K)$ has $(e_\alpha)_{\alpha<\gamma}$ as a transfinite bimonotone basis.

Fix X with a transfinite basis $(x_\alpha)_{\alpha<\gamma}$. The support $\operatorname{supp} x$ of $x \in X$ is the set $\{\alpha < \gamma : x_\alpha^*(x) \neq 0\}$. For a given interval $I \subseteq \gamma$, let $X_I = P_I X$, and for $\alpha < \gamma$, let $X_\alpha = X_{[0,\alpha)}$. For $x, y \in X$ finitely supported, we write $x < y$ to denote that $\max \operatorname{supp} x < \min \operatorname{supp} y$.

A sequence $(y_\alpha)_{\alpha<\xi}$ is called a *transfinite block subsequence* of $(x_\alpha)_{\alpha<\gamma}$ if and only if for all $\alpha < \xi$, y_α is finitely supported and for all $\alpha < \beta < \xi$, $y_\alpha < y_\beta$. Notice that a transfinite block subsequence of a transfinite basis is always a transfinite basis of its closed linear span.

Fix two Banach spaces X and Y. A bounded operator $T : X \to Y$ is an isomorphism iff TX is closed and T is one-to-one. T called *strictly singular* if it is not an isomorphism when restricted to any infinite dimensional closed subspace of X (i.e., for all infinite dimensional closed $X' \hookrightarrow X$, either TX' is not closed or $T|X'$ is not $1-1$). This is equivalent to saying that for all $Y \hookrightarrow X$ and $\varepsilon > 0$, there is an infinite dimensional closed subspace Y' of Y such that $\|T|Y'\| \leq \varepsilon$.

It is well known that most of the structure of the infinite dimensional closed subspaces of a separable Banach space X with a basis $(x_n)_n$ is described by its block sequences. Namely that for every infinite dimensional closed subspace Y of X and every $\varepsilon > 0$ there exists a normalized sequence in Y and a block sequence $(w_n)_n$ of $(x_n)_n$ which are $1 + \varepsilon$-equivalent. The method used for the proof of this result is called the *gliding hump* argument. This result is not extendable in the case of the transfinite block sequences. For example, consider a biorthogonal basis $(x_\alpha)_{\alpha<\omega\cdot2}$ of a Hilbert space and set Y the subspace generated by the sequence $(x_n + x_{\omega+n})_n$.

We now describe how block sequences are connected to the subspaces in the transfinite case.

Proposition A.3. *Let $(x_\alpha)_{\alpha<\gamma}$ be a transfinite basis of X and Y an infinite dimensional closed subspace X. Then there exists a $\lambda \leq \gamma$ and a closed subspace Z of Y such that*

1. *$P_\lambda : Z \to X_\lambda$ is an isomorphism.*

2. *For every $\varepsilon > 0$ there exists a semi-normalized block sequence $(w_n)_n$ in X_λ and a normalized sequence $(z_n)_n$ in Z such that $\sum_n \|P_\lambda z_n - w_n\| < \varepsilon$.*

3. *There exists a subspace Z' of Z isomorphic to a block subspace of X.*

4. *If we additionally assume that Y has a Schauder basis $(y_n)_n$, then the sequence $(z_n)_n$ in 2. can be chosen to be a block sequence of $(y_n)_n$.*

Proof. We assume that $(x_\alpha)_{\alpha<\gamma}$ is a bimonotone basis. Let

$$\beta_0 = \min\{\beta : P_\beta : Y \to X_\beta \text{ is not strictly singular}\}. \qquad (A.1)$$

Let us show that $\lambda = \beta_0$ is the required ordinal. Notice that β_0 has to be necessarily a limit ordinal. Since P_{β_0} is not strictly singular on Y, there exists a subspace Z of Y such that $P_{\beta_0} : Z \to X_{\beta_0}$ is an isomorphism. On the other hand for every $\gamma < \beta_0$, $P_\gamma : Y \to X_\gamma$ is strictly singular hence for every $\varepsilon > 0$ and every subspace Z' of Z there exists $W \hookrightarrow Z'$ such that $\|P_\gamma|W\| < \varepsilon$. Now we are ready to apply a modified gliding hump argument to obtain $(z_n)_n$, $(w_n)_n$ as they are required in 2. Indeed for a given ε we choose $(\varepsilon_n)_n$ such that $\varepsilon_n > 0$, $\sum \varepsilon_n < \varepsilon/4$. We choose a normalized $z_1 \in Z$. Since β_0 is a limit ordinal, there must exist $\gamma_1 < \beta_0$ such that $\|P_{[\gamma_1,\beta_0)}z_1\| < \varepsilon_1$. Hence setting $w_1 = P_{\gamma_1}z_1$ we have that $\|w_1 - P_{\beta_0}z_1\| < \varepsilon_1$. Since $P_{\gamma_1} : Z \to X_{\gamma_1}$ is strictly singular there exists a normalized $z_2 \in Z$ and $\|P_{\gamma_1}z_2\| < \varepsilon_2$. Choose $\gamma_2 > \gamma_1$ such that $\|P_{[\gamma_2,\beta_0)}z_2\| < \varepsilon_2$ and set $w_2 = P_{[\gamma_1,\gamma_2)}z_2$. Observe that $\|P_{\beta_0}z_2 - w_2\| < 2\varepsilon_2$ and $w_1 < w_2$. Continuing in this manner we obtain $(z_n)_n$ and $(w_n)_n$ such that for all n, $\|P_{\beta_0}z_n - w_n\| \leq 2\varepsilon_n$, hence

$$\sum_n \|P_{\beta_0}z_n - w_n\| \leq \varepsilon/2. \qquad (A.2)$$

Since we assume that the transfinite basis $(x_\alpha)_{\alpha<\gamma}$ is bimonotone, (A.2) implies that $(P_{\beta_0}z_n)_n$ and $(w_n)_n$ are equivalent.

The desired property 3. follows from 2, while 4. results of a careful choice of $(z_n)_n$ in 2. $\qquad \square$

As we have mentioned in the introduction the manner that block subspaces saturate the subspaces of X is weaker than the corresponding result for spaces X with a basis $(x_n)_n$. In the next proposition we provide a sufficient condition which ensures the complete extension of the result from the Schauder bases to transfinite Schauder bases fulfilling the additional condition.

Proposition A.4. Let $(x_\alpha)_{\alpha<\gamma}$ be a transfinite basis of X. Assume that for every I, J disjoint intervals of γ the spaces X_I and X_J are totally incomparable. Then for every Y closed infinite dimensional subspace of X and every $\varepsilon > 0$ there exist $(y_n)_n$, $(z_n)_n$ normalized sequences such that $(y_n)_n \subset Y$, $(z_n)_n$ block sequence of $(z_\alpha)_{\alpha<\gamma}$ and $\sum_n \|y_n - z_n\| < \varepsilon$.

Proof. From Proposition A.3 there exists a subspace Z of Y and $\lambda \leq \gamma$ such that $P_\lambda : Z \to X_\lambda$ is an isomorphism. Assume that $\lambda < \gamma$ and set $I = [1, \lambda)$ and $J = [\lambda, \gamma)$. Then $P_J : Z \to X_J$ is a strictly singular operator. Hence we may find (w_n), (z_n) as in Proposition A.3 (2) such that $\sum_n \|P_J(z_n)\| < \varepsilon$ which yields that $\sum_n \|z_n - w_n\| < 2\varepsilon$. $\square$

Definition A.5. A transfinite basis $(x_\alpha)_{\alpha<\gamma}$ is called *shrinking* iff for all $(\alpha_n)_n \uparrow$, $(x_{\alpha_n})_n$ is shrinking in the usual sense (i.e., $(x_{\alpha_n}^*)$ generates in norm the dual of the closed span of $(x_{\alpha_n})_n$).

It is called *boundedly complete* iff for all $(\alpha_n)_n \uparrow$, $(x_{\alpha_n})_n$ is boundedly complete in the usual sense (i.e., for all sequence of scalars $(\lambda_n)_n$, if there is some $C > 0$ such that for all n, $\|\sum_{i=1}^n \lambda_i x_{\alpha_i}\| \leq C$, then $\sum_i \lambda_i x_{\alpha_i}$ converges in norm).

The following result is the extension of the well-known James' characterization of reflexivity [35] in the general setting of a Banach space with a transfinite basis.

Proposition A.6. Let $(x_\alpha)_{\alpha<\gamma}$ be a transfinite basis of X. Then X is reflexive iff $(x_\alpha)_{\alpha<\gamma}$ is shrinking and boundedly complete.

Proof. The direct implication is consequence of the James' characterization. The opposite requires the following two claims. $\square$

Claim 1. If $(x_\alpha)_{\alpha<\gamma}$ is shrinking then the biorthogonal basis $(x_\alpha^*)_{\alpha<\gamma}$ generates in the norm topology the dual space X^*.

Proof of the claim. Assume the contrary. Then there exists $x^* \in X^*$ not in the closed linear span Y of $(x_\alpha^*)_{\alpha<\gamma}$. Set $\beta_0 = \min\{\beta \leq \gamma : P_\beta^* x^* \notin Y\}$. Then $P_{\beta_0}^* x^* \notin Y$ but for all $\gamma < \beta_0$, $P_\gamma^* x^* \in Y$. Therefore there exists an increasing sequence of successive disjoint intervals $I_1 < I_2 < \cdots < I_n < \cdots < \beta_0$ and $\varepsilon > 0$ such that for each $n \in \mathbb{N}$, $P_{I_n}^* x^* \in Y$ and $\|P_{I_n}^* x^*\| \geq \varepsilon$. Observe that if $x^* \in X^*$, $x^* = w^* - \sum_{\alpha<\gamma} \mu_\alpha x_\alpha^*$, where for each $\alpha < \gamma$, $\mu_\alpha = x^*(x_\alpha)$. Moreover if I is an interval of γ such that $P_I^* x^* \in Y$ and $\varepsilon' > 0$, then there is a finite subset $F_{\varepsilon'}$ of I such that $\|y_{\varepsilon'}^* - x^*\| < \varepsilon$, where

$$y_{\varepsilon'}^* = w^* - \sum_{\alpha\in\gamma\setminus I} \mu_\alpha x_\alpha^* + \sum_{\alpha\in F_{\varepsilon'}} \mu_\alpha x_\alpha^*.$$

Using this observation we inductively select finite sets $F_1 \subseteq I_1, \ldots, F_n \subseteq I_n$ such that setting

$$y_n^* = \sum_{i=1}^n \sum_{\alpha\in F_i} \mu_\alpha x_\alpha^* + P_{\beta_0\setminus\bigcup_{i=1}^n I_n} x^*, \tag{A.3}$$

we have that

$$\|P_{\beta_0}^* x^* - y_n^*\| < \varepsilon_n < \frac{\varepsilon}{4}. \tag{A.4}$$

Set $y^* = w^* - \lim_n y_n^*$ and (A.3) and (A.4) yield that $\operatorname{supp} y_n^* \subseteq \bigcup_n F_n$ and also $\|P_{F_n}^* y^*\| > \varepsilon/2$. Since each F_n is a finite set we can enumerate $\bigcup_n F_n$ as $(\alpha_n)_n \uparrow$ and clearly y^* yields that the sequence $(x_{\alpha_n})_n$ is not a shrinking Schauder basis, a contradiction. $\qquad\square$

Claim 2. If $(x_\alpha)_{\alpha<\gamma}$ is boundedly complete then for every $x^{**} \in X^{**}$, the series $\sum_{\alpha<\gamma} x^{**}(x_\alpha^*)x_\alpha$ converges in norm.

Proof of the claim. Suppose the contrary and fix $x^{**} \in X^{**}$ but not in X. The proof is similar to the previous one. For each $\alpha < \gamma$, let $\lambda_\alpha = x^{**}x_\alpha^*$ and let

$$\beta_0 = \min\{\beta < \gamma : P_\beta^{**}x^{**} \notin X\}.$$

Using a similar argument we can choose an increasing sequence $(F_n)_n$ of finite subsets of γ such that $w^* - \sum_{\alpha \in \bigcup_n F_n} \lambda_\alpha x_\alpha^*$ exists and for every n, $\|\sum_{\alpha \in F_n} \lambda_\alpha x_\alpha\| > \varepsilon > 0$. This yields that the sequence $(x_\alpha)_{\alpha \in \bigcup_n F_n}$ is not boundedly complete, a contradiction. $\qquad\square$

Appendix B

The Proof of the Finite Representability of J_{T_0}

The goal of this part is to prove the basic inequality and show the finite interval representability of the James-like space J_{T_0} in $\mathfrak{X}_{\omega_1}$ (Theorem VII.9). Reaching these two goals involves a similar sort of problems and for this reason we introduce a general theory applicable to both cases and hopefully to many other cases to come.

General theory

The theory deals with a block sequence of vectors $(x_k)_{k=1}^n$, a sequence of scalars $(b_k)_{k=1}^n$, and a functional $f \in K_{\omega_1}$, and tries to estimate $|f(\sum_{k=1}^n b_k x_k)|$ in terms of $|g(\sum_{k=1}^n b_k e_k)|$ for an appropriately chosen functional g of an auxiliary Tsirelson-like space X with basis $(e_i)_i$. The natural approach is to start with a tree-analysis $(f_t)_{t \in \mathcal{T}}$ of f, and to try to replace the functional f_t at each node $t \in \mathcal{T}$ by a functional g_t in the norming set of the auxiliary space, and in doing this to try to copy, as much as possible, the given tree-analysis $(f_t)_{t \in \mathcal{T}}$. Not all nodes $t \in \mathcal{T}$ have the same importance in this process. It turns out that the crucial replacements $f_t \mapsto g_t$ are made for t belonging to some sets $\mathcal{A} \subseteq \mathcal{T}$ such that $(f_t)_{t \in \mathcal{A}}$ is in some sense responsible for the estimation of the action of the whole functional f on each of the vectors x_k. These are the *maximal antichains* of $\mathcal{T}$ defined below. Observe that some of the replacements $f_t \mapsto g_t$ are necessary before this procedure has a chance to work. Suppose for example the replacements are made in an auxiliary mixed Tsirelson space X where a particular $(\mathcal{A}_{n_{j_0}}, \frac{1}{m_{j_0}})$-operation is not allowed. Then, every time we find a node $t \in \mathcal{T}$ such that the corresponding f_t has weight $w(f_t) = m_{j_0}$, the replacement g_t has to be something avoiding this operation, i.e., we cannot put the combination $g_t = (1/w(f_t)) \sum_{s \in S_t} g_s$. This sort of nodes are the ones that we call "catchers" below, because their own tree analyses $(f_s)_{s \succeq t}$ cannot be taken into account.

Antichains and arrays of antichains

Recall that every $f \in K_{\omega_1}$ has a tree-analysis $(f_t)_{t \in \mathcal{T}}$ such that: For every $t \in \mathcal{T}$, (a) if $u \succeq t$, then $\operatorname{ran} f_u \subseteq \operatorname{ran} f_t$, and (b) if f_t is of type I, then $f_t = (1/w(f_t)) \sum_{s \in S_t} f_s$.

Recall that $A \subseteq \mathcal{T}$ is called an *antichain* if for every $t \neq t' \in A$, neither $t \preceq t'$ nor $t' \preceq t$. Given $t, t' \in \mathcal{T}$, we define $t \wedge t' = \max\{v \in \mathcal{T} : v \preceq t, t'\}$. Notice that $A \subseteq \mathcal{T}$ is an antichain iff $t \wedge t' \not\succeq t, t'$ for every $t \neq t' \in \mathcal{T}$.

Definition B.1. Fix a tree-analysis $(f_t)_{t \in \mathcal{T}}$ of f as above. Given a finitely supported vector x, a set $\mathcal{A} \subseteq \mathcal{T}$ is called a *regular antichain* for x and $(f_t)_{t \in \mathcal{T}}$ if:

(a.1) For every $t \in \mathcal{A}$, f_t is not of type II.

(a.2) $f_{t_1 \wedge t_2}$ is of type II for every $t_1 \neq t_2 \in \mathcal{A}$, and

(a.3) $\operatorname{ran} f_t \cap \operatorname{ran} x \neq \emptyset$, for every $t \in \mathcal{A}$.

$\mathcal{A}$ is a *maximal antichain* for x if in addition $\mathcal{A}$ satisfies

(a.4) For every $t \in \mathcal{T}$, if $\operatorname{supp} f_t \cap \operatorname{ran} x \neq \emptyset$, then there is some $u \in \mathcal{A}$ comparable with t.

Let $(x_k)_{k=1}^n$ be a block sequence, and let $\mathcal{A} = (\mathcal{A}_k)_{k=1}^n$ be such that each $\mathcal{A}_k$ is a regular antichain for the vector x_k and the tree-analysis $(f_t)_{t \in \mathcal{T}}$. For a given $t \in \mathcal{T}$, we define

$$D_t^{\mathcal{A}} = \bigcup_{u \succeq t}\{k \in [1, n] : u \in \mathcal{A}_k\}, \quad E_t^{\mathcal{A}} = D_t^{\mathcal{A}} \setminus (\bigcup_{s \in S_t} D_s^{\mathcal{A}}).$$

Whenever there is no possible confusion we simply write D_t and E_t to denote $D_t^{\mathcal{A}}$ and $E_t^{\mathcal{A}}$, respectively.

$\mathcal{A} = (\mathcal{A}_k)_{k=1}^n$ is called a *(maximal) regular array* for $(x_k)_{k=1}^n$ and $(f_t)_{t \in \mathcal{T}}$ if each $\mathcal{A}_k$ is a (maximal) regular antichain for x_k and $(f_t)_{t \in \mathcal{T}}$, and in addition

(a.5) for every $t \in \bigcup_k \mathcal{A}_k$ such that f_t is of type I, either t is a *catcher*, i.e., $D_s = \emptyset$ for every $s \in S_t$, or for every $k \in E_t$, t is *splitter* of x_k, i.e., for every $k \in E_t$ there are at least $s_1 \neq s_2 \in S_t$ such that $\operatorname{ran} f_{s_i} \cap \operatorname{ran} x_k \neq \emptyset$.

We denote by $\mathrm{S}(\mathcal{A})$ and $\mathrm{C}(\mathcal{A})$ the set of splitter nodes and catcher nodes of $\mathcal{A}$, respectively. Notice that if $t_i \in \mathcal{A}_{k_i}$ $(i = 1, 2)$ are catcher nodes, then they are incomparable, and that $\mathcal{A}_k = \mathrm{S}(\mathcal{A}) \cup \mathrm{C}(\mathcal{A})$.

Note that if no f_t $(t \in \mathcal{T})$ is of type II then $\#\mathcal{A}_k \leq 1$ for all k, and so the tree-analysis below becomes much simpler.

Definition B.2. *(The functor $\mathcal{A}(x, \mathrm{C})$.)* Given a block vector x and $\mathrm{C} \subseteq \mathcal{T}$ consisting of nodes of type I, let $\mathcal{A}(x, \mathrm{C})$ be the set of nodes $t \in \mathcal{T}$ such that

(A.1) f_t is not of type II.

(A.2) $\operatorname{ran} f_t \cap \operatorname{ran} x \neq \emptyset$.

(A.3) For every $s \preceq t$ if $s \in S_u$ and f_u is of type I, then for every $s' \in S_u \setminus \{s\}$ we have that $\operatorname{ran} f_{s'} \cap \operatorname{ran} x = \emptyset$.

(A.4) If f_t is of type I and $t \notin C$, then t is a splitter of x.

(A.5) for every $u \not\succeq t$, $u \notin C$.

Proposition B.3. *Let $\mathcal{A} = \mathcal{A}(x, C)$ be a maximal regular antichain such that $\{t \in \mathcal{A} \setminus C : f_t$ of type I$\} \subseteq S(\mathcal{A})$. Moreover, if $(x_k)_{k=1}^n$ is a block sequence, then the corresponding $\boldsymbol{\mathcal{A}} = (\mathcal{A}(x_k, C))_{k=1}^n$ is a maximal regular array such that*

(a) $\{t \in \bigcup_k \mathcal{A}_k \setminus C : f_t$ of type I$\} \subseteq S(\boldsymbol{\mathcal{A}})$, and

(b) $C \subseteq C(\boldsymbol{\mathcal{A}})$ and for every $t \in C$, E_t is an interval of integers.

Proof. Fix $t \neq t' \in \mathcal{A}_k$. That $f_{t \wedge t'}$ is of type II follows from the fact that if $u \not\succeq t$, then $u \notin C$, by (A.5), hence if f_u is of type I, then (A.3) implies that u is not splitter of x. We show the maximality of $\mathcal{A}$: Fix $t \in \mathcal{T}$ such that $\operatorname{supp} f_t \cap \operatorname{ran} x \neq \emptyset$. Let $t_0 \succeq t$ be such that f_{t_0} is of type 0 and $\operatorname{supp} f_{t_0} \subseteq \operatorname{ran} w_k$, and set $b = [0, t_0] = \{v \in \mathcal{T} : v \preceq t_0\}$ which is a $\preceq$-well ordered set, and $t \in b$. We distinguish two cases: Suppose first that $b \cap C = \emptyset$. Let $u_0 = \min\{u \in b : v$ satisfies (A.1), (A.4)$\}$. Notice that u_0 exists since t_0 satisfies (A.1) and (A.4). The minimality of u_0 shows that u_0 satisfies (A.3), hence $u_0 \in \mathcal{A}$. Suppose now that $b \cap C \neq \emptyset$, and set $v_0 = \min b \cap C$. It is not difficult to show that $u_0 = \max\{u \preceq v_0 : u$ satisfies (A.1), (A.4)$\}$ is in $\mathcal{A}$ (notice that v_0 satisfies (A.1) and (A.4), hence u_0 is well defined.)

Repeating this procedure for each vector in a given a block sequence $(x_k)_{k=1}^n$, one gets that the array $(\mathcal{A}(x_k, C))_{k=1}^n$ is maximal and regular. Finally suppose that $t \in C$ and suppose that $k_1 < k_2 < k_3$ with $k_1, k_3 \in E_t$. It is routine to check that t satisfies (A.1)–(A.5) for x_{k_2}, hence it follows that $k_2 \in E_t$. $\qquad\square$

Proposition B.4. *Suppose that $\boldsymbol{\mathcal{A}} = (\mathcal{A}_k)_{k=1}^n$ is a regular array for a block sequence $(x_k)_{k=1}^n$ and $(f_t)_{t \in \mathcal{T}}$. Then:*

(b.0) *If $t \in \mathcal{A}_k$ is splitter or if f_t is of type 0, then $\operatorname{supp} f_t \cap \operatorname{ran} x_k \neq \emptyset$.*

(b.1) *If f_t is of type I, then $\{D_s\}_{s \in S_t} \cup \{\{k\} : k \in E_t\}$ is a block family, and if t is splitter, then $\#E_t \leq \#S_t - 1$.*

Suppose that in addition $\boldsymbol{\mathcal{A}} = (\mathcal{A}_k)_{k=1}^n$ is maximal for $(x_k)_{k=1}^n$.

(b.2) *Fix $t \in \mathcal{A}_k$ and fix $u \not\succeq s \preceq t$ with f_u of type I and $s \in S_t$. Then for every $s' \in S_u \setminus \{s\}$ $\operatorname{ran} f_{s'} \cap \operatorname{ran} x_k = \emptyset$.*

(b.3) *Suppose that f_t is of type II, $k \in D_t$ and $s \in S_t$. If $\operatorname{supp} f_t \cap \operatorname{ran} x_k \neq \emptyset$, then $k \in D_s$.*

Proof. (b.0) If f_t is of type 0, the conclusion is clear. If t is a splitter, let $s_1 \neq s_2 \in S_t$ be such that $f_{s_1} < f_{s_2}$ and $\operatorname{ran} f_{s_1} \cap \operatorname{ran} x_k$, $\operatorname{ran} f_{s_2} \cap \operatorname{ran} x_k \neq \emptyset$. Then $\max \operatorname{supp} f_{s_1} \in \operatorname{ran} x_k$.

(b.1) For the first part, if t is catcher, there is nothing to prove, so we assume t is splitter. First we show that $\{D_s\}_{s \in S_t} \cup \{\{k\} : k \in E_t\}$ is a disjoint family. If

$k \in E_t \cap D_s$ for some $s \in S_t$, then there is some $u \succeq s$ with $u \in \mathcal{A}_k$. But $t \in \mathcal{A}_k$ and $t \not\succeq u$, a contradiction. Suppose that $k \in D_s \cap D_{s'}$ with $s \neq s' \in S_t$. Then there are $u, u' \in \mathcal{A}_k$ such that $u \succeq s$, $u' \succeq s'$. Hence $u \wedge u' = t$ but f_t is of type I, contradicting (a.2). For the second part, suppose that $k_1 < k_2 < k_3$ are such that $k_1, k_3 \in D_s$ for some $s \in S_t$. This implies that $\operatorname{ran} x_{k_1} \cap \operatorname{ran} f_s, \operatorname{ran} x_{k_3} \cap \operatorname{ran} f_s \neq \emptyset$, and hence $\operatorname{ran} x_{k_2} \subseteq \operatorname{ran} f_s$. This implies that $\operatorname{ran} x_{k_1} \cap \operatorname{ran} f_{s'} = \emptyset$ for every $s' \in S_t \setminus \{s\}$. Since t is splitter, $k_2 \notin E_t$, and, by (a.3), $k_2 \notin D_{s'}$ for every $s' \in S_t \setminus \{s\}$.

Let $S_t = \{s_1 < \cdots < s_d\}$ be ordered such that $f_{s_i} < f_{s_j}$ whenever $i < j$. For $k \in E_t$, the set $H_k = \{i \in [1, d] : \operatorname{ran} x_k \cap \operatorname{ran} f_{s_i} \neq \emptyset\}$ has at least two elements. We claim that the mapping $k \mapsto \max H_k \in \{2, \ldots, d\}$ is one-to-one. To see this note that for $k < k'$ we obtain that $H_k \cap H_{k'} = \{\max H_k\}$ if $\max H_k = \max H_{k'}$, and $H_k < H_{k'}$ otherwise.

(b.2) Fix $s' \in S_t \setminus \{s\}$, and suppose that $\operatorname{ran} f_{s'} \cap \operatorname{ran} x_k \neq \emptyset$. Since $\operatorname{ran} f_s \cap \operatorname{ran} x_k \neq \emptyset$, we get that $\operatorname{supp} f_{s'} \cap \operatorname{ran} x_k \neq \emptyset$. By maximality of $\mathcal{A}_k$, there is $t' \in \mathcal{A}_k$ comparable with s'. Since $\mathcal{A}_k$ is an antichain, we get that $t' \succeq s'$, and hence $t \wedge t' = u$. But f_u is of type I, a contradiction.

(b.3) This follows using (a.4) and (a.1), (a.2).   □

Assignments, filtrations, and their relationships

Definition B.5. Given a block sequence $(x_k)_{k=1}^n$, and a regular array $\mathcal{A} = (\mathcal{A}_k)_{k=1}^n$ for $(x_k)_{k=1}^n$, a sequence $(g_{k,t}^{\mathcal{A}})_{t \in \mathcal{A}_k, k} \subseteq c_{00}(\mathbb{N})$ is called an $\mathcal{A}$-*assignment* provided that $\operatorname{supp} g_{k,t} \subseteq \{k\}$ for every k and $t \in \mathcal{A}_k$. The property (b.1) ensures that every $\mathcal{A}$-assignment $(g_{k,t}^{\mathcal{A}})_{t \in \mathcal{A}_k, k}$ naturally *filters down* to the whole tree $(G_{k,t}^{\mathcal{A}})_{t \in \mathcal{T}}$ as follows: If $k \notin D_t^{\mathcal{A}}$, then $G_{k,t}^{\mathcal{A}} = 0$, and if $t \in \mathcal{A}_k$, then $G_{k,t}^{\mathcal{A}} = g_{k,t}^{\mathcal{A}}$. Suppose that $k \in D_t^{\mathcal{A}} \setminus D_s^{\mathcal{A}}$. If f_t is of type I, then we define recursively $G_{k,t}^{\mathcal{A}} = (1/w(f_t)) G_{k,s}^{\mathcal{A}}$, where $s \in S_t$ is the unique $s = s(k, t) \in S_t$ such that $k \in D_s^{\mathcal{A}}$ (by (b.1)). If f_t is of type II, $f_t = \sum_{s \in S_t} \lambda_s f_s$, then we simply set $G_{k,t}^{\mathcal{A}} = \sum_{s \in S_t} \lambda_s G_{k,s}^{\mathcal{A}}$. For $t \in \mathcal{T}$, let

$$G_t^{\mathcal{A}} = \sum_{k \in D_t^{\mathcal{A}}} G_{k,t}^{\mathcal{A}}.$$

We call $(G_t^{\mathcal{A}})_{t \in \mathcal{T}}$ the *filtration* of $(g_{k,t}^{\mathcal{A}})_{t \in \mathcal{A}_k, k}$. Whenever there is no possible confusion, we write $g_{k,t}$, $G_{k,t}$ and G_t instead of the respective $g_{k,t}^{\mathcal{A}}$, $G_{k,t}^{\mathcal{A}}$ and $G_t^{\mathcal{A}}$.

Proposition B.6.

(c.1) *Fix $t \in \mathcal{T}$. For every k, $\operatorname{supp} g_{k,t} \subseteq \{k\}$. Hence $\operatorname{supp} g_t \subseteq D_t$.*

(c.2) *If f_t is not of type II, then $G_t = \sum_{k \in E_t} g_{k,t} + (1/w(f_t)) \sum_{s \in S_t} G_s$.*

(c.3) *If f_t is of type II, $f_t = \sum_{s \in S_t} \lambda_s f_s$, then $G_t = \sum_{s \in S_t} \lambda_s G_s$.*

Proof. (c.1) is clear.

(c.2) If f_t is of type 0, this is clear. Suppose that f_t is of type I. Then by definition

$$G_t = \sum_{k \in E_t} G_{k,t} + \sum_{k \in D_t \setminus F_t} G_{k,t} = \sum_{k \in E_t} g_{k,t} + \sum_{s \in S_t} \sum_{k \in D_s} G_{k,t} -$$
$$= \sum_{k \in E_t} g_{k,t} + \sum_{s \in S_t} \tfrac{1}{w(f_t)} \sum_{k \in D_s} G_{k,s} = \sum_{k \in E_t} g_{k,t} + \tfrac{1}{w(f_t)} \sum_{s \in S_t} G_s.$$

$$\text{(B.1)}$$

(c.3) Suppose that f_t is of type II, i.e., $f_t = \sum_{s \in S_t} \lambda_s f_s$, and suppose that $k \in D_t$. Then, by (c.1), $G_t(e_k) = G_{t,k}(e_k) = \sum_{s \in S_t} \lambda_s G_{k,s}(e_k) = (\sum_{s \in S_t} \lambda_s G_s)(e_k)$. If $k \notin D_t$, then $G_t(e_k) = 0$, and $\sum_{s \in S_t} \lambda_s G_s(e_k) = 0$. $\qquad\square$

Definition B.7. *(Canonical Assignment)* Suppose that $\mathcal{A} = (\mathcal{A}_k)_k$ is a regular array for $(x_k)_{k=1}^n$ and $(f_t)_{t \in \mathcal{T}}$. Let $f_{k,t} = f_t(x_k)e_k^*$ for $k \in [1, n]$ and $t \in \mathcal{A}_k$. This is the $\mathcal{A}$-*canonical assignment.*

Remark B.8. Note that if the array $\mathcal{A}$ is maximal, then filtering down the canonical assignment we get $f_t(w_k) = F_{k,t}(e_k)$, for every $t \in \mathcal{T}$, and $k \in D_t$: If $k \in E_t$, this is just by definition. Suppose $k \notin F_t$. If f_t is of type I, then $F_{k,t}(e_k) = (1/w(f_t))F_{k,s}(e_k)$, where $s \in S_t$ is unique such that $k \in D_s$. By the maximality of $\mathcal{A}_k$, we get that supp $f_{s'} \cap \operatorname{ran} w_k = \emptyset$ for every $s' \in S_t \setminus \{s\}$ (by (b.2)), hence $f_t(x_k) = (1/w(f_t))f_s(x_k) = (1/w(f_t))F_{k,s}(e_k) = F_{k,t}(e_k)$, by inductive hypothesis. If $f_t = \sum_{s \in S_t} \lambda_s f_s$ is of type II, then by maximality of $\mathcal{A}_k$, $f_t(x_k) = \sum_{s \in S_t, k \in D_s} \lambda_s f_s(x_k) = \sum_{s \in S_t, k \in D_s} \lambda_s F_{k,t}(e_k) = F_{k,t}(e_k)$, the last equality because $F_{k,u} = 0$ if $k \notin D_u$.

We obtain that $f_t(\sum_{k \in D_t} b_k x_k) = F_t(\sum_{k \in D_t} b_k e_k) = F_t(\sum_{k=1}^n b_k e_k)$ for every sequence of scalars $(b_k)_{k=1}^n$. The last equality follows from supp $G_t \subseteq D_t$. In particular, $f(\sum_{k=1}^n b_k x_k) = G_\emptyset(\sum_{k=1}^n b_k e_k)$, since $D_\emptyset = \{k : \operatorname{supp} f \cap \operatorname{ran} x_k \neq \emptyset\}$, by maximality of $\mathcal{A}$.

Proposition B.9. *Suppose that* $\mathcal{A} = (\mathcal{A}_k)_{k=1}^n$ *is a regular array (not necessarily maximal) for* $(x_k)_{k=1}^n$ *and* $(f_t)_{t \in \mathcal{T}}$*. Fix scalars* $(b_k)_{k=1}^n$*,* $(c_k)_{k=1}^n$ *and suppose that* $(g_{k,t})_{t \in \mathcal{A}_k, k}$*,* $(h_{k,t})_{t \in \mathcal{A}_k, k}$ *are* $\mathcal{A}$*-assignments.*

(1) *If for every* $t \in \mathcal{A}_k$ $g_{k,t}(b_k e_k) \leq h_{k,t}(c_k e_k)$*, then for every* $t \in \mathcal{T}$ $G_{k,t}(b_k e_k) \leq H_{k,t}(c_k e_k)$*.*

(2) $\|G_{k,u}(e_k)\|_\infty \leq \max\{\|g_{k,t}\|_\infty : t \in \mathcal{A}_k\}$*, for every* $u \in \mathcal{T}$*.*

(3) *If for every* $t \in \bigcup_{k=1}^n \mathcal{A}_k$ $\sum_{k \in E_t} g_{k,t}(b_k e_k) \leq \sum_{k \in E_t} h_{k,t}(c_k e_k)$*, then for every* $t \in \mathcal{T}$*,*

$$G_t\Big(\sum_{k \in D_t} b_k e_k\Big) \leq H_t\Big(\sum_{k \in D_t} c_k e_k\Big).$$

(4) $\|G_u\|_\infty \leq \|\sum_{t \in \mathcal{A}_k, k} g_{k,t}\|_\infty$ *for every* $u \in \mathcal{T}$*.*

Proof. This follows from Proposition B.6. $\qquad\square$

Two successive filtrations

In some applications of the theory one needs to do the process of assignment and filtration twice starting with different arrays of antichains. To see this, suppose that $\mathcal{C}$ and $\mathcal{D}$ are regular arrays for $(x_k)_{k=1}^n$ and $(f_t)_{t\in\mathcal{T}}$. Then we can naturally define a $\mathcal{D}$-assignment $(g_{k,t}^{\mathcal{D}})_{t\in\mathcal{D}_k,k}$ by taking the filtration $g_{k,t}^{\mathcal{D}} = G_{k,t}^{\mathcal{C}}$. For this to work, one needs the following special relationship between $\mathcal{C}$ and $\mathcal{D}$.

Definition B.10. We write $\mathcal{C} \not\prec \mathcal{D}$ if for every k, every $c \in \mathcal{C}_k$ and every $d \in \mathcal{D}_k$, we have that $c \not\prec d$. A $\mathcal{C}$-assignment $(g_{k,t}^{\mathcal{C}})_{k\in\mathcal{C}_k,k}$ is called *coherent* provided that $g_{k,t}^{\mathcal{C}} = 0$ whenever $f_t(w_k) = 0$.

Proposition B.11. *Suppose that $\mathcal{C} \not\prec \mathcal{D}$ are two regular arrays for $(x_k)_{k=1}^n$ and $(f_t)_{t\in\mathcal{T}}$, and suppose that $\mathcal{D}$ is in addition maximal. Fix a coherent $\mathcal{C}$-assignment $(g_{k,t}^{\mathcal{C}})_{t\in\mathcal{C}_k,k}$. Then:*

(a) *For every $k \in D_t^{\mathcal{C}} \cap D_t^{\mathcal{D}}$, $G_{k,t}^{\mathcal{C}} = G_{k,t}^{\mathcal{D}}$.*

(b) $g_{\emptyset}^{\mathcal{C}} = g_{\emptyset}^{\mathcal{D}}$.

Proof. (a) If $k \in E_t^{\mathcal{D}}$, this is just by definition. Suppose f_t is of type I, and suppose that $k \in D_s^{\mathcal{D}}$, for some $s \in S_t$. Then $G_{k,t}^{\mathcal{D}} = (1/w(f_t))G_{k,s}^{\mathcal{D}}$. Since $\mathcal{D}$ is a maximal regular array, by Proposition B.4 (b.2), $\mathrm{ran}\, f_{s'} \cap \mathrm{ran}\, w_k = \emptyset$ for every $s' \in S_t \setminus \{s\}$. If $k \in D_s^{\mathcal{C}}$, then we are done by the inductive hypothesis. So, suppose $k \in E_t^{\mathcal{C}}$, i.e., $t \in \mathcal{C}_k$. Hence, $t \not\succeq u$ for some $u \in \mathcal{D}_k$ (because $k \in D_s^{\mathcal{D}}$), contradicting our assumption that $\mathcal{C} \not\prec \mathcal{D}$. If $f_t = \sum_{s\in S_t} \lambda_s f_s$ is a sub-convex combination, then $G_{k,t}^{\mathcal{D}} = \sum_{s\in S_t, k\in D_s^{\mathcal{D}}} G_{k,s}^{\mathcal{D}} = \sum_{s\in S_t, k\in D_s^{\mathcal{D}} \cap D_s^{\mathcal{C}}} G_{k,s}^{\mathcal{D}} = \sum_{s\in S_t, k\in D_s^{\mathcal{D}} \cap D_s^{\mathcal{C}}} G_{k,s}^{\mathcal{C}} = G_{k,t}^{\mathcal{C}}$. To see the last equality note that if $k \notin D_s^{\mathcal{D}}$, then, by maximality of $\mathcal{D}$, $\mathrm{supp}\, f_u \cap \mathrm{ran}\, w_k = \emptyset$ for every $u \succeq s$, so, by coherence of the assignment, $G_{k,t}^{\mathcal{C}} = 0$; if $k \notin D_s^{\mathcal{C}}$, then $k \notin D_u^{\mathcal{C}}$ for all $u \succeq s$, and so $g_{k,u}^{\mathcal{C}} = 0$ for all $u \succeq s$ $u \in \mathcal{C}_k$, giving us $G_{k,s}^{\mathcal{D}} = 0$.

(b) Note now that $g_{\emptyset}^{\mathcal{D}} = \sum_{k\in D_{\emptyset}^{\mathcal{D}}} g_{k,\emptyset}^{\mathcal{D}} = \sum_{k\in D_{\emptyset}^{\mathcal{C}} \cap D_{\emptyset}^{\mathcal{D}}} g_{k,\emptyset}^{\mathcal{D}} = g_{\emptyset}^{\mathcal{C}}$. For if $k \in D_{\emptyset}^{\mathcal{C}} \setminus D_{\emptyset}^{\mathcal{D}}$, then by maximality of $\mathcal{D}$, for all $u \in \mathcal{T}$, $\mathrm{supp}\, f_u \cap \mathrm{ran}\, w_k = \emptyset$, hence, by coherence of the $\mathcal{C}$-assignment $g_{k,u}^{\mathcal{C}} = 0$ for all u, and hence $g_{k,\emptyset}^{\mathcal{D}} = 0$; if $k \in D_{\emptyset}^{\mathcal{D}} \setminus D_{\emptyset}^{\mathcal{C}}$, then every $\mathcal{C}_k = \emptyset$, and so $g_{k,\emptyset}^{\mathcal{C}} = 0$. $\square$

Let us now give the two main applications of this general theory of tree-analyses.

The proof of the basic inequality

Let W be the minimal subset of $c_{00}(\mathbb{N})$ containing $\{\pm e_k^*\}_k$, and closed under $((\mathcal{A}_{4n_j}, \frac{1}{m_j}))$ operations. Fix a (C, ε)-RIS $(x_k)_{k=1}^n$, and fix $(j_k)_{k=1}^n$ witnessing that $(x_k)_{k=1}^n$ is a (C, ε)-RIS, i.e., a) $\|x_k\| \leq C$, b) $|\mathrm{supp}\, x_k| \leq m_{j_{k+1}} \varepsilon$ and c) for all type I functionals ϕ of K with $w(\phi) < m_{j_k}$, $|\phi(x_k)| \leq C/w(\phi)$. Fix a sequence

$(b_k)_{k=1}^n$ of scalars, $\max_k |b_k| \leq 1$, and $f \in K_{\omega_1}$. Let $(f_t)_{t \in \mathcal{T}}$ be a tree-analysis of f. Consider the maximal regular array $\mathcal{A} = (\mathcal{A}(x_k, C))_{k=1}^n$, where C is the set of nodes t such that f_t is of type I and $w(f_t) = m_{j_0}$.

We introduce the following two $\mathcal{A}$-assignments $(g_{k,t})_{t \in \mathcal{A}_k, k}$, and $(r_{k,t})_{t \in \mathcal{A}_k, k}$. Fix k and $t \in \mathcal{A}_k$. If t_t is of type 0, then we set $g_{k,t} = e_k^*$ and $r_{k,t} = 0$. Suppose that t is of type I, and $w(f_t) \neq m_{j_0}$. Let $l_t = \min\{k \in E_t : w(f_t) \leq m_{j_k}\}$ if it exists, and $l_t = \infty$ otherwise. If $k > l_t$, then we set $g_{k,t} = (1/w(f_t))e_k^*$ and $r_{k,t} = 0$. If $k < k_t$, then we set $g_{k,t} = 0$ and $r_{k,t} = \varepsilon e_k^*$. If $k = l_t$, then we set $g_{k,t} = e_k^*$ and $r_{k,t} = 0$. Suppose now that $w(f_t) = m_{j_0}$. Notice that E_t is an interval. Set $k_t = \max\{l \in D_t : |b_l| = \|(b_i)_{i \in E_t}\|_\infty\}$. If $k = k_t$ then we set $g_{k,t} = e_k^*$, and $r_{k,t} = \varepsilon e_k$; if not, then we set $g_{k,t} = 0$, and $r_{k,t} = \varepsilon e_k^*$. Let $(G_t)_{t \in \mathcal{T}}$ and $(R_t)_{t \in \mathcal{T}}$ be the corresponding filtrations.

Claim (D). Fix $t \in \mathcal{T}$. Then: (d.1) $\|R_t\|_\infty \leq \varepsilon$.
(d.2) $|f_t(\sum_{k \in D_t} b_k x_k)| \leq C(G_t + R_t)(\sum_{k \in D_t} |b_k| e_k)$.
(d.3) For every t for which f_t is of type I, either $G_t \in \text{conv}\{h \in W : w(h) = w(f_t)\}$ or $G_t = e_k^* + h_t$ for some $k \notin \text{supp}\, h_t$ and $h_t \in \text{conv}\{h \in W : w(h) = w(f_t)\}$.

Proof of the claim. (d.1) follows from Proposition B.9, and (d.2) follows also from the same proposition applied to the canonical $\mathcal{A}$-assignment, the assignment $(C(g_{k,t} + h_{k,t}))_{t \in \mathcal{A}_k, k}$, and the sequences of scalars $(b_k)_k$ and $(|b_k|)_k$.

(d.3) If $w(f_t) = m_{j_0}$, then l is a catcher and $G_t = \sum_{h \subset E_t} g_{k,t} = e_{k_t}^* \in W$. Suppose that t is of type I, $w(f_t) \neq m_{j_0}$. By (c.2) and the particular $\mathcal{A}$-assignment, we know that either $G_t = (1/w(f_t))(\sum_{k \in E_t, k > l_t} e_k^* + \sum_{s \in S_t} G_s)$ or $G_t = e_{l_t}^* + h_t$, where $h_t = (1/w(f_t))(\sum_{k \in E_t, k > l_t} e_k^* + \sum_{s \in S_t} G_s)$. Assume this last case holds. *Subcase 1a.* For every $s \in S_t$ the functional f_s is not of type II. From the inductive hypothesis, we have that for every $s \in S_t$, $G_s = h_s$ or $G_s = e_{l_s}^* + h_s$, $h_s \in W$. For $s \in S_t$ such that $G_s = e_{l_s}^* + h_s$, set $I_s^1 = \{n \in \mathbb{N} : n < l_s\}$ and $I_s^2 = \{n \in \mathbb{N} : n > l_s\}$. We set $h_s^1 = I_s^1 h_s$, $h_s^2 = I_s^2 h_s$. Then, for every $s \in S_t$ the functionals h_s^1, $e_{l_s}^*$, and h_s^2 are successive and belong to W. By (b.1), for $s \neq s' \in S_t$ the corresponding functionals together with $\{e_k^*\}_{k \in E_t, k > l_t}$ form a block family, and we obtain that

$$\#\{e_k^*\}_{k \in E_t, k > l_t} + \#\{e_{l_s}^* : s \in S_t\} + \#\{h_s^1 : s \in S_t\} + \#\{h_s^2 : s \in S_t\} \leq 4\#S_t.$$
(B.2)

Therefore, $(1/w(f_t))(\sum_{k \in E_t, k > l_t} e_k^* + \sum_{s \in S_t} G_s) \in W$.
Subcase 1b. There are $s \in S_t$ for which f_s is of type II. Let B_1 be the set of immediate successors s of t such that f_s is of type II, and $B_2 = S_t \setminus S_1$. Observe that every sub-convex combination $f_s = \sum_{u \in S_s} r_u f_u$ satisfies that f_u is of type I. We may assume, allowing repetitions if needed, that for every $s \in S_t$ such that f_s is of type II, $f_s = (1/k) \sum_{q=1}^k f_{s,q}$, where each $f_{s,q} \in \{f_u : u \in S_s\}$. For each $q = 1, 2, \ldots, k$ we set $h_t^q = (1/m_j)(\sum_{l \in E_t, l > l_t} e_l^* + \sum_{s \in B_1} G_s + \sum_{s \in B_2} G_{s,q})$, where $G_{s,q} = G_u$ for $u \in S_s$ such that $f_{s,q} = f_u$. A similar argument as in the previous subcase shows that $h_t^q \in W$ with $w(h_t^q) = m_j$ for $q = 1, 2, \ldots, k$ and $h_t = (1/k) \sum_{q=1}^k h_t^q$, as required. $\qquad\square$

The particular case $t = \emptyset$, the root of $\mathcal{T}$, gives us the conclusion of the basic inequality.

Remark B.12. Note that a finer assignment using the same array of antichains will actually give us the conclusion of the basic inequality for a bit smaller auxiliary space $T[(\mathcal{A}_{2n_j}, \frac{1}{m_j})_j]$.

The proof of the finite interval representability of J_{T_0}

The general scheme of the proof is quite similar to the proof of the basic inequality though the input block sequence of vectors is slightly differently chosen. Notice however that the finite interval representability involve two inequalities needed for showing that the representing operator as well as its inverse are uniformly bounded. Thus, while in the proof of the Basic Inequality we could afford to go the auxiliary space $T[(\mathcal{A}_{4n_j}, \frac{1}{m_j})_j]$ this is no longer possible in this case. In other words, we need to improve on the counting (B.2). It is exactly for this reason that we introduce below two arrays of antichains and use two successive filtrations as explained above in Subsection B.

Fix a transfinite block sequence $(x_\alpha)_{\alpha<\gamma}$, $n \in \mathbb{N}$, a sequence $I_1 \leq I_2 \leq \cdots \leq I_n$ of successive, not necessarily distinct, infinite intervals of γ, and $\varepsilon > 0$. Let j_0 be such that $m_{2j_0+1} > 100n/\varepsilon$ and set $l = n_{2j_0+1}/m_{2j_0+1}$. Find a $(1, j_0)$-dependent sequence $(z_1, \psi_1, \ldots, z_{n_{2j_0+1}}, \psi_{n_{2j_0+1}})$ such that (a) $\mathrm{ran}\,\psi_i \subseteq \mathrm{ran}\,z_i$ for every $i = 1, \ldots, n_{2j_0+1}$ and (b) $(z_k)_{k=(i-1)l+1}^{il} \subseteq \langle x_\alpha \rangle_{\alpha \in I_i}$ for every $i = 1, \ldots, n$. Let

$$\phi = \tfrac{1}{m_{2j_0+1}} \sum_{i=1}^{n_{2j_0+1}} \psi_i,$$

and for each $k = 1, \ldots, n$ we set

$$w_k = \tfrac{m_{2j_0+1}}{l} \sum_{i=(k-1)l+1}^{kl} z_i \quad \text{and} \quad \phi_k = \tfrac{1}{m_{2j_0+1}} \sum_{i=(k-1)l+1}^{kl} \psi_i \in K_{\omega_1}.$$

Proposition B.13. *Fix $k = 1, \ldots, n$. Then*

(1) $\mathrm{ran}\,\phi_k \subseteq \mathrm{ran}\,w_k$, $\phi_k w_k = 1$ *and* $1 \leq \|w_k\| \leq 24$.

(2) *For every $f \in K_{\omega_1}$ of type I with $w(f) > m_{2j_0+1}$, $|f(w_k)| \leq 1/m_{2j_0+1}^2$.*

(3) *Let $f \in K_{\omega_1}$ be of type I, $f = (1/w(f)) \sum_{i=1}^d f_i$ with $w(f) = m_{2j+1}$ for $j < j_0$ and $d \leq n_{2j+1}$. Let $d_0 = \max\{i \leq d : w(f_i) < m_{2j_0+1}\}$, and set $f_\mathrm{L} = 1/m_{2j+1} \sum_{i=1}^{d_0-1} f_i$ and $f_\mathrm{R} = 1/m_{2j+1} \sum_{i=d_0+1}^d f_i$. Then $|f_\mathrm{L}(w_k)| \leq 1/m_{2j_0+1}^2$ and $|f_\mathrm{R}(w_k)| \leq 1/m_{2j_0+1}$.*

(4) *Let $f = (1/w(f)) \sum_{i=1}^d f_i$ with $w(f) = m_{2j_0+1}$ and $d \leq n_{2j_0+1}$ be such that $\#\{i \in [1,d] : w(f_i) = w(\psi_i) \text{ and } \mathrm{supp}\,z_i \cap \mathrm{supp}\,f_i \neq \emptyset\} \leq 2$. Then, $|f(w_k)| \leq 1/m_{2j_0+1}^2$.*

Proof. First of all, note that $(z_i)_{i=(k-1)l+1}^{kl}$ is a $(12, 1/n_{2j_0+1})$-RIS. Note also that (1) and (2) follow from Proposition II.19.

(3) By the properties of special sequences,

$$\# \bigcup_{i=1}^{d_0-1} \operatorname{supp} f_i \leq w(f_{d_0}) < m_{2j_0+1}. \tag{B.3}$$

So, $|f_{\mathrm{L}}(w_k)| \leq \|f_0\|_{\ell_1}\|w_k\|_\infty \leq m_{2j_0+1}^3/n_{2j_0+1} \leq 1/m_{2j_0+1}^2$. Let us now estimate $|f_{\mathrm{R}}(w_k)|$. To save on notation we only estimate for $k=1$. Set

$$F_0 = \{r \in [1,l] : \#(\{i \in [d_0+1, d] : \operatorname{ran} z_r \cap \operatorname{supp} f_i \neq \emptyset\}) \geq 2\}, \quad F_1 = [1,l] \setminus F_0.$$

Notice that $|F_0| \leq n_{2j+1} - 1$. For $i = 0, 1$ let $w^i = (m_{j_1}/l)\sum_{k \in F_i} z_k$. Since $f_{\mathrm{R}} \in K_{\omega_1}$ and since $(z_k)_k$ is a $(12, 1/n_{2j_0+1})$-RIS, we have that

$$|f_{\mathrm{R}}(w^0)| \leq \|w^0\| \leq \tfrac{m_{2j_0+1}}{l}\sum_{k \in F_0} \|z_k\| \leq \tfrac{m_{2j_0+1}}{l} 6n_{2j+1}. \tag{B.4}$$

To estimate $|f_{\mathrm{R}}(w^1)|$ we use the basic inequality. For each $i = d_0+1, \ldots, d$, let

$$H_i = \{k \in F_1 : \operatorname{ran} z_k \cap \operatorname{supp} f_i \neq \emptyset\}.$$

Note that $\{H_i\}_i$ is a partition of F_1 and is a block family. For $i = d_0+1, \ldots, d$, we set $w^{1,i} = m_{j_1}/l\sum_{k \in H_i} z_k$. Clearly $w^1 = w^{1,d_0+1} + \cdots + w^{1,d}$ and hence

$$|f_{\mathrm{R}}(w^1)| \leq \sum_{i=d_0+1}^{d} |f_{\mathrm{R}}(w^{1,i})| = \tfrac{1}{m_{2j+1}}\sum_{i=d_0+1}^{d} |f_i(w^{1,i})|. \tag{B.5}$$

Let us estimate now $|f_i(w^{1,i})|$, for $i = d_0+1, \ldots, d$. For a fixed such i, applying again the basic inequality, we obtain $|f_i(w^{1,i})| \leq 12(g_1^i + g_2^i)(m_{2j_0+1}/l\sum_{k \in H_i} e_k)$, where in the worst case, $g_1^i = h^i + e_{k_i}^*$, with $h^i \in W$, and $h^i \in \operatorname{conv}_{\mathbb{Q}}\{h \in W : w(h) = w(f_i)\}$. Since the auxiliary space is 1-unconditional, by Proposition II.9, $|h^i((m_{2j_0+1}/l)\sum_{k \in H_i} e_k)| \leq m_{2j_0+1}/w(f_i)$. Note that $\|g_2^i\|_\infty \leq 1/n_{2j_0+1}$. Putting all these inequalities together we get

$$|f_{\mathrm{R}}(w^1)| \leq \tfrac{12}{m_{2j+1}}\left(\sum_{i=d_0+1}^{d} \tfrac{m_{2j_0+1}}{w(f_i)} + \tfrac{m_{2j_0+1}n_{2j+1}}{l} + \tfrac{m_{2j_0+1}}{n_{2j_0+1}}\right)$$
$$\leq \tfrac{12}{m_{2j+1}}\left(\sum_{i=d_0+1}^{d} \tfrac{m_{2j_0+1}}{w(f_i)} + \tfrac{m_{2j_0+1}^2 n_{2j+1}}{n_{2j_0+1}} + \tfrac{m_{2j_0+1}}{n_{2j_0+1}}\right). \tag{B.6}$$

Using (B.4) and (B.6) we obtain

$$|f_{\mathrm{R}}(w_1)| \leq \tfrac{12m_{2j_0+1}}{m_{2j+1}}\left(\tfrac{2n_{2j+1}m_{2j_0+1}}{n_{2j_0+1}} + \tfrac{1}{n_{2j_0+1}} + \sum_{i=d_0+1}^{d} \tfrac{1}{w(f_i)}\right) \leq \tfrac{1}{m_{2j_0+1}}. \tag{B.7}$$

(4) Let $E = \{i \in [1,d] : w(f_i) = w(\psi_i) \text{ and } \operatorname{supp} z_i \cap \operatorname{supp} f_i \neq \emptyset\}$. By our assumptions, $\#E \leq 2$. For $i \in [(k-1)l, kl] \setminus E$ the properties of the dependent sequences yield that $|f(z_i)| \leq 1/n_{2j_0+1}$. Hence, $|f(w_k)| \leq 2 \cdot 24m_{2j_0+1}/l + m_{2j_0+1}/n_{2j_0+1} \leq 1/m_{2j_0+1}^2$. $\qquad\square$

Lemma B.14. $\|\sum_{k=1}^{n} b_k w_k\| \leq 121 \|\sum_{k=1}^{n} b_k v_k\|_{J_{T_0}}$ *for every choice of scalars* $(b_k)_{k=1}^n$.

Proof. Fix a sequence $(b_k)_{k=1}^n$ of scalars with $\max_k |b_k| \leq 1$, an $f \in K_{\omega_1}$, and its tree $(f_t)_{t \in \mathcal{T}}$.

Antichains. A node $t \in \mathcal{T}$ is called *relevant* if (1) $w(f_t) \leq m_{2j_0+1}$, and (2) if $u \not\geq t$ is its immediate predecessor, if f_u is of type I, and if $w(f_u) = m_{2j+1} < m_{2j_0+1}$, then $t = s(u) = \max\{s \in S_u : w(f_s) < m_{2j_0+1}\}$, where the maximum is taken according to the block ordering $S_u = \{s_1 < \cdots < s_d\}$. Let C be the set of nodes t which are either non-relevant, or such that f_t is of type I and $w(f_t) = m_{2j_0+1}$. Let $\boldsymbol{\mathcal{B}} = (\mathcal{B}_k)_{k=1}^n$ where $\mathcal{B}_k = \mathcal{A}(w_k, \mathrm{C})$ for $k = 1, \ldots, n$ (see (A.1)–(A.5) in Proposition B.3 above). For each k, let $\mathcal{B}_k^{\mathrm{unc}} = \mathrm{S}(\mathcal{B}_k) \setminus \mathrm{C}$ be the set of splitters that are not in C, $\mathcal{B}_k^{\mathrm{cnd}} = \mathcal{B}_k \cap \mathrm{C}$, and $\mathcal{B}_k^{\mathrm{at}} = \mathcal{B}_k \setminus (\mathcal{B}_k^{\mathrm{unc}} \cup \mathcal{B}_k^{\mathrm{cnd}})$.

Fix $u \in \mathcal{B}_k^{\mathrm{unc}}$, and observe that u is splitter of x_k for every $k \in E_u$. List all $s \in S_u$ such that $\mathrm{ran}\, f_s \cap \mathrm{ran}\, w_k \neq \emptyset$,$\{s_{k,1}, \ldots, s_{k,d}\}$ ordered according to the block ordering $f_{s_{k,1}} < \cdots < f_{s_{k,d}}$. We set now

$$w_{k,u}^{\mathrm{in}} = w_k \,|[\min \mathrm{supp}\, w_k, \max \mathrm{supp}\, f_{s_{k,1}}]$$
$$w_{k,u}^{\mathrm{fin}} = w_k - w_{k,t}^{\mathrm{in}}.$$

For $\star \in \{\mathrm{in}, \mathrm{fin}\}$, let $\mathcal{B}_{k,u}^{\star} = \mathcal{A}(w_{k,u}^{\star}, \mathrm{C}^{nr})$, where C^{nr} is the set of non-relevant nodes of $\mathcal{T}$. Set $\mathcal{B}_k^{\star} = \bigcup_{u \in \mathcal{B}_k^{\mathrm{unc}}} \mathcal{B}_{k,u}^{\star}$. Observe that $\boldsymbol{\mathcal{B}}^{\star} = (\mathcal{B}_k^{\star})_{k=1}^n$ is a regular (not necessarily maximal) arrow for $(w_k)_{k=1}^n$ and $(f_t)_{t \in \mathcal{T}}$, whenever $\star \in \{\mathrm{in}, \mathrm{fin}, \mathrm{cnd}, \mathrm{at}\}$.

Assignments and filtrations. Consider the following $\boldsymbol{\mathcal{B}}^{\star}$-assignments $(g_{k,t}^{\star})_{k \in \mathcal{B}_k^{\star}, k}$ where $\star \in \{\mathrm{in}, \mathrm{fin}, \mathrm{cnd}\}$, and $(r_{k,t}^{\star})_{k \in \mathcal{B}_k^{\star}, k}$ where $\star \in \{\mathrm{in}, \mathrm{fin}, \mathrm{cnd}, \mathrm{at}\}$: Fix k, and $t \in \bigcup_{\star \in \{\mathrm{in}, \mathrm{fin}, \mathrm{cnd}, \mathrm{at}\}} \mathcal{B}_k^{\star}$.

(a) Suppose that f_t is of type 0. Then we set $r_{k,t}^{\mathrm{at}} = (1/m_{2j_0+1})e_k^*$ if $t \in \mathcal{B}_k^{\mathrm{at}}$, and we set $g_{k,t}^{\star} = 0$ and $r_{k,t}^{\star} = (1/m_{2j_0+1})e_k^*$, if $t \in \mathcal{B}_k^{\star}$ for some $\star \in \{\mathrm{in}, \mathrm{fin}\}$.

(b) Suppose that t is non-relevant. Then clearly $t \notin \mathcal{B}_k^{\mathrm{at}}$. Fix $\star \in \{\mathrm{in}, \mathrm{fin}, \mathrm{cnd}\}$. We set $g_{k,t}^{\star} = 0$ in all cases. Suppose that $w(f_t) > m_{2j_0+1}$. Then we set $r_{k,t}^{\star} = (1/m_{2j_0+1})e_k^*$ for $\star \in \{\mathrm{in}, \mathrm{fin}\}$, and $r_{k,t}^{\mathrm{cnd}} = (\mathrm{sgn}(b_k)/m_{2j_0+1})e_k^*$. Finally, if $t \neq s(u)$, where u is the immediate predecessor of t (see the definition of relevant node), then we set $r_{k,t}^{\star} = \|f_t(w_k)\|e_k^*$ for $\star \in \{\mathrm{in}, \mathrm{fin}\}$ and $r_{k,t}^{\mathrm{cnd}} = \mathrm{sgn}(b_k)\|f_t(w_k)\|e_k^*$.

(c) Suppose now that t is relevant.

(c.1) $w(f_t) = m_{2j_0+1}$. If $t \in \mathcal{B}_k^{\star}$, for $\star \in \{\mathrm{in}, \mathrm{fin}\}$, then we set $g_{k,t}^{\star} = (1/w(f_t))e_k^*$ and $r_{k,t}^{\star} = 0$. Suppose that $t \in \mathcal{B}_k^{\mathrm{cnd}}$. Suppose that $f_t = \pm I(1/m_{2j_0+1})\sum_{i=1}^{n_{2j_0+1}} g_i$, where $I \subseteq \omega_1$ is an interval, and $\Phi = (g_1, \ldots, g_{n_{2j_0+1}})$ is a $2j_0 + 1$-special sequence. Set $\Psi = (\psi_1, \ldots, \psi_{n_{2j_0+1}})$. Consider $I_t = \{i \in [1, \kappa_{\Phi,\Psi} - 1] : Ig_i \neq 0\} = [k(t,1), k(t,2)]$, and let $\varepsilon_t = \mathrm{sgn}(\sum_{k=k(t,1)+1}^{k(t,2)-1} b_k)$. If $k = k(t,i)$ for $i = 1, 2$, then we set $g_{k,t}^{\mathrm{cnd}} = \mathrm{sgn}(b_{k(t,i)})e_{k(t,i)}^*$ and $r_{k,t}^{\mathrm{cnd}} = 0$. We set $g_{k,t}^{\mathrm{cnd}, \mathcal{B}_k^{\mathrm{cnd}}} = \varepsilon_t e_k^*$ and $r_{k,t}^{\mathrm{cnd}, \mathcal{B}_k^{\mathrm{cnd}}} = 0$ if $k \in (k(t,1), k(t,2))$. We set $g_{k,t}^{\mathrm{cnd}} = 0$, $r_{k,t}^{\mathrm{cnd}} = \mathrm{sgn}(b_k)(1/m_{2j_0+1})e_k^*$ otherwise.

(c.2) Suppose that $w(f_t) \neq m_{2j_0+1}$. Then $t \in \mathcal{B}_k^{\star}$, for some $\star \in \{\mathrm{in}, \mathrm{fin}\}$. Set $g_{k,t}^{\star} = (1/w(f_t))e_k^*$ and $r_{k,t}^{\star} = 0$ for all cases, except for $w(f_t) = m_{2j+1} < m_{2j_0+1}$.

In this case, we observe that since t is splitting there are at least two immediate successor $s_1 \neq s_2 \in S_t$ such that $\operatorname{ran} w_{k,u}^\star \cap \operatorname{ran} f_{s_i} \neq \emptyset$ $(i = 1, 2)$ for some $u \in \mathcal{B}_k^\star$. This implies that there is at most one $k \in E_t^{\mathcal{B}^\star}$ such that $\operatorname{ran} f_{s(t)} \cap \operatorname{ran} w_{k,v}^\wedge \neq \emptyset$ for $v \in \mathcal{B}_k^{\mathrm{unc}}$, and $t \in \mathcal{B}_{k,v}^\star$. Then we set $g_{k,t}^\star = (1/m_{2j_0+1})e_k^\star$ and $r_{k,t}^\star = 0$ if k is this one, and $g_{k,t}^\star = 0$ and $r_{k,t}^\star = (1/m_{2j_0+1})e_k^\star$ otherwise.

Let $(G_t^\star)_{t \in \mathcal{T}}$, $(R_t^\star)_{t \in \mathcal{T}}$ be the corresponding filtrations. Recall that given a regular array $\mathcal{A} = (\mathcal{A}_k)_k$ for $(x_k)_k$ and $(f_t)_{t \in \mathcal{T}}$ the canonical $\mathcal{A}$-assignment $(f_{k,t}^{\mathcal{A}})_{t \in \mathcal{A}_k, k}$ is defined by $f_{k,t}^{\mathcal{A}} = f(x_k)e_k^\star$. It was shown in Remark B.7 that if in addition $\mathcal{A}$ is maximal, then for every $(a_k)_{k=1}^n$ and every $t \in T$, $F^{\mathcal{A}}(\sum_{k \in D_t^{\mathcal{A}}} a_k e_k) = f_t(\sum_{k \in D_t^{\mathcal{A}}} a_k w_k)$.

Claim. *Fix $t \in \mathcal{T}$, and for $\star \in \{\mathrm{in}, \mathrm{fin}, \mathrm{cnd}, \mathrm{at}\}$ let $D_t^\star = D_t^{\mathcal{B}^\star}$. Then:*

(e.1) $|F_t^\star(\sum_{k \in D_t^\star} b_k e_k)| \leq 24(G_t^\star + R_t^\star)(\sum_{k \in D_t^\star} |b_k| e_k)$ *for $\star \in \{\mathrm{in}, \mathrm{fin}\}$.*

(e.2) $|F_t^{\mathrm{cnd}}(\sum_{k \in D_t^{\mathrm{cnd}}} b_k e_k)| \leq 24(G_t^{\mathrm{cnd}} + R_t^{\mathrm{cnd}})(\sum_{k \in D_t^{\mathrm{cnd}}} b_k e_k)$.

(e.3) $|F_t^{\mathrm{at}}(\sum_{k \in D_t^{\mathrm{at}}}) b_k e_k| \leq 24|R_t^{\mathrm{at}}(\sum_{k \in D_t^{\mathrm{at}}}) b_k e_k|$.

(e.4) $G_t^\star \in W(T_0)$ *for $\star \in \{\mathrm{in}, \mathrm{fin}\}$, and $G_t^{\mathrm{cnd}} \in 3W(J_{T_0})$.*

(e.5) $\|R_t^{\mathrm{at}}\|_\infty \leq 1/m_{2j_0+1}$. *For $\star \in \{\mathrm{in}, \mathrm{fin}, \mathrm{cnd}\}$, either t is non-relevant, $w(f_t) < m_{2j_0+1}$ and $G_k^\star = \sum_{k \in E_t^{\mathcal{B}^\star}} \|f_t(w_k)\| e_k^\star$ or $\|R_t^\star\|_\infty \leq 1/m_{2j_0+1}$.*

Proof of the claim. (e.1)–(e.3) are immediate applications of Proposition B.9.

(e.4) Most of the cases follows immediately by definition of the corresponding assignments. We hint the non-trivial ones: Suppose that t is relevant. If $w(f_t) = m_{2j_0+1}$, then $D_t^{\mathrm{cnd}} = E_t^{\mathrm{cnd}}$, and the corresponding assignment gives that $G_t^{\mathrm{cnd}} = \lambda_1 e_{k(t,1)}^\star + \lambda_2 e_{k(t,2)}^\star + \varepsilon_t \sum_{k \in E_t^{\mathrm{cnd}} \cap (k(t,1), k(t,2))} e_k^\star \in 3W(J_{T_0})$, where $\lambda_i = \mathrm{sgn}(b_{k(t,i)}) \chi_{E_t^{\mathrm{cnd}}}(k(t,i))$, for $i = 1, 2$, and where χ_E denotes the characteristic function of E. Fix $\star \in \{\mathrm{in}, \mathrm{fin}\}$. We claim that $\#D_t^\star \leq 1$: Suppose not, and say that $k_1 < k_2 \in D_t^\star$. Then since $w(f_t) = m_{2j_0+1}$ there are $u_i \in \mathcal{B}_{k_i}^{\mathrm{unc}}$ and $s_i \in \mathcal{B}_{k_i,u_i}^\star$ $(i = 1, 2)$ such that $u_1, u_2 \npreceq t \preceq s_1, s_2$. If $\star = \mathrm{in}$, then since $\operatorname{ran} f_t \subseteq \operatorname{ran} f_{s(k_1, u_1)}$, it follows that $\operatorname{ran} f_t < \operatorname{ran} w_{k_2}$, and since $\operatorname{ran} f_{s_2} \subseteq \operatorname{ran} f_t$, we obtain that $\operatorname{ran} f_{s_2} \cap \operatorname{ran} w_{k_2} = \emptyset$, contradicting the fact that $s_2 \in \mathcal{B}_{k_2,u_2}^{\mathrm{in}}$. If $\star = \mathrm{fin}$, in a similar manner we obtain that $\operatorname{ran} w_{k_1} \cap \operatorname{ran} f_{s_1} = \emptyset$, a contradiction. Hence, either $G_t^\star = 0$, or $G_t^\star = (1/m_{2j_0+1})e_k^\star$, certainly in $W(T_0)$ considered as sub-convex combinations.

Suppose now that $w(f_t) \neq m_{2j_0+1}$. There are three subcases to consider: If $w(f_t) > m_{2j_0+1}$, then t is non-relevant, hence a catcher node, and $G_t^\star = \sum_{k \in E_t^\star} g_{k,t}^\star = 0$. If $w(f_t) = m_{2j}$ with $j \leq j_0$, then the inductive hypothesis gives that $G_t^{\mathrm{cnd}} \in 3W(J_{T_0})$ (since $E_t^{\mathrm{cnd}} = \emptyset$). Fix $\star \in \{\mathrm{in}, \mathrm{fin}\}$. Observe that for every $k \in E_t^\star$ there is $s \in S_t$ such that $\operatorname{ran} f_s \subseteq \operatorname{ran} w_k$, in which case $D_s^\star = \emptyset$, and so $\#E_t^\star + \#\{s \in S_t : G_s^{\mathrm{cnd}} \neq 0\} \leq \#S_t$, and then, $G_t^\star = (1/w(f_t))(\sum_{k \in E_t^\star} e_k^\star + \sum_{s \in S_t} G_s^\star) \in W(T_0)$.

If $w(f_t) = m_{2j+1} < m_{2j_0+1}$, then using that there is at most one immediate successor $s(t)$ of t which is relevant we obtain that either $G_t^{\mathrm{cnd}} = 0$, or $G_t^{\mathrm{cnd}} = (1/m_{2j})G_{s(t)}^{\mathrm{cnd}}$, and for $\star \in \{\mathrm{in, fin}\}$, either $G_t^\star = (1/m_{2j+1})e_k^*$, or $G_t^\star = (1/m_{2j+1})G_{s(t)}^\star$.

(e.5) $\|R_t^{\mathrm{at}}\|_\infty \le 1/m_{2j_0+1}$ follows from Proposition B.9, since this is so for the corresponding assignment of which R_t^{at} is a filtration. Suppose that $\star \in \{\mathrm{in, fin, cnd}\}$. The proof is by backwards induction over t. Again we concentrate on non-trivial cases. Suppose that f_t is of type I and t is relevant. Then if $w(f_t) = m_{2j}$ with $j \le j_0$, the desired result follows from the definition of the corresponding assignments, and inductive hypothesis. Suppose that $w(f_t) = m_{2j+1}$ with $j < j_0$. Then $R_t^\star = \sum_{k \in E_t^\star} r_{k,t}^\star + (1/w(f_t)) \sum_{s \in S_t} R_s^\star$. By the definition of the assignments, $\|r_{k,t}^\star\|_\infty \le 1/m_{2j_0+1}$ for every $k \in E_t^\star$. Observe that all $s \in S_t$, except possibly one, $s(t)$, are non-relevant and that $r_{k,s}^\star = \|f_s(w_k)\|e_k^*$ for every $k \in E_s^\star = D_s^\star$. Hence, for every $s \in S_t \setminus \{s(t)\}$, $\|(1/w(f_t))R_s^\star\|_\infty = \max\{(1/w(f_t))\|f_s(w_k)\| : k \in E_s^\star\} \le 1/m_{2j+1}$; the last inequality follows from Proposition B.13. By inductive hypothesis $\|R_{s(t)}^\star\|_\infty \le 1/m_{2j_0+1}$, so we are done.

Suppose that t is non-relevant. The case $w(f_t) > m_{2j_0+1}$ is immediate. Suppose that $w(f_t) = m_{2j}$ and $t \ne s(u)$, where u is the immediate predecessor of t (see the definition of relevant node). Notice that t is a catcher, so $E_t^\star = D_t^\star$, and $R_t^\star = \sum_{k \in E_t^\star} \|f_t(w_k)\|e_k^*$, as desired. $\qquad\square$

We are now ready to finish the proof using part B of the general theory above. Notice that for each $\star \in \{\mathrm{in, fin, cnd, at}\}$, $\mathcal{B}^\star \not\prec \mathcal{B}$, and that the canonical assignments of $\mathcal{B}^\star$ are coherent. Let $(h_{k,t}^\star)_{t \in \mathcal{B}_k, k}$ be the assignments induced by the canonical $\mathcal{B}^\star$-assignments $(f_{k,t}^\star)_{t \in \mathcal{B}_k^\star, k}$, for $\star \in \{\mathrm{in, fin, cnd, at}\}$.

Claim. *For every $t \in T$, $H_t^{\mathrm{in}} + H_t^{\mathrm{fin}} + F_t^{\mathrm{cnd}} + F_t^{\mathrm{at}} = F_t^{\mathcal{B}}$, the canonical assignment of $\mathcal{B}$.*

Proof of the claim. We show that for every $t \in B_k$, $h_{k,t}^{\mathrm{in}} + h_{k,t}^{\mathrm{fin}} + h_{k,t}^{\mathrm{cnd}} + h_{k,t}^{\mathrm{at}} = f_t(w_k)e_k^*$. The only non trivial case is if $t \in \mathcal{B}_k^{\mathrm{unc}}$. Notice that since $B_{k,t}^\star$ is a maximal antichain for $w_{k,t}^\star$ and $(f_s)_{s \succeq t}$, we obtain that $h_{k,t}^\star = F_{k,t}^\star = f_t(w_{k,t}^\star)e_k^*$. Hence, $f_{k,t}^{\mathrm{in}} + f_{k,t}^{\mathrm{fin}} = (f_t(w_{k,t}^{\mathrm{in}}) + f_t(w_{k,t}^{\mathrm{fin}}))e_k^* = f_t(w_k)e_k^*$, and $h_{k,t}^\star = 0$ for $\star \in \{\mathrm{cnd, at}\}$. $\qquad\square$

Finally, by Proposition B.11, $H_\emptyset^\star = F_\emptyset^\star$, for $\star \in \{\mathrm{in, fin, cnd, at}\}$. Hence,

$$|f(\textstyle\sum_{k=1}^n b_k w_k)| = |F_\emptyset^{\mathcal{B}}(\textstyle\sum_{k=1}^n b_k e_k)| \le \textstyle\sum_{\star \in \{\mathrm{in, fin, cnd, at}\}} |F_\emptyset^\star(\textstyle\sum_{k \in D_\emptyset^\star} b_k e_k)| \le$$

$$\le 24(5\|\textstyle\sum_{k=1}^n b_k e_k\|_{J_{T_0}} + 4\|\textstyle\sum_{k=1}^n b_k e_k\|_\infty) \le 120\|\textstyle\sum_{k=1}^n e_k\|_{J_{T_0}} + \varepsilon.$$

$$\tag{B.8}$$

Corollary B.15. *The natural isomorphism $F : \langle w_1, \ldots, w_n \rangle \to \langle v_1, \ldots, v_n \rangle$ defined by $F(w_i) = v_i$ satisfies that $\|F\| \le 1$ and $\|F^{-1}\| \le 120 + \varepsilon$. Consequently, J_{T_0} is finite interval representable on the basis $(e_\alpha)_{\alpha < \omega_1}$ of $\mathfrak{X}_{\omega_1}$ with a constant $C < 121$.*

Proof. Proposition VII.11 shows that $\|F\| \le 1$; the other inequality follows from Lemma B.14. $\qquad\square$

Bibliography

[1] D. E. Alspach, S. A. Argyros, *Complexity of weakly null sequences*, Dissertationes Mathematicae, 321, 1992.

[2] A. D. Andrew, W. L. Green, *On James' quasireflexive Banach space as a Banach algebra*, Canad. J. Math. **32** (1980), no. 5, 1080–1101.

[3] S. A. Argyros, *A universal property of Hereditarily Indecomposable Banach spaces*, Proc. Amer. Math. Soc., **129** (2001), 3231–3239.

[4] S.A. Argyros, A. Arvanitakis, A.Tolias, *Strictly singular extensions and HI spaces with no reflexive subspaces*, in preparation,

[5] S. A. Argyros, I. Deliyanni, *Banach spaces of the type of Tsirelson*, preprint, 1992.

[6] S. A. Argyros, I. Deliyanni, *Examples of asymptotic l_1 Banach spaces*, Trans. Amer. Math. Soc. **349** (1997), no. 3, 973–995.

[7] S. A. Argyros, I. Deliyanni, *On the structure of the Diagonal strictly singular operators in HI spaces*, in preparation,

[8] S. A. Argyros, I. Deliyanni, D. N. Kutzarova, A. Manoussakis, *Modified mixed Tsirelson spaces*, J. Funct. Anal. **159** (1998), no. 1, 43–109.

[9] S. A. Argyros, I. Deliyanni, A. Manoussakis, *Distortion and spreading models in mixed Tsirelson spaces*, Studia Math. **157** (3) 2003, 199–236.

[10] S. A. Argyros, V. Felouzis, *Interpolating Hereditarily Indecomposable Banach spaces*, Journal Amer. Math. Soc., **13** (2000), 243–294.

[11] S.A. Argyros, G. Godefroy, H. Rosenthal, *Descriptive set theory and Banach spaces*, Handbook of the geometry of Banach spaces, Vol. 2, 1007–1069, North-Holland, Amsterdam, 2003.

[12] S. A. Argyros, J. Lopez-Abad, S. Todorcevic, *A class of Banach spaces with no unconditional basic sequence*, C. R. Acad. Sci. Paris. Ser. **I 337** (2003), 43–48.

[13] S. A. Argyros, J. Lopez-Abad, S. Todorcevic, *A class of Banach spaces with few non strictly singular operators*, J. of Func. Anal. (to appear).

[14] S. A. Argyros, A. Manoussakis, *An Indecomposable and Unconditionally sat-urated Banach space*, Studia Math. **159** (1), (2003), 1–32.

[15] S. A. Argyros, A. Manoussakis, *A sequentially unconditional Banach space with few operators*, Proc. LMS (to appear).

[16] S. A. Argyros, A. Tolias, *Methods in the theory of hereditarily indecomposable Banach spaces*, Mem. Amer. Math. Soc. 170 (2004), no. 806,

[17] S. A. Argyros, A. Tolias, *Indecomposability and unconditionality in duality*, Geom. Funct. Anal. 14 (2004), no. 2, 247–282.

[18] S. F. Bellenot, *Tsirelson superspaces and ℓ_p*, Journal of Funct. Anal., 69, No 2, 1986, 207–228.

[19] S. Bellenot, R. Haydon, E. Odell, *Quasi-Reflexive and tree spaces*, Contemporary Mathematics, **85** (1989), 19–43.

[20] J. Bernués, I. Deliyanni, *Families of finite subsets of* N *of low complexity and Tsirelson type spaces*, Math. Nachr. **222** (2001), 15–29.

[21] P. G. Casazza, T. Shura, *Tsirelson's space*, Lecture Notes in Math. 1363, Springer Verlag, 1989.

[22] I. Deliyanni, A. Manoussakis, *Asymptotic ℓ_p HI-spaces*, in preparation.

[23] G.A. Edgar, *A long James space*, Proc. Conf. on Measure Theory, Lecture Notes in Math. **794**, (1980) 31–37.

[24] M. Fabian, P. Habala, P. Hajek,V. Montesinos Santaluca, J. Pelant, V. Zizler, *Functional analysis and infinite-dimensional geometry.*, CMS Books in Mathematics 8, Springer-Verlag, New York, 2001.

[25] V. Ferenczi, *Operators on subspaces of Hereditarily Indecomposable Banach spaces*, Bull. London Math. Soc. **29** (1997), no. 3, 338–344.

[26] V. Ferenczi, *Hereditarily finitely decomposable Banach spaces*, Studia Math. **123** (1997), no. 2, 135–149.

[27] V. Ferenczi, *A uniformly convex Hereditarily Indecomposable Banach space*, Israel J. Math., **102** (1997), 199–225.

[28] T. Figiel, W. B. Johnson, *A uniformly convex Banach space which contains no ℓ_p*, Compositio Math. 29, 1974, 179–190.

[29] W. T. Gowers, *A Banach space not containing c_0, ℓ_1 or a reflexive subspace*, Transactions of the AMS 344, No 1, 1994, 407–420.

[30] W. T. Gowers, *An Infinite Ramsey Theorem and Some Banach-Space Dichotomies*, Ann. of Math. (2) **156**, (2002), 797–833.

[31] W. T. Gowers, *A solution to Banach's hyperplane problem*, Bull. London. Math. Soc., **28**, (1996), 297–304.

[32] W. T. Gowers, *Ramsey methods in Banach spaces*, Handbook of the geometry of Banach spaces, Vol. 2, 1071–1097, North-Holland, Amsterdam, 2003

[33] W. T. Gowers, B. Maurey, *The unconditional basic sequence problem*, Journal of AMS 6, 1993, 851–874.

[34] W. T. Gowers, B. Maurey, *Banach spaces with small spaces of operators*, Math. Ann. 307 (1997), no. 4, 543–568.

[35] R. C. James, *Bases and reflexivity of Banach spaces*. Ann. of Math. (2) **52**, (1950). 518–527.

[36] R. C. James, *A non-reflexive Banach space isometric with its second conjugate space*, Proc. Nat. Acad. Sci. U. S. A. 37, (1951). 174–177.

[37] R. C. James, *Uniformly non-square Banach spaces*, Ann. of Math. 80, 1964, 542–550.

[38] R. C. James, *A separable somewhat reflexive Banach space with nonseparable dual*, Bull. Amer. Math. Soc. 80 (1974), 738–743.

[39] W. B. Johnson,*A reflexive Banach space which is not sufficiently Euclidean* Studia Math. 55 (1976), no. 2, 201–205.

[40] J. Lindenstrauss *On nonseparable reflexive Banach spaces*. Bull. Amer. Math. Soc. **72** (1966) 967–970.

[41] J. Lindenstrauss, L. Tzafriri, *Classical Banach spaces I*, Springer-Verlag Vol. 92, (1977)

[42] J. Lopez-Abad, A. Manoussakis, *On Tsirelson type spaces* (in preparation).

[43] A. Manoussakis, *On the structure of a certain class of mixed Tsirelson spaces*, Positivity **5** (2001), no. 3, 193–238.

[44] A. Manoussakis, *Some remarks on spreading models and mixed Tsirelson spaces*, Proc. Amer. Math. Soc. 131 (2003), 2515–2525.

[45] B. Maurey, *A remark about distortion*, Operator Theory: Advances and Applications 77, 1995, 131–142.

[46] B. Maurey, *Banach spaces with few operators*, Handbook of the geometry of Banach spaces, Vol. 2, 1247–1297, North-Holland, Amsterdam, 2003

[47] B. Maurey, V. D. Milman, N. Tomczak-Jaegermann, *Asymptotic infinite dimensional theory of Banach spaces*, Operator Theory: Advances and Applications 77, 1995, 149–175.

[48] B. Maurey, H. Rosenthal, *Normalized weakly null sequences with no unconditional subsequence*, Studia Math. 61, 1977, 77–98.

[49] V. D. Milman, *The geometric theory of Banach spaces, part II: Geometry of the unit sphere*, Math. Surveys 26, 1971, 79–163.

[50] V. D. Milman, N. Tomczak-Jaegermann, *Asymptotic ℓ_p spaces and bounded distortions*, Banach spaces, Contemp. Math. 144, 1993, 173–196.

[51] E. Odell, *On subspaces, asymptotic structure, and distortion of Banach spaces; connections with logic*, Analysis and logic (Mons, 1997), 189–267, London Math. Soc. Lecture Note Ser., 262, Cambridge Univ. Press, Cambridge, 2002.

[52] E. Odell, Th. Schlumprecht, *The distortion problem*, Acta Math. 173, 1994, 259–281.

[53] E. Odell, Th. Schlumprecht, *Distortion and Stabilized Structure in Banach spaces; New Geometric Phenomena for Banach and Hilbert Spaces*, Proc. International Congress of mathematicians, vol. 1,2,(Zürich, Switzerland 1994), Birkhäuser Verlag, Basel, 1995, 955–965.

[54] J. Rainwater, *Weak convergence of bounded sequences*, Proc. Amer. Math. Soc. 14 (1963), 999.

[55] Th. Schlumprecht, *An arbitrarily distortable Banach space*, Israel J. Math. 76, 1991, 81–95.

[56] J. Schreier, *Ein Gegenbeispiel zur Theorie der schwachen Konvergenz*, Studia Math. 2, 1930, 58–62.

[57] S. Shelah, *A Banach space with few operators*, Israel J. Math. 30 (1978), no. 1-2, 181-191.

[58] S. Shelah, J. Steprāns, *A Banach space on which there are few operators*, Proc. Amer. Math. Soc. 104 (1988), no. 1, 101–105.

[59] I. Singer, *Bases in Banach spaces. II*, Editura Academiei Republicii Socialiste România, Bucharest, 1981.

[60] S. Todorcevic, *Partitioning pairs of countable ordinals*, Acta Math. 159 (1987), no. 3-4, 261–294.

[61] S. Todorcevic, *Coherent sequences*, Handbook of set theory, to appear.

[62] N. Tomczak-Jaegermann, *Banach spaces of type p have arbitrarily distortable subspaces*, GAFA, 6 (1996), 1074–1082.

[63] B. S. Tsirelson, *Not every Banach space contains ℓ_p or c_0*, Funct. Anal. Appl. 8 1974, 138–141.

[64] H. M. Wark, *A non-separable reflexive Banach space on which there are few operators*, J. London Math. Soc. (2) 64 (2001), no. 3, 675–689.

[65] M. Zippin, On perfectly homogeneous bases in Banach spaces, *Israel J. of Math.* 4 (1966), p. 265–272

Part B

High-Dimensional
Ramsey Theory and
Banach Space Geometry

Stevo Todorcevic

Part II

Introduction

One of the first remarkable applications of Ramsey theory to Banach space theory
dates back to the mid 1970s and is due to Brunel and Sucheston. They were
able to isolate the notion of a spreading model (e_n) of a basic sequence (x_n) in
a given Banach space X. The spreading model (e_n) is a basic sequence in some
other Banach space Y. It is finitely block equivalent to (x_n) but it typically has
considerably more properties especially if one is willing to pass to its infinite block
subsequences. For example, (e_n) has a block subsequence which is unconditional.
Working from an unconditional spreading model (e_n) one can go further and find
a basic sequence (f_n) in some Banach space Z such that (f_n) is finitely block
representable in (e_n) and such that (f_n) is equivalent to the standard basis of
some ℓ_p $(1 \le p < \infty)$ or c_0, a well-known result of J. L. Krivine. It follows that
for any basic sequence (x_n) in some Banach space X one of the classical spaces ℓ_p
$(1 \le p < \infty)$ or c_0 is finitely block representable in (x_n). This in particular gives
precise connection between the local theory of general Banach spaces and the local
theory of classical Banach spaces, as well as a new proof of Dvoretzky's theorem
that ℓ_2 is finitely represented in every infinite dimensional Banach space. Starting
from this set of results we have split the rest of the lectures into two parts, those
dealing with problems of finding subsequences of a given sequence with some
desired properties (like being unconditional, for example) and those of finding
block subsequences with desired properties. In Ramsey theory this corresponds to
concentrating on different "Ramsey spaces" so we tried to explain this in some
details. The discovery of the first Ramsey space $\mathbb{N}^{[\infty]}$ of all infinite increasing
sequences of natural numbers has in fact been started in the early 1960s from
purely utilitarian reasons of developing a theory of well-quasi-orderings, a theory
that culminated in Laver's proof of Fraïssé's conjecture stating that the class
of countable linear orderings is well-quasi-ordered under embeddability. We have
presented parts of this beautiful theory of Nash–Williams not only because of its
own right but also because of the importance of his notions of 'fronts' and 'barriers'
in the theory of Tsirelson norms. This is an area of functional analysis that has
seen a considerable growth in the last 10 to 15 years. Some of this deep theory is
presented in the parallel set of lecture notes of Spiros A. Argyros which form the
first part of this volume and which we warmly recommend to the reader.

In the early 1970s Nash–Williams theory has been reformulated and considerably strengthened by the work of Galvin, Prikry, Silver and specially Ellentuck by introducing the topological Ramsey theory. In hindsight the first application of this new topological Ramsey theory in the Banach space theory is Rosenthal's ℓ_1-theorem stating that a bounded infinite sequence (x_n) of elements of some Banach space contains either an infinite subsequence (x_{n_k}) which is weakly Cauchy or an infinite subsequence (x_{n_k}) such that for some positive constants $\delta < \varepsilon$,

$$\delta \sum_{k=1}^{m} |\lambda_k| \leq \| \sum_{k=1}^{m} \lambda_k x_{n_k} \| \leq \varepsilon \sum_{k=1}^{m} |\lambda_k|$$

for all choices of integers $m \geq 1$ and scalars $\lambda_1, \ldots, \lambda_k$. The proof of Rosenthal's ℓ_1-theorem which used the topological Ramsey theory has actually been found by Farahat [27] shortly after Rosenthal's original proof [78]. In fact, it was Farahat's proof which has introduced topological Ramsey theory to this area of Banach space theory. Out of the numerous results proved in the wake of Farahat's proof of Rosenthal's ℓ_1-theorem one could not go without mentioning Elton–Odell's beautiful result that every infinite dimensional normed space X contains an infinite normalized sequence $(x_n)_{n=0}^{\infty}$ of vectors such that for some $\varepsilon > 0$ and all $m \neq n$, $\|x_m - x_n\| > 1 + \varepsilon$. Diestel's book [20] contains an exposition of several other results about sequences in Banach spaces obtained using the topological Ramsey space $\mathbb{N}^{[\infty]}$ of infinite subsets of $\mathbb{N}$.

The Ramsey theory of $\mathbb{N}^{[\infty]}$ is particularly well suited for the study of sequence — convergence and summability of series in Banach spaces, or more generally, in topological abelian groups. We have presented this in Sections II.9 and II.10. For example, we show how Nash–Williams' version of this theory enters naturally into the proof of Rosenthal's theorem stating every weakly null sequence (x_n) in some Banach space X contains either a subsequence (x_{n_k}) all of whose further subsequences are Cesàro summable, or a subsequence (x_{n_k}) whose spreading model is isomorphic to ℓ_1. In Section II.10 we show that a simple application of the topological Ramsey theorem for $\mathbb{N}^{[\infty]}$ gives a simple and uniform way for deducing several basic principles of functional analysis such as principles of uniform boundedness and automatic continuity, Schur's ℓ_1-theorem, Orlicz–Pettis theorem, Vitali–Hahn–Saks theorem, etc.

While the Ramsey space $\mathbb{N}^{[\infty]}$ has numerous applications to Banach space theory the block Ramsey spaces such as the space $\mathrm{FIN}^{[\infty]}$ of all infinite block sequences of finite sets seem to be more relevant to deeper problems of this theory. We have chosen to exemplify this on the problem of finding an unconditional sequence of a given weakly null sequence (x_n) of some Banach space X. The Ramsey theory of $\mathbb{N}^{[\infty]}$ via an unpublished result of Rosenthal shows that every weakly null sequence in $\ell_\infty(\Gamma)$ consisting of characteristic functions of subsets of Γ contains an unconditional subsequence. In these notes (see II.5), we shall extend this result to all weakly null sequences in $l_\infty(\Gamma)$ that are in some sense separated away from 0. There is however a bound on how far one can go in this direction.

The famous example of Maurey–Rosenthal shows that this is no longer true even in the case of weakly null sequences of the function space $C(\omega^{\omega^2} + 1)$. However, it still makes sense asking for an unconditional *block-subsequence* of a given weakly null sequence. The search for a block Ramsey theorem applicable to this problem from the Banach space theory lasted for almost twenty years. The culmination of this search is the famous Gowers' dichotomy which although it does not depend on any previous result about known Ramsey spaces, such as $\mathbb{N}^{[\infty]}$ or $\mathrm{FIN}^{[\infty]}$, it uses the basic idea of the 'combinatorial forcing' present in essentially all results establishing the existence of various Ramsey spaces. We have taken some effort in pointing explicitly any place where combinatorial forcing is used not only to point out the similarities in various arguments but to also indicate the existence of some sort of method that might be applicable to many other cases.

We have also taken some effort to point out the difference between the classical notion of a Ramsey subset of a given Ramsey space and Gowers' notion of strategically and approximately Ramsey sets of infinite normalized block sequences of vectors of a given Banach space E. For example, in the last sections of these notes we show that while classical Ramsey property is always closed under the operation of complementation, Gowers' notion of strategic or approximate Ramsey property in general is not preserved by taking complements. This is done by transferring the strategic or approximate Ramsey property of a given class of sets of normalized block sequences to the class of their continuous images. So, for example, if the Banach space E contains no isomorphic copy of c_0 then there is a rather automatic way to transfer the strategic Ramsey property of analytic sets of infinite block sequences of E to the strategic Ramsey property of G_δ sets of infinite block sequences of E. This explains why the full strength of Gowers' analytic Ramsey theorem for Banach spaces (Theorem 4.1 of [35]) is really contained in its G_δ-case and why all known applications of this theorem involve only simple sets of block sequences.

There is an aspect of Ramsey theory that calls for applications. This is a part of Ramsey theory that tries to classify 'canonical' equivalence relations on a given Ramsey space. For example in the case of the Ramsey space $\mathbb{N}^{[\infty]}$ one is interested in the behavior of equivalence relations (equivalently, maps from $\mathbb{N}^{[\infty]}$ into some other space) on $\mathbb{N}^{[\infty]}$, not globally but relative to the possibility of going to an arbitrary chosen sub-cube $M^{[\infty]}$ over an infinite subset M of $\mathbb{N}$. We have chosen to present a result of Pudlak and Rödl that essentially deals with the classification of all Borel equivalence relations on $\mathbb{N}^{[\infty]}$ with at most countably many classes. We have also included some block Ramsey analogues such as the beautiful result of A. D. Taylor which identifies the list of five equivalence relations on FIN and the results of J. Lopez-Abad which does the same for the generalizations FIN_k of FIN.

Finally, we mention that this set of notes is a reworked version of the material that was originally prepared for the Advanced Course on Ramsey Methods in Analysis at the Centre de Recerca Matemàtica during January 2004. We have tried to present some of the aspects of Ramsey theory that have shown to be useful in some areas of functional analysis but undoubtedly have missed many others. Even the small selection when fully presented would fill an amount of lecturing of a much greater length than that given given to us. So it was necessary to compromise and give only the key concepts and ideas of the theory. The length restriction is also responsible for our compromise of being rather brief and at times superficial regarding the credits and other historical discussions. We choose this opportunity to apologize to the numerous authors that have built this subject and hope that the rich source of literature that we include at the end of these notes will indirectly remedy this flaw since any reader interested in more details will have no problems of finding a more complete picture there. During the preparation of this set of notes I have profited from remarks of many participants of the Advanced Course. Special thanks go to Neus Castells and Jordi Lopez Abad who helped not only with mathematical contents of these notes but also with the technical aspects of preparing these notes. Naturally, all the remaining omissions and errors are my sole responsibility.

Chapter I

Finite-Dimensional Ramsey Theory: Finite Representability of Banach Spaces

I.1 Finite-Dimensional Ramsey Theorem

Consider a set S with possibly some structure such as for example a partial semigroup structure, where the associative relation is defined only for x and y respecting some order. Then given a positive integer k, Ramsey-theoretic results at this level are typically not about colorings of the full k-power S^k but rather some restricted version $S^{[k]} \subseteq S^k$ that respects the partial semigroup structure on S. In the case $S = \mathbb{N}$ with the usual order, $S^{[k]}$ is the set of all k-element subsets of $\mathbb{N}$, and similarly for any other ordered set $(S, <)$.

Theorem I.1.1 (Ramsey). *For every positive integer k and for every finite coloring of the symmetric power $\mathbb{N}^{[k]}$ there is an infinite set $M \subseteq \mathbb{N}$ such that $M^{[k]}$ is monochromatic.*

Proof. Choose a nonprincipal ultrafilter $\mathcal{U}$ on $\mathbb{N}$. It is very convenient to think about $\mathcal{U}$ as a *quantifier* over $\mathbb{N}$ in the sense that a formula of the form $(\mathcal{U}n)\varphi(n)$ is interpreted as saying that the set $\{n : \varphi(n)\}$ belongs to $\mathcal{U}$. The corresponding quantifier respects all propositional connectives in the sense that for every choices of formulas $\varphi(x)$ and $\psi(x)$ we have:

(a) $\neg(\mathcal{U}n)\varphi(n) \longleftrightarrow (\mathcal{U}n)\,\neg\varphi(n)$,

(b) $(\mathcal{U}n)\varphi(n)\,\&\,(\mathcal{U}n)\psi(n) \longleftrightarrow (\mathcal{U}n)\,(\varphi(n)\,\&\,\psi(n))$,

(c) $(\mathcal{U}n)\varphi(n) \vee (\mathcal{U}n)\psi(n) \longleftrightarrow (\mathcal{U}n)\,(\varphi(n) \vee \psi(n))$.

Given a positive integer k and an ultrafilter $\mathcal{U}$ we can define its k-power $\mathcal{U}^k$ by deciding when a subset X of $\mathbb{N}^{[k]}$ belongs to $\mathcal{U}^k$ as follows:

$$X \in \mathcal{U}^k \text{ iff } (\mathcal{U}n_0)(\mathcal{U}n_1)\cdots(\mathcal{U}n_{k-1})X(n_0, n_1, \ldots, n_{k-1})$$

where, of course, $X(n_0, n_1, \ldots, n_{k-1})$ has the meaning that the k-element set $\{n_0, n_1, \ldots, n_{k-1}\}$ is written in its $<$-increasing order and that it belongs to X. Then the properties (a), (b), and (c) of the quantifier $(\mathcal{U}n)$ for this particular formula translate to the following facts about $\mathcal{U}^k$:

(a') $X \notin \mathcal{U}^k$ iff $(\mathbb{N}^{[k]}\setminus X) \in \mathcal{U}^k$.
(b') $X, Y \in \mathcal{U}^k$ iff $X \cap Y \in \mathcal{U}^k$.
(c') $X \in \mathcal{U}^k$ or $Y \in \mathcal{U}^k$ iff $X \cup Y \in \mathcal{U}^k$.

It follows that $\mathcal{U}^k$ is indeed a nonprincipal ultrafilter on $\mathbb{N}^{[k]}$. Hence, given a finite coloring of $\mathbb{N}^{[k]}$ we can find a color P such that $P \in \mathcal{U}^k$.

Given a set $s \subseteq \mathbb{N}$ of size $l < k$, let

$$P_s = \{t \in \mathbb{N}^{[k-l]} : \max(s) < \min(t) \ \& \ s \cup t \in P\}.$$

Note the following properties of these sets:
(0) $P_\emptyset = P \in \mathcal{U}^k$
(1) $\{s \in P^{[l]} : P_s \in \mathcal{U}^{k-l}\} \in \mathcal{U}^l$ for $0 \le l < k$.
    Let
$$T = \{s \in \mathbb{N}^{[\le k]} : P_s \in \mathcal{U}^{[k-|s|]}\},$$

where in case $s \in \mathbb{N}^{[k]}$, we choose the convention that "$P_s \in \mathcal{U}^0$" means "$s \in P$". Note that:
(2) $\emptyset \in T$
(3) $s \in T \cap \mathbb{N}^{[<k]}$ implies $T_s = \{n : s \cup \{n\} \in T\} \in \mathcal{U}$.

Using (2) and (3) one builds an infinite strictly increasing sequence $(m_j) \subseteq \mathbb{N}$ such that for every $j \in \mathbb{N}$, and every $s \subseteq \{m_0, \ldots, m_{j-1}\}$ of size $< k$,

$$T_s \supseteq \{m_j, m_{j+1}, \ldots\}.$$

In other words, we have an infinite set $M = \{m_j : j \in \mathbb{N}\}$ such that $M^{[k]} \subseteq P$, as required. $\square$

The previous proof suggests the following definition.

Definition I.1.2. An ultrafilter $\mathcal{U}$ on $\mathbb{N}$ is *Ramsey* if it is nonprincipal and has the property that for every positive integer k its k-power $\mathcal{U}^k$ is generated by sets of the form $M^{[k]}$ for $M \in \mathcal{U}$.

Corollary I.1.3. *A nonprincipal ultrafilter $\mathcal{U}$ on $\mathbb{N}$ is Ramsey iff for every positive integer k and every finite coloring of $\mathbb{N}^{[k]}$ there is $M \in \mathcal{U}$ such that $M^{[k]}$ is monochromatic.*

Lemma I.1.4. *A nonprincipal ultrafilter $\mathcal{U}$ on $\mathbb{N}$ is Ramsey iff for every sequence $\{M_i\}_{i=0}^\infty \subseteq \mathcal{U}$ there exists $M \in \mathcal{U}$ such that $j \in M_i$ for all $i < j$ in M.*

Proof. Given a sequence $\{M_i\}_{i=0}^{\infty}$ of elements of $\mathcal{U}$ define $c : \mathbb{N}^{[2]} \to \{0,1\}$ by letting $c(\{i,j\}) = 0$ iff $j \in M_i$. If $M \in \mathcal{U}$ is such that $c \upharpoonright M^{[2]}$ is constant, then the constant value has to be 0, giving us the conclusion $j \in M_i$ whenever $i < j \in M$. For the converse implication, given $c : \mathbb{N}^{[2]} \to \{0,1\}$, for each i let $\varepsilon(i) \in \{0,1\}$ be such that $M_i = \{j > i : c(i,j) = \varepsilon(i)\} \in \mathcal{U}$. Get $M \in \mathcal{U}$ such that $j \in M_i$ whenever $i < j \in M$. Shrinking M we may assume that $\varepsilon \upharpoonright M$ is constant, so $c \upharpoonright M^{[2]}$ is also constant. $\qquad\square$

Lemma I.1.5. *A nonprincipal ultrafilter $\mathcal{U}$ on $\mathbb{N}$ is Ramsey iff for every $f : \mathbb{N} \to \mathbb{N}$ there exists an $M \in \mathcal{U}$ such that $f \upharpoonright M$ is either $1-1$ or constant.*

Proof. Given f, define a coloring $c : \mathbb{N}^{[2]} \to \{0,1\}$ by letting $c(i,j) = 0$ iff $f(i) = f(j)$. Find $M \in \mathcal{U}$ with monochromatic $M^{[2]}$. Then f is either constant or one-to-one on M. We prove the converse implication using the characterisation given in the previous Lemma. Let $\{M_i\}_{i=0}^{\infty} \subseteq \mathcal{U}$ be a given sequence which we may assume to be decreasing and with empty intersection. Define $f : \mathbb{N} \to \mathbb{N}$ by $f(i) = \min\{j : i \in M_j\}$. Then there is $M \in \mathcal{U}$ on which f is one-to-one. Choose a strictly increasing infinite sequence (n_k) in $\mathbb{N}$ such that for all $i,j,k \in \mathbb{N}$, if $f(i) = j$, then $i \leq n_k$ implies $j < n_{k+1}$ and $j \leq n_k$ implies $i < n_{k+1}$. Applying the selection property to the function that collapses each interval $[n_k, n_{k+1})$ to a point, we obtain an $N \in \mathcal{U}$ which takes at most one point from each of the intervals. Going to a subset of N, we may assume that N does not intersect two consecutive intervals $[n_k, n_{k+1})$. Then $P = M \cap N$ is a member of $\mathcal{U}$ such that $j \in M_i$ whenever $i < j \in P$. $\qquad\square$

Exercise I.1.6. Show that if $\mathcal{U}$ is a Ramsey ultrafilter on $\mathbb{N}$ then for every $(x_n) \in c_0$ there is $M \in \mathcal{U}$ such that $\sum_{n \in M} |x_n| < \infty$.

Note that for a given ultrafilter $\mathcal{U}$ on some set S and a sequence $(x_s)_{s \in S}$ of elements of some compact space X there is a unique member x of X such that

$$\{s \in S : x_s \in \mathrm{U}\} \in \mathcal{U}$$

for every open neighborhood U of x. We shall denote this fact by the formula

$$\lim_{s \to \mathcal{U}} x_s = x.$$

Note also that the set βS of all ultrafilters on S has a natural compact Hausdorff topology generated by basis

$$\{\mathcal{U} \in \beta S : A \in \mathcal{U}\} \quad (A \subseteq S).$$

Exercise I.1.7. Show that βS is the Stone-Čech compactification of the discrete space S, or in other words, that for every $f : S \to X$ where X is a compact Hausdorff space there is a continuous map $\beta f : \beta S \to X$ such that $\beta f \upharpoonright S = f$.

Exercise I.1.8. Use Ramsey's Theorem to show that every bounded infinite sequence $(x_n)_{n=0}^{\infty}$ of real numbers has an infinite subsequence $(x_{n_k})_{k=0}^{\infty}$ such that both $(x_{n_k})_{k=0}^{\infty}$ and $(x_{n_{k+1}-n_k})_{k=0}^{\infty}$ are convergent.

I.2 Spreading Models of Banach Spaces

Recall the notion of a *Schauder basis* of an infinite-dimensional Banach space X, a sequence $(e_i) \subseteq X$ with the property that for every $x \in X$ there is a unique sequence (a_i) of scalars such that

$$x = \sum_{i=0}^{\infty} a_i e_i.$$

Given a Schauder basis (e_i) of X the sequence $(P_n)_{n=1}^{\infty}$ of *canonical projections* is defined by $P_n(x) = \sum_{i=0}^{n-1} a_i e_i$ where $x = \sum_{i=0}^{\infty} a_i e_i$. The sequence $(\|P_n\|)_{n=1}^{\infty}$ of norms is uniformly bounded and the corresponding least upper bound is the *basis constant* of (e_i). By going to an equivalent norm of X this constant can be made to be equal 1. One also implicitly assumes that $\|e_i\| = 1$ for all i.

Exercise I.2.1. Show that if (e_i) is a Schauder basis of a Banach space X with constant $C > 0$, and if $x = \sum_{i=0}^{\infty} a_i e_i$ is a vector from X of norm at most 1 then $|a_i| \leq 2C$ for all i.

Definition I.2.2. A sequence $(e_i) \subseteq X$ is an *unconditional basic sequence* if there is $C < \infty$ such that for all finite sequence $(a_i)_{i=0}^{n-1}$ of scalars and $(\varepsilon_i)_{i=0}^{n-1} \in \{\pm 1\}^n$,

$$\| \sum_{i=0}^{n-1} \varepsilon_i a_i e_i \| \leq C \| \sum_{i=0}^{n-1} a_i e_i \|.$$

The smallest C satisfying this is called the *unconditional basis constant of* (e_i).

Exercise I.2.3. Show that $(e_i) \subseteq X$ is an unconditional basic sequence if for every x in the closed linear span of (e_i) there exists a sequence (a_i) of scalars such that

$$x = \sum_{i=0}^{\infty} a_{\pi(i)} e_{\pi(i)}$$

for every permutation π of $\mathbb{N}$.

Exercise I.2.4. Show that the standard bases of ℓ_p $(1 \leq p < \infty)$ and c_0 are 1-unconditional.

One of the first uses of Ramsey Theory in the study of Banach spaces is based on the following concept.

Definition I.2.5. Given a normalized basic sequence (x_n) in some Banach space X, its *spreading model* is a normalized basic sequence (y_n) in possibly another Banach space Y such that for every integer $k \geq 0$ there is an integer $l \geq 0$ such that

$$| \, \| \sum_{i=0}^{k} a_{n_i} x_{n_i} \| - \| \sum_{i=0}^{k} a_{n_i} y_{n_i} \| \, | < \frac{1}{k}$$

for every sequence of integers $l \leq n_0 < \cdots < n_k$ and for every sequence $(a_{n_i})_{i \leq k}$ of scalars of modulus at most 1.

Note that while a spreading model for a given normalized basic sequence (x_n) needs not to exist, it is unique whenever it does.

Recall that two basic sequences (x_n) and (y_n) in possibly different Banach spaces X and Y are *C-equivalent* for some constant $C > 0$ whenever

$$(1/C)\| \sum_{n=0}^{\infty} a_n x_n \| \leq \| \sum_{n=0}^{\infty} a_n y_n \| \leq C\| \sum_{n=0}^{\infty} a_n x_n \|$$

for every choice of scalars (a_n) of modulus at most 1.

Definition I.2.6. A basic sequence (x_n) in some Banach space X is *C-spreading* for some $C > 0$ if it is C-equivalent to all of its infinite subsequences.

Example I.2.7. The canonical basic sequences of c_0 and ℓ_p $(p \geq 1)$ are 1-spreading.

Note the following reformulation of the existence of a 1-spreading model.

Definition I.2.8. A basic sequence (x_n) in some Banach space X is *asymptotically spreading* if there is a 1-spreading basic sequence (y_n) in a possibly different Banach space Y such that for all k, $\varepsilon > 0$ there is $n \in \mathbb{N}$ such that

$$(x_{n_i})_{i=0}^{k} \sim_{1+\varepsilon} (y_i)_{i=0}^{k} {}^1$$

for every choice of integers $n < n_0 < \cdots < n_k$.

The following Lemma gives the intended meaning to this definition in regard to Definition I.2.5.

Lemma I.2.9. *A basic sequence (x_n) is asymptotically spreading iff it admits a 1-spreading model.* $\qquad\qquad\square$

Theorem I.2.10 (Brunel–Sucheston). *Every normalized basic sequence (x_n) in some Banach space X has a subsequence which is asymptotically spreading.*

Proof. Let $C > 0$ be the basic constant of (x_n). For every integer $k > 0$ we fix a finite $(\frac{1}{2k})$-net D_k in $[-2C, 2C]^k$ in the metric determined by the supremum norm. We assume moreover that D_k is closed under scalar multiplication by $\pm\ell/k$ $(0 \leq \ell \leq k)$. Then for every $k > 0$ and $d = (d_0, \ldots, d_{k-1}) \in D_k$ we have a map $f_d : \mathbb{N}^{[k]} \longrightarrow \mathbb{R}^+$ defined by

$$f_d(\{n_0, n_1, \ldots, n_{k-1}\}_<) = \| \sum_{i=0}^{k-1} d_i x_{n_i} \|.$$

---

[1] $(u_i) \sim_C (v_i)$ is our notation for C-equivalence defined above.

Note that $f_d(\{n_0, n_1, \ldots, n_{k-1}\}) \leq \sum_{i=0}^{k-1} |d_i|$ for every $\{n_0, n_1, \ldots, n_{k-1}\} \in \mathbb{N}^{[k]}$. In other words, the range of f_d is bounded in $\mathbb{R}$ for all $d \in D_k$ and $k > 0$. It follows that

$$\lim_{s \to \mathcal{U}^k} f_d(s) = r_d$$

is finite for all $\ell > 0$ and $d \in D_k$. To simplify the notation we assume that $\mathcal{U}$ is, in fact, a Ramsey ultrafilter, so by Corollary 1.3 we can find $M_k \in \mathcal{U}$ ($k > 0$) such that

$$|f_d(s) - r_d| < \frac{1}{2k}.$$

for all $k > 0$, $d \in D_k$ and $s \in M_k^{[k]}$. We assume that the sets M_k are decreasing. Pick infinite $M = (n_i)_{i \in \mathbb{N}}$ (in fact in $\mathcal{U}$) such that $n_j \in M_{n_i}$ whenever $i < j$.

Define a norm $\| \cdot \|$ on c_{00} by letting

$$\| \sum_{i=0}^{k-1} d_i e_i \| = r_d \quad (k > 0,\ d = (d_0, \ldots, d_{k-1}) \in D_k).$$

Note that indeed this set of equalities defines a norm since for example for a scalar λ for which $\lambda d = (\lambda d_0, \ldots, \lambda d_{k-1})$ belongs to D_k, we have

$$\| \sum_{i=0}^{k-1} \lambda d_i e_i \| = \lambda r_d.$$

Note that in this norm the basis (e_i) of c_{00} is 1-spreading, i.e., $\| \sum_{i=0}^{k-1} \lambda a_i e_{n_i} \| = \| \sum_{i=0}^{k-1} \lambda a_i e_i \|$ for every choice of $n_0 < \cdots < n_{k-1}$, and $k > 0$. By the choice of $M = (n_j)$,

$$| \| \sum_{i=0}^{k-1} d_i e_i \| - \| \sum_{i=0}^{k-1} d_i x_{n_{j_i}} \| | < \frac{1}{k}$$

for every $k > 0$ and every choice of $j_0 < j_1 < \cdots < j_{k-1}$ as long as $j_0 \geq k$. Since (D_k) is a sequence of finer and finer nets, this sequence of estimates are sufficient to conclude that $((e_i), \| \cdot \|)$ is a spreading model of the subsequence (x_{n_j}) of our given sequence (x_n). $\square$

The following fact shows the going to a block-subsequence of the spreading model (e_n) one can improve its behavior.

Lemma I.2.11. *If* $(e_n)_{n=0}^{\infty}$ *is spreading then* $(e_{2n} - e_{2n+1})_{n=0}^{\infty}$ *is 1-unconditional.*

Proof. Let $f_i = e_{2i} - e_{2i+1}$, $i = 0, 1, \ldots$. We need to verify that for every pair of finite sets of integers $I \subseteq J$ and every choice a_j ($j \in J$) of scalars of modulus at most 1,

$$\| \sum_{i \in I} a_i f_i \| \leq \| \sum_{j \in J} a_j f_j \|.$$

Pick an integer $n > 0$. Choose a block sequence s_j $(j \in J)$ of finite subsets of $\mathbb{N}$ (i.e., $s_i < s_j$ whenever $i < j$ in J) such that $|s_j| = 2$ for $j \in I$ and $|s_j| = n + 1$ for $j \in J \setminus I$. For $j \in J$ and $k < |s_j|$ let $s_j(k)$ denote the kth member of s_j according to the increasing enumeration of s_j. Then by the 1-spreading property of (e_i) and for all $k < n$,

$$\| \sum_{j \in J} a_j f_j \| = \| \sum_{i \in I} a_i (e_{s_i(1)} - e_{s_i(0)}) + \sum_{j \in J \setminus I} a_j (e_{s_j(k+1)} - e_{s_j(k)}) \|.$$

Summing up these n inequalities and dividing by n, we get

$$\| \sum_{j \in J} a_j f_j \| \geq \frac{1}{n} \| n \sum_{i \in I} a_i (e_{s_i(1)} - e_{s_i(0)}) + \sum_{k=0}^{n-1} \sum_{j \in J \setminus I} a_j (e_{s_j(k+1)} - e_{s_j(k)}) \|$$

$$\geq \| \sum_{i \in I} a_i (e_{s_i(1)} - e_{s_i(0)}) \| - \frac{1}{n} \sum_{j \in J \setminus I} |a_j| \, \| (e_{s_j(n)} - e_{s_j(0)}) \|$$

$$\geq \| \sum_{i \in I} a_i f_i \| + \frac{2|J \setminus I|}{n}.$$

Since $\frac{2|J \setminus I|}{n} \to 0$ as $n \to \infty$ this gives us the desired inequality. $\qquad \square$

The following result gives us a sufficient condition for the spreading model (e_n) to be unconditional.

Theorem I.2.12 (Brunel–Sucheston). *Every normalized weakly null basic sequence* (x_n) *in some Banach space X has a subsequence with a 1-spreading unconditional model.* $\qquad \square$

Proof. On the basis of the previous proof, it suffices to show that the spreading model (e_i) of the subsequence (x_{n_i}) of (x_n) is 1-unconditional under the additional assumption that (x_n) is weakly null. In other words we have to prove that

$$\| \sum_{i \neq i_0}^{k} a_i e_i \| \leq \| \sum_{i=0}^{k} a_i e_i \|$$

for all $0 \leq i_0 \leq k$ and all choices of scalars a_i $(i \leq k)$ with $a_{i_0} \neq 0$. Fix $\epsilon > 0$. Find n such that

$$\| \| \sum_{i \neq i_0}^{k} a_i e_i \| - \| \sum_{i \neq i_0}^{k} a_i x_{n_i} \| \| < \epsilon$$

$$\| \| \sum_{i=0}^{k} a_i e_i \| - \| \sum_{i=0}^{k} a_i x_{n_i} \| \| < \epsilon$$

for all choices of integers $n_k > n_{k-1} > \ldots > n_0 \geq n$. Since (x_{n_i}) is weakly null by

Mazur's theorem (see II.3.35 below) there is a sequence of integers $k_l > k_{l-1} > \cdots > k_0 \geq n_k$. and a sequence λ_i $(i \leq l)$ of positive scalars which add to 1 such that the convex-combination $y = \sum_{i \neq i_0}^{l} \lambda_i x_{n_i}$ has norm at most $\frac{\epsilon}{|a_{i_0}|}$. Then for every $j \in [k_0, k_l]$,

$$\|\sum_{i=0}^{k} a_i e_i\| \geq \|\sum_{i=0}^{i_0-1} a_i x_{n_i} + a_{i_0} x_j + \sum_{i=i_0+1}^{k} a_i x_{n_i}\| - \epsilon$$

Note that this inequality remains valid if we replace x_j by any convex combination of vectors satisfying this inequality, and so in particular if we replace x_j by y. Hence,

$$\|\sum_{i=0}^{k} a_i e_i\| \geq \|\sum_{i=0}^{i_0-1} a_i x_{n_i} + a_{i_0} y + \sum_{i=i_0+1}^{k} a_i x_{n_i}\| - \epsilon$$

$$\geq \|\sum_{i \neq i_0}^{k} a_i x_{n_i}\| + |a_{i_0}|\|y\| - \epsilon \geq \|\sum_{i \neq i_0}^{k} a_i x_{n_i}\| - 2\epsilon.$$

Since $\epsilon > 0$ was arbitrary this finishes the proof. $\qquad\qquad\square$

We shall see below that going to *block subsequences* of the spreading model one can always achieve the unconditionality. This will be the subject matter of our next section but let us finish the present section by pointing out some standard examples that can be used to test these new notions.

Example I.2.13. The *Schreier space* S is the completion of $(c_{00}, \|\cdot\|_S)$, where the norm is given by

$$\|x\|_S = \sup \left\{ \sum_{n \in E} |x(n)| : E \subseteq \mathbb{N} \text{ and } |E| \leq \min(E) + 1 \right\}$$

Note that while ℓ_1 does not embed into S the spreading model of its Schauder basis is isomorphic to ℓ_1. This is essentially the first nontrivial example of what is usually called *asymptotic ℓ_1 space*.

The second example of an asymptotic ℓ_1 space is the following well-known space.

Example I.2.14. The *Tsirelson space* is the completion of $(c_{00}, \|\cdot\|_T)$, where $\|\cdot\|_T$ is implicitly given by a system of equations

$$\|x\|_T = \|x\|_\infty \vee \sup \left\{ \frac{1}{2} \sum_{\ell=0}^{n-1} \|E_\ell x\| : n \leq E_0 < E_1 < \cdots < E_{n-1} \right\},$$

where for a subset E of $\mathbb{N}$ and $x \in c_{00}$ by Ex we denote the member of c_{00} which is equal to x on E and 0 otherwise. Notice that the standard basis (e_n) of c_{00} is a Schauder basis of T.

Exercise I.2.15. (1) Prove that for every $\varepsilon > 0$ there is a vector $x = \sum_n a_n e_n \in T$ such that $\varepsilon \|x\|_{\ell_1} > \|x\|_T$. Give explicitly such a vector x.
(2) Prove the statement (1) where (e_n) is replaced by an arbitrary normalized sequence (x_n) of finitely supported vectors of T such that for every n max supp $x_n <$ min supp x_{n+1}[2]. Use this to show that ℓ_1 does not embed into T.
(3) Verify that Schreier's as well as Tsirelson's space have spreading models isomorphic to ℓ_1.

I.3 Finite Representability of Banach Spaces

The construction of the spreading model presented in the previous section suggests considering ultrapowers $X^I/\mathcal{U}$ of Banach spaces X. Thus, for a Banach space X, an (infinite) index set I and a (nonprincipal) ultrafilter $\mathcal{U}$ on I we let

$$X^I/\mathcal{U} = \{(x_i)_{i \in I} \in X^I : \lim_{i \to \mathcal{U}} \|x_i\| \le \infty\}/ \sim,$$

where we put $(x_i)_{i \in I} \sim (y_i)_{i \in I}$ if $\lim_{i \to \mathcal{U}} \|x_i - y_i\| = 0$, and where we equip $X^I/\mathcal{U}$ with the norm

$$\|[(x_i)_{i \in I}]\| = \lim_{i \to \mathcal{U}} \|x_i\|.$$

Definition I.3.1. A Banach space X is *finitely representable* in a Banach space Y if for every $\varepsilon > 0$ and every finite-dimensional subspace X_0 of X there is a finite-dimensional subspace Y_0 of Y and an isomorphism $T : X_0 \to Y_0$ such that $\|T\|\|T^{-1}\| < 1 + \epsilon$.

Note that if (e_n) is a spreading model of some sequence (x_n) in some Banach space X then any space which is finitely representable in $\overline{(e_n)}$ is also finitely representable in X.

Theorem I.3.2. *The ultrapower $X^I/\mathcal{U}$ is a Banach space which contains a natural isometric copy of X and which is finitely representable in X.*

Proof. To see that $X^I/\mathcal{U}$ is complete, consider a Cauchy sequence $(v^n)_{n=0}^\infty$ of elements of $X^I/\mathcal{U}$. Going to a subsequence of $(v^n)_{n=0}^\infty$ we can choose a decreasing sequence $I = I_0 \supseteq \cdots \supseteq I_n \supseteq \ldots$ of elements of $\mathcal{U}$ and representatives $(x_i^n)_{i \in I} \in v^n$ $(n = 0, 1 \ldots)$ such that
(1) $\|x_i^m - x_i^n\| \le 2^m$ for all $i \in I_n$
(2) $x_i^{m+1} = x_i^m$ for $i \in I_m \setminus I_{m+1}$.
Then for every $i \in I$ the sequence $(x_i^n)_{n=0}^\infty$ is Cauchy in X so we can assign to it a limit x_i. Then $[(x_i)_{i \in I}]$ is the limit of the sequence $(v^n)_{n=0}^\infty$ in $X^I/\mathcal{U}$.

Identifying $x \in X$ with the constant sequence $(x)_{i \in I}$ gives us an isometric embedding of X into $X^I/\mathcal{U}$.

---

[2]such a sequence (x_n) is usually called a *block subsequence* of (e_n)

To show that $X^I/\mathcal{U}$ is finitely representable in X, choose a finite-dimensional subspace V_0 of $X^I/\mathcal{U}$ spanned by vectors $v^0, \ldots, v^n$ of norm 1. Choose $C > 0$ such that

$$C \sum_{k=0}^{n} |a_k| \le \| \sum_{k=0}^{n} a_k v^k \| \le \sum_{k=0}^{n} |a_k|$$

for every choice of scalars $(a_i)_{i=0}^n$. Let $v^k = [(x_i^k)_{i \in I}]$ for $k \le n$.

Fix $\epsilon > 0$. Choose a finite $(C \cdot \epsilon)$-net D in the unit ball of ℓ_1^{n+1} and for $d \in D$ and $k \le n$ let d_k be the kth coordinate of d relative to the basis of ℓ_1^{n+1}. Then

$$(\forall d \in D)(\mathcal{U}_i)C \cdot \sum_{k=0}^{n} |d_k| \le \| \sum_{k=0}^{n} d_k x_i^k \| \le \sum_{k=0}^{n} |a_k|.$$

Since D is finite we can find a single set $J \in \mathcal{U}$ such that for all $i \in J$ and $d \in D_n$,

$$C \cdot \sum_{k=0}^{n} |d_k| \le \| \sum_{k=0}^{n} d_k x_i^k \| \le \sum_{k=0}^{n} |a_k|.$$

Fix $i \in J$ and let X_0 be the linear span of $(x_i^k)_{k=0}^n$, and let $T : V_0 \to X_0$ be determined by $T(v^k) = x_i^k$. Then one straightforwardly checks that

$$\|T\| \cdot \|T^{-1}\| \le 1 + 3\epsilon \qquad \square$$

The following result shows that the notion of ultrapower captures the notion of finite representability in Banach spaces in a very precise sense.

Theorem I.3.3. *If a Banach space X is finitely representable in some other Banach space Y then there is an ultrapower $Y^I/\mathcal{U}$ of Y which contains an isometric copy of X.*

Proof. Let I be the collection of all pairs (V, ϵ) where V is a finite-dimensional subspace of X and $\epsilon > 0$. We order I by letting $(U, \epsilon) \le (V, \delta)$ if $U \subseteq V$ and $\epsilon \ge \delta$. Note that $(I, \le)$ is upward directed, so we can find an ultrafilter $\mathcal{U}$ on I such that

$$\{(V, \delta) \in I : (V, \delta) \ge (U, \epsilon)\} \in \mathcal{U}$$

for all $(V, \epsilon) \in I$.

By the hypothesis of the theorem, for every finite-dimensional subspace $V \subseteq X$ and $\epsilon > 0$ we can choose an isomorphism $T_{(V,\epsilon)} : V \to U$ where U is some finite-dimensional subspace of Y. This determines $T : X \to Y^I/\mathcal{U}$ by

$$T(x) = (T_{(V,\epsilon)}(x))_{(V,\epsilon) \in I},$$

where we let $T_{(V,\epsilon)}(x) = 0$ if $x \notin V$. Then T is a norm-preserving linear operator from X into $Y^I/\mathcal{U}$. $\square$

The ultrapower technique leads to a natural realization of the spreading model of a given basic sequence (x_n) in some Banach space X. To see this take a nonprincipal ultrafilter $\mathcal{U}$ on $\mathbb{N}$ and consider its finite Fubini powers $\mathcal{U}^k$ $(k \geq 1)$. Note that there is a natural identification

$$X^{\mathbb{N}^{[1]}}/\mathcal{U}^1 \subseteq X^{\mathbb{N}^{[2]}}/\mathcal{U}^2 \subseteq \cdots \subseteq X^{\mathbb{N}^{[k]}}/\mathcal{U}^k \subseteq \ldots$$

via the sequence of linear isometric embeddings

$$\varphi : X^{\mathbb{N}^{[k]}}/\mathcal{U}^k \to X^{\mathbb{N}^{[l]}}/\mathcal{U}^l (k \leq l)$$

defined by

$$\varphi([(y_s)_{s \in \mathbb{N}^{[k]}}]) = [(z_t)_{t \in \mathbb{N}^{[l]}}]$$

where $z_t = y_{t \restriction k}$. Each ultrapower $X^{\mathbb{N}^{[k]}}/\mathcal{U}^k$ has a distinguished vector e_k determined by our starting sequence $(x_n)_{n=0}^{\infty} \subseteq X$ as

$$e_k = [(y_s)_{s \in \mathbb{N}^{[k]}}],$$

where $y_s = x_{\max(s)}$. Then for every $x \in X$, every integer $n > 0$ and every sequence a_k $(k \leq n)$ of scalars we have the equality

$$\left\| x + \sum_{k=0}^{l-1} a_k e_k \right\|_l = \lim_{s \to \mathcal{U}^l} \left\| x + \sum_{n \in s} a_{|s \cap n|} x_n \right\|$$

where for $s \in \mathbb{N}^{[l]}$ and $n \in s$, we let $s \cap n = \{m \in s : m < n\}$ and where on the left-hand side $x \in X$ is identified with the constant sequence $[(x)_{s \in \mathbb{N}^{[l]}}] \in X^{\mathbb{N}^{[l]}}/\mathcal{U}^l$. It follows that $(e_k)_{k=0}^{\infty}$ is a spreading model of a subsequence of (x_n) found via the simple diagonal procedure presented above during the course of the proof of Theorem I.2.10.

The spreading sequence (e_k) has some interesting *block-sequences*, i.e., sequences $(e_{2n} - e_{2n+1})_{n=0}^{\infty}$ considered above in Lemma I.2.11. In fact one can find block-subsequences of (e_n) with stronger properties.

Lemma I.3.4. *Every spreading normalized basic sequence $(e_n)_{n=0}^{\infty}$ has a block-subsequence whose spreading model $(f_k)_{k=0}^{\infty}$ is 1-equivalent to $(\epsilon_n f_n)_{n=0}^{\infty}$ for any choice of $(\epsilon_n)_{n=0}^{\infty}$ of signs.*

Proof. By Lemma I.2.11 we may assume that $(e_n)_{n=0}^{\infty}$ is 1-unconditional. We distinguish two cases
Case 1. $\| \sum_{i=0}^{l} e_i \| \to \infty$ as $l \to \infty$. Choose finite sets F_k $(k = 0, 1, \ldots)$ such that for every k
(a) $\| \sum_{i=0}^{|F_k|-1} e_i \| \geq k^2$, and
(b) $\max F_k + 1 < \min F_{k+1}$.
    For $k = 0, 1, \ldots$, let

$$f_k = \frac{\sum_{i \in F_k} \sigma_k(i) e_i}{\| \sum_{i=0}^{|F_k|-1} e_i \|},$$

where $\sigma_k : F_k \to \{-1, 1\}$ is the alternating sign-assignment starting with 1. We want to show that the spreading model of (f_k) has the desired property. To see this, it suffices to show that for every $n \le k_0 < \cdots < k_{n-1}$, every sequence of signs $(\epsilon_k)_{k=0}^n$ and every choice of scalars $(a_i)_{i=0}^{n-1}$,

$$\Big|\,\|\sum_{i=0}^{n-1} a_i f_{k_i}\| - \|\sum_{k=0}^{n-1} \epsilon_i a_i f_{k_i}\|\,\Big| \le \frac{2n \max_{i=0}^{n-1} |a_i|}{k_0^2}. \tag{I.1}$$

Fix $n \le k_0 < \cdots < k_{n-1}$, a sequence of signs $(\epsilon_k)_{k=0}^n$ and a sequence of scalars $(a_i)_{i=0}^{n-1}$. For $i \le n - 1$, let

$$g_{k_i} = \frac{\sum_{i \in G_k} \tau_k(i) e_i}{\|\sum_{i=0}^{|G_k|-1} e_i\|},$$

where $G_i = (F_i \setminus \{\min F_i\}) \cup \{\max F_i + 1\}$, and $\tau_i : G_i \to \{-1, 1\}$ is the alternating sign-assignment starting with 1. Notice that
(c) for every $i \le n - 1$ $|F_i| = |G_i|$,
(d) for every $i < n - 1$, $G_i < F_{i+1}$, and
(e) for every $i \le n - 1$ $f_{k_i} + g_{k_i} = (1/\|\sum_{i=0}^{|F_{k_i}|-1} e_i\|)(e_{l_i} + (-1)^{|F_i|+1} e_{r_i+1})$, where $l_i = \min F_i$ and $r_i = \max F_i$.

Let $I^+ = \{i < n : \epsilon_k = 1\}$ and $I^- = \{i < n : \epsilon_k = -1\}$. Note that since (e_i) is spreading we have that

$$\|\sum_{i=0}^{n-1} \epsilon_i a_i f_{k_i}\| = \|\sum_{i \in I^+} a_i f_{k_i} - \sum_{i \in I^-} a_i g_{k_i}\|. \tag{I.2}$$

So, using that (e_n) is 1-unconditional, and equation (I.2)

$$\Big|\,\|\sum_{i=0}^{n-1} a_i f_{k_i}\| - \|\sum_{i=0}^{n-1} \epsilon_i a_i f_{k_i}\|\,\Big| = \Big|\,\|\sum_{i=0}^{n-1} a_i f_{k_i}\| - \|\sum_{i \in I^+} a_i f_{k_i} - \sum_{i \in I^-} a_i g_{k_i}\|\,\Big| \le$$

$$\le \|\sum_{i \in I^-} a_i (f_{k_i} + g_{k_i})\| \le$$

$$\le \frac{\max_{i \le n-1} |a_k|}{k_0^2} \|\sum_{i \in I^-} (e_{l_i} + (-1)^{|F_i|+1} e_{r_i+1})\| \le$$

$$\le \frac{\max_{i \le n-1} |a_k|}{k_0^2} 2n, \tag{I.3}$$

as desired.

Case 2. $\sup_l \|\sum_{i=0}^l e_i\| < \infty$. Then (e_i) is equivalent to the standard basis of c_0, so by James' theorem for each $l > 0$ we can find a block subsequence $(f_k^l)_{k=0}^\infty$ of (e_i) which is $(1 + 1/l)$-equivalent to the standard basis of c_0. So the spreading model $(f_i)_{i=0}^\infty$ of $(f_k^k)_{k=0}^\infty$ is 1-equivalent to the standard basis of c_0 and is therefore equivalent to $(\epsilon_i f_i)_{i=0}^\infty$ for every choice $(\epsilon_i)_{i=0}^\infty$ of signs. $\square$

To summarize what has been proved upto this point we need a definition quite analogous to Definition I.3.1 above.

Definition I.3.5. Suppose X and Y are Banach spaces with Schauder bases (e_n) and (f_n), respectively. We say that X is finitely block representable in Y if for every $\epsilon > 0$ and every finite-dimensional block subspace X_0 of X there is a finite-dimensional block subspace Y_0 of Y and an isomorphism $T : X_0 \to Y_0$ such that $\|T\|\|T^{-1}\| < 1 + \epsilon$.

Clearly, if (e_i) is a spreading model of a subsequence of some basic sequence (x_n) in some Banach space X, then the space generated by (e_i) is finitely block representable in $\overline{(x_n)}$. So the following result summarizes the results obtained so far.

Theorem I.3.6. *Suppose (x_n) is a normalized basic sequence in some Banach space X. Then there is a Banach space Y with a Schauder basis (y_n) which is finitely block representable in $\overline{(x_n)}$ and has the property*

$$\| \sum_{i=0}^{n} a_i y_i \| = \| \sum_{i=0}^{n} \sigma_i a_i y_{k_i} \|$$

for every n, every sequence $k_0 < k_1 < \cdots < k_n$ of integers, every sequence $a_0 < a_1 < \cdots < a_n$ of scalars, and every sequence $\sigma_0, \sigma_1, \ldots, \sigma_n$ of signs. $\square$

Corollary I.3.7. *For every positive integer n and $\epsilon > 0$ there is an integer m such that every basic sequence $(e_i)_{i=0}^{m}$ in some Banach space X contains a block subsequence $(f_i)_{i=0}^{n}$ such that*

$$\| \sum_{i=0}^{n} \sigma_i a_i f_i \| \leq (1+\epsilon) \| \sum_{i=0}^{n} a_i f_i \|$$

for every choice $a_0, a_1, \ldots, a_n$ of scalars and $\sigma_0, \sigma_1, \ldots, \sigma_n$ of signs. $\square$

Starting from the basic sequence (y_n) of Theorem I.3.6 one can find its block-subsequence whose spreading model is equivalent to the basis of one of the standard sequence space ℓ_p $(1 \leq p < \infty)$ or c_0. This is the content of the following result.

Theorem I.3.8 (Krivine). *Suppose (e_i) is a basic sequence in some Banach space such that*

$$\| \sum_{i=0}^{n} a_i e_i \| = \| \sum_{i=0}^{n} \sigma_i a_i e_{k_i} \|$$

for every $n \geq 0$, sequence of integers $0 \leq k_0 < k_1 < \cdots < k_n$, sequence $a_0, a_1, \ldots, a_n$ of scalars and sequence $\sigma_0, \sigma_1, \ldots, \sigma_n$ of signs. Then the unit vector basis of either ℓ_p for some $1 \leq p < \infty$ or of c_0 is finitely block representable in (e_i).

Proof. We need a convenient way to "double" a given finitely supported vector of (e_i) as well as to "triple" it. In other words to have bounded operators $x \mapsto D(x)$ and $x \mapsto T(x)$ which commute and which perform these two operations, respectively. A natural way to formalize this is to re-enumerate (e_i) as $(e_i)_{i \in \mathbb{Q}}$ and define D and T on the span X_1 of $(e_i)_{i \in \mathbb{Q} \cap [0,1)}$ by letting $D(e_q) = e_{q/2} + e_{(q+1)/2}$ and $T(e_q) = e_{q/3} + e_{(q+1)/3} + e_{(q+2)/3}$. Note that this indeed defines bounded linear operators on X_1 which commute. In fact $\|D\| \leq 2$ and $\|T\| \leq 3$. Taking an ultrapower over a block subsequence (x_n) of (e_i) such that for some $1 \leq \lambda \leq 2$, the sequence $\|D(x_n) - \lambda x_n\|$ converges to 0 one finds a sequence $(\tilde{y}_n)$ in the ultrapower and $1 \leq \mu \leq 3$ such that $\|T(x_n) - \mu \tilde{y}_n\|$ converges to 0. So, for some normalized block subsequence (z_n) of (x_n) we have that $\|D(z_n) - \lambda z_n\| \to 0$ and $\|D(z_n) - \mu z_n\| \to 0$. For $n, k \geq 0$ let $z_{n\,k}$ be the copy of the vector z_n when $\mathbb{Q} \cap [0,1)$ is shifted to $\mathbb{Q} \cap [k, k+1)$ in the natural way. Using (f_k) to denote the standard basis of c_{00} we use $(z_{n\,k})$ to define a norm on c_{00} by

$$\| \sum_{k=0}^{l} a_k f_k \| = \lim_{n \to \mathcal{U}} \| \sum_{k=0}^{l} a_k z_{n\,k} \|.$$

Let Y be the completion of c_{00} relative to this norm. Then for every pair x and y of finitely supported vectors of Y and index k such that $x < k, k+1, k+2 < y$, we have

$$\|x + f_k + f_{k+1} + y\| = \|x + \lambda f_k + y\|$$

and

$$\|x + f_k + f_{k+1} + f_{k+2} + y\| = \|x + \mu f_k + y\|.$$

It follows that for every pair of integers m and n,

$$\| \sum_{k=1}^{2^m 3^n} f_k \| = \lambda^m \mu^n \tag{I.4}$$

More generally,

$$\| \sum_{k=1}^{l} e_k \| = \| \sum_{i=1}^{j} \lambda^{m_i} \mu^{n_i} e_i \| \tag{I.5}$$

whenever $l = \sum_{i=1}^{j} 2^{m_i} 3^{n_i}$. So if $\lambda = 1$ or $\mu = 1$ we would get that $\sum_{k=1}^{2^m} f_k = 1$ or $\sum_{k=1}^{3^n} f_k = 1$, respectively, so the basis (f_k) of Y is 1-equivalent to the basis of c_0. So we may concentrate on the case when $\lambda, \mu > 1$. Note that by (I.4) the function $\varphi(2^m 3^{-n}) = \lambda^m \mu^{-n}$ is a nondecreasing multiplicative function, and since $\{2^n 3^{-n} : m, n \in \mathbb{N}\}$ is dense in $\mathbb{R}^+$, it extends to a nondecreasing multiplicative function $\varphi : \mathbb{R}^+ \to \mathbb{R}^+$. So there is $1 \leq p < \infty$ such that $\varphi(v) = v^{1/p}$. In particular, $\lambda = 2^{1/p}$ and $\mu = 3^{1/p}$.

Consider now the function $\psi(l) = \| \sum_{k=1}^{l} f_k \|$. Then taking $m_i = m$ and $n_i = n$ ($1 \leq i \leq j$) in (I.5), we obtain that $\psi(j \cdot 2^m \cdot 3^n) = \lambda^m \cdot 3^n \cdot \psi(j) = (2^m \cdot 3^n)^{\frac{1}{p}} \cdot \psi(j)$.

Since ψ is nondecreasing and subadditive and since $\{2^n 3^{-n} : m, n \in \mathbb{N}\}$ is dense in $\mathbb{R}^+$ one easily concludes that, in fact, $\psi(j) = j^{1/p}$ for all $j \in \mathbb{N}$. Using this and (I.5) again we see that if for each $i \leq j$ we have scalars a_i of the form $\lambda^{m_i} \mu^{n_i}$, then for $l = \sum_{i=1}^{j} 2^{m_i} 3^{n_i}$ we have that

$$\| \sum_{i=1}^{j} a_i f_i \| = \| \sum_{k=1}^{l} f_k \| = l^{\frac{1}{p}} = (\sum_{i=1}^{j} a_i^p)^{\frac{1}{p}}.$$

Since the sequence normalizations of sequences $(a_1, \ldots, a_j)$ of this form are dense in the unit sphere of ℓ_p^j we have this equality for every choice of scalars. □

Corollary I.3.9. *For every basic sequence (x_n) in some Banach space X the standard basis of some ℓ_p $(1 \leq p < \infty)$ or of c_0 is finitely block representable in the closed linear span of (x_n).* □

It is worth stating also a finite form of Corollary I.3.9 proved via the usual compactness argument.

Corollary I.3.10. *For every integer n and constants $0 < \epsilon$, $C < \infty$ there is an integer $m = m(n, \epsilon, C)$ such that if X is an m-dimensional Banach space with normalized basis $(x_j)_{j=1}^{m}$ and constant C there is a $p \subset [1, \infty]$ and a block subsequence $(f_i)_{i=1}^{n}$ of $(x_j)_{j=1}^{m}$ which is $(1 + \epsilon)$-equivalent to the standard basis of ℓ_p^n.* □

We finish this section by mentioning another finite-dimensional result.

Theorem I.3.11 (Odell–Rosenthal–Schlumprecht). *Suppose h is a uniformly continuous real-valued function defined on the unit sphere of some Banach space X with Schauder basis (x_n). Then there is $\lambda \in \mathbb{R}$ such that for every $\epsilon > 0$ and every integer $n > 0$ there is a block subsequence $(f_i)_{i=0}^{n}$ such that $|h(y) - \lambda| < \epsilon$ for all $y \in \overline{(f_i)_{i=0}^{n}}$ of norm 1.* □

It turns out that an infinite-dimensional analogue of this result is false even for Banach spaces of the form ℓ_p $(1 \leq p < \infty)$, though it is true for the Banach space c_0. We return to this point in another section of these notes.

Chapter II

Ramsey Theory of Finite and Infinite Sequences

II.1 The Theory of Well-Quasi-Ordered Sets

Infinite-dimensional Ramsey theory is a branch of Ramsey theory initiated in the early 1960s by Nash–Williams in his attempts to extend a well-known theorem of Kruskal that the finite trees are well-quasi-ordered (w.q.o.) under the inf-preserving embeddings (see Theorem II.1.9 below). Recall that a *quasi-order* (q.o.) is a set Q with a binary transitive and reflexive relation $\leq$. We write $x < y$ if $x \leq y$ and $y \not\leq x$, we write $x \mid y$ if $x \not\leq y$ and $y \not\leq x$, and we write $x \equiv y$ if $x \leq y$ and $y \leq x$. A quasi-ordered set Q is *well-quasi-ordered* (w.q.o.) if there are no infinite descending sequences

$$x_0 > x_1 > \cdots > x_n > \cdots$$

and no infinite antichains

$$x_m \mid x_n \qquad (m \neq n).$$

Lemma II.1.1. *A quasi-ordered set Q is w.q.o. iff for every infinite sequence $(x_n) \subseteq Q$ there exists $m < n$ such that $x_m \leq x_n$ iff for every infinite sequence $(x_n) \subseteq Q$ there exists an infinite sequence $n_0 < n_1 < \cdots < n_k < \cdots$ of integers such that $x_{n_i} \leq x_{n_j}$ whenever $i < j$.*

Proof. Use the 2-dimensional Ramsey theorem. $\square$

Corollary II.1.2. *If Q_0 and Q_1 are w.q.o., then so is their cartesian product $Q_0 \times Q_1$.* $\square$

Lemma II.1.3. *Suppose (x_n) is a sequence of elements of a q.o. set Q. Then there is an infinite subsequence (x_{n_k}) such that either*

(1) $x_{n_k} \mid x_{n_\ell}$ whenever $k \neq \ell$, or

(2) $x_{n_k} \leq x_{n_\ell}$ whenever $k < \ell$, or

(3) $x_{n_k} > x_{n_\ell}$ whenever $k < \ell$.

Proof. Apply the 2-dimensional Ramsey theorem to the 3-coloring given by (1), (2) and (3).   ☐

A quasiordered set Q is *well-founded* if it contains no infinite sequence (x_n) such that $x_m > x_n$ whenever $m < n$.

Corollary II.1.4. *If a well-founded q.o. set Q is not w.q.o., then it contains an infinite sequence (x_n) such that $x_m \not\leq x_n$ whenever $m < n$.*   ☐

An infinite sequence (x_n) of elements of some q.o. set Q is a *bad array* if $x_m \not\leq x_n$ holds for all $m < n$.

Definition II.1.5 (Nash–Williams). A bad array (x_n) of a q.o. set Q is *minimal bad* if there is no bad array (y_n) of Q such that:

(1) $\forall m \exists n\, y_m \leq x_n$,

(2) $\exists m \exists n\, y_m < x_n$.

Lemma II.1.6. *Suppose Q is a well-founded but not well-quasi-ordered set. Then there is a minimal bad array (x_n) in Q.*

Proof. By Corollary II.1.4 we can start with a bad array (x_n^0) of Q and assume that its first term x_0^0 is minimal first term of all possible bad arrays. Choose now a bad array (x_n^1) such that x_1^1 is minimal second term among all bad arrays (y_n) such that $y_o = x_0^0$. Then choose a bad array $(x_n^2) \leq (x_n^1)$ such that x_2^2 is minimal third term among all bad arrays (y_n) such that $y_0 = x_0^0$ and $y_1 = x_1^1$, and so on. We claim that the diagonal sequence (x_n^n) is a minimal bad array of Q. For suppose there is a bad array (y_i) such that $\forall i \exists n\, y_i \leq x_n^n$ and $\exists i \exists n\, y_i < x_n^n$. Let m be the minimal n such that for some i, $y_i < x_n^n$. Going to a subsequence of (y_i) we may assume that in fact $y_0 < x_m^m$ and that for all i there is $n \geq m$ such that $y_i < x_n^n$. Then the sequence $x_0^0, \ldots, x_{n-1}^{n-1}, y_0, y_1, y_2, \ldots$ is a bad array, contradicting the choice of x_n^n.   ☐

Corollary II.1.7. *If (x_n) is a minimal bad array of a quasi-ordered set Q then $\{q \in Q : (\exists i) q < x_i\}$ is w.q.o.*   ☐

Given a q.o. set Q we let the set $Q^{<\infty}$ of all finite sequences of elements of Q be ordered by: $(x_0, \ldots, x_m) \leq (y_0, \ldots, y_m)$ if there exists $k_0 < k_1 < \cdots < k_m \leq n$ such that

$$x_i \leq y_{k_i} \text{ for all } i \leq m.$$

Theorem II.1.8 (Higman). *If Q is w.q.o. then so is $Q^{<\infty}$.*

Proof. First of all note that $Q^{<\infty}$ is well-founded. So if $Q^{<\infty}$ is not w.q.o., then by Lemma II.1.6 we have a minimal bad array $(t^n)_{n=0}^\infty$ of members of $Q^{<\infty}$. Note that a subsequence of $(t^n)_{n=0}^\infty$ is also a bad array of $Q^{<\infty}$, so by Lemma II.1.1 we may assume that if

$$t^n = (t_0^n, \ldots, t_{k_n}^n) \quad (n = 0, 1, 2, \ldots),$$

then

$$t_0^m \leq t_0^n \text{ whenever } m \leq n.$$

Note that in particular $k_n \geq 1$ for all n. So if we let $s^n = (t_1^n, \ldots, t_{k_n}^n)$, we get an infinite sequence $(s^n)_{n=0}^\infty$ of nonempty members of $Q^{<\infty}$. Note that (1) and (2) of Definition II.1.5 are satisfied for $(x_n) = (t^n)$ and $(y_n) = (s^n)$, so we conclude that (s^n) is not a bad array of $Q^{<\infty}$. Thus there exists $m < n$ such that $s^m \leq s^n$. Combining this with the fact that $t_0^m \leq t_0^n$ we conclude that $t^m \leq t^n$, contradicting the assumption that (t^n) is a bad array. This finishes the proof. $\qquad\square$

Recall that a *tree* is a partially ordered set $(T, \leq_T)$ with the property that $\{x \in T : x \leq_T y\}$ is totally ordered for all $y \in T$. A tree T is *rooted* if it has a minimal element, $\text{root}(T)$. Given a quasi-ordered set Q, a *Q-tree* or a *tree labeled by Q* is a pair (T, f) where T is a rooted tree and $f : T \to Q$. Given two Q-trees (S, f) and (T, g) we put $(S, f) \leq (T, g)$ if there is a $\Phi : S \to T$ such that for all $x, y \in S$:
(1) $x \leq_S y$ iff $\Phi(x) \leq_T \Phi(y)$
(2) $\Phi(x \wedge y) = \Phi(x) \wedge \Phi(y)$
(3) $f(x) \leq_Q g(\Phi(x))$.

Theorem II.1.9 (Kruskal). *If Q is w.q.o., then so is the class of all finite Q-trees.*

Proof. If the class of Q-trees is not w.q.o., then by Lemma II.1.6 we could choose a minimal bad array (T_i, f_i) $(i = 0, 1, 2, \ldots)$ of Q-trees. Going to a subsequence we may assume that

$$f_n(\text{root}(T_n)) \leq_Q f_m(\text{root}(T_m)) \text{ where } n < m.$$

Let J be the class of all Q-trees of the form (T, f) for which we can find n and an immediate successor x of $\text{root}(T_n)$ such that

$$T = \{y \in T_n : x \leq_T y\} \text{ and } f = f_n \restriction T.$$

By II.1.7 the class J is w.q.o. and by Higman's theorem so is $J^{<\infty}$. Thus we can find $n < m$ such that the element of $J^{<\infty}$ corresponding to the immediate successors of $\text{root}(T_n)$ is $\leq$ the element of $J^{<\infty}$ corresponding to the immediate successors of $\text{root}(T_m)$. It follows that $(T_n, f_n) \leq (T_m, f_m)$ contradicting our original assumption that the sequence $\{(T_i, f_i)\}$ is bad. $\qquad\square$

Unfortunately Higman's Theorem is not true in the infinite dimension as the following example shows:

Example II.1.10 (R. Rado). Let $Q_1 = \{(i,j) : i < j \in \mathbb{N}\}$ quasiordered by

$$(i,j) \leq (k,l) \text{ iff either } i = k \text{ and } j \leq l, \text{ or } j < k.$$

Then Q_1^∞ is not w.q.o.   □

In fact Rado's example is a minimal quasi-ordered set with this property.

Theorem II.1.11 (R. Rado). *Suppose Q is w.q.o. while its power Q^ω is not w.q.o. Then Q contains an isomorphic copy of the quasi-ordered set Q_1.*

Proof. Suppose $x_n : \omega \to Q$ $(n = 0, 1, 2, \dots)$ is a bad array of members of Q^ω. By our assumptions Q is w.q.o., so for each n one can find an integer k_n such that for each $k \geq k_n$ there are infinitely many l with the property that $x_n(k) \leq x_n(l)$. Using Higman's theorem and then Ramsey's theorem and going to a subsequence we may assume that $x_n \upharpoonright k_n \leq x_m \upharpoonright k_m$ whenever $n \leq m$. It follows that we can cut off the initial segment $x_n \upharpoonright k_n$ from each of the x_n and still obtain a bad array. So to save on notation we assume that $k_n = 0$ for all n. Then for each $n \leq m$ there is an integer k_{nm} such that $x_n(k_{nm}) \not\leq x_m(k)$ for all $k \in \omega$. So in particular we have that

$$x_n(k_{nm}) \not\leq x_m(k_{mp}) \text{ whenever } n < m < p.$$

Color a triple $\{n, m, p\}_<$ red or blue depending whether or not $x_n(k_{nm}) \leq x_n(k_{mp})$. Also color a quadruple $\{n, m, p, q\}_<$ green or yellow depending whether or not $x_n(k_{nm}) \leq x_p(k_{pq})$. Let M be an infinite subset of ω with $M^{[3]}$ and $M^{[4]}$ monochromatic. Then $\{x_n(k_{nm}) : n, m \in M, n < m\}$ is an isomorphic copy of Q_1 in Q.   □

Exercise II.1.12. Show that if Q is w.q.o., then its power Q^ω is well-founded.

It was this difficulty with the notion of w.q.o. that has motivated the following remarkable definition.

Definition II.1.13 (Nash–Williams). A quasi-ordered set Q is *better-quasi-ordered (b.q.o.)* if for every continuous map $f \colon \mathbb{N}^\infty \longrightarrow Q^1$ there exists $X \in \mathbb{N}^\infty$ such that

$$f(X) \leq f(_*X),$$

where $_*X$ denotes the shift of X, i.e., $_*X(n) = X(n+1)$.

This is in fact a modern reformulation of Nash–Williams' original definition. In practice it is more convenient to work with the set $\mathbb{N}^{[\infty]}$ of all infinite strictly increasing sequences of integers rather than with $\mathbb{N}^\infty$ since in this case we can in particular identify sequences with the corresponding subsets of $\mathbb{N}$. Latter advances of this area of Ramsey theory have shown that this definition does not change if we quantify over Borel functions rather than continuous ones. In the following section we shall examine this remarkable definition moving towards its original form as given by Nash–Williams.

Exercise II.1.14. Show that every w.q.o. is b.q.o.

Exercise II.1.15. Show that the product of two b.q.o.'s is a b.q.o.

---

[1]Q is taken here with its discrete topology.

II.2   Nash–Williams' Theory of Fronts and Barriers

The purpose of this section is to study properties of families $\mathcal{F}$ that can always be achieved by taking restrictions of the form

$$\mathcal{F} \upharpoonright M = \{s \in \mathcal{F} : s \subseteq M\} = \mathcal{F} \cap \mathcal{P}(M),$$

where M is an infinite subset of $\mathbb{N}$. Note that there is another form of restriction that one can take:

$$\mathcal{F}[M] = \{s \cap M : s \in \mathcal{F}\}.$$

One can also take two possible kind of closures of $\mathcal{F}$, the *downwards closure*

$$\widehat{\mathcal{F}} = \{t : t \subseteq s \text{ for some } s \in \mathcal{F}\},$$

as well as *topological closure* $\overline{\mathcal{F}}$ in the Cantor space $2^{\mathbb{N}}$ when sets are identified with their characteristic functions. One of the point behind the notion of *barrier* that we study in this section is that we have the equality

$$\overline{\mathcal{F} \upharpoonright M} = \widehat{\mathcal{F}} \upharpoonright M = \mathcal{F}[M]$$

whenever $\mathcal{F}$ is a barrier on some superset of M, so this way one avoids the possible traps caused by mixing up between these two kinds of restrictions and the two kinds of closures. So let us introduce the key notions of this section.

Definition II.2.1. A family $\mathcal{B}$ of finite subsets of $\mathbb{N}$ is called a *front* if

(1) $s \not\sqsubseteq t$[2] whenever $s \neq t \in \mathcal{B}$.

(2) $\bigcup \mathcal{B}$ is infinite and for every infinite $M \subseteq \bigcup \mathcal{B}$ there exists $s \in \mathcal{B}$ such that $s \sqsubseteq M$.

If $\mathcal{B}$ instead of (1) has the following stronger condition:

(1') $s \not\subseteq t$ whenever $s \neq t \in \mathcal{B}$.

then $\mathcal{B}$ is called a *barrier on* $\bigcup \mathcal{B}$.

   To connect this notion with the Definition II.1.13 note that whenever we have a front $\mathcal{B}$ such that $\bigcup \mathcal{B} = \mathbb{N}$ and a mapping

$$f : \mathcal{B} \longrightarrow Q,$$

we can extend f to a continuous map

$$\hat{f} : \mathbb{N}^{[\infty]} \longrightarrow Q$$

by letting $\hat{f}(M) = f(s)$ where $s \in \mathcal{B}$ is unique with the property that $s \sqsubseteq M$. Clearly every continuous map $g : \mathbb{N}^{[\infty]} \longrightarrow Q$ (with discrete topology) has the form $\hat{f}$ for some $f : \mathcal{B} \longrightarrow Q$.

---

[2]$s \sqsubseteq t$ means that s is an initial segment of t.

Exercise II.2.2. Prove this last assertion.

Definition II.2.3. (i) A family $\mathcal{F}$ of finite subsets of $\mathbb{N}$ is *thin* if $s \not\sqsubseteq t$ for every pair s, t of distinct members of $\mathcal{F}$.
(ii) A family $\mathcal{F}$ of finite subsets of $\mathbb{N}$ is *Sperner* if $s \not\subseteq t$ for every pair $s \neq t \in \mathcal{F}$.
(iii) A family $\mathcal{F}$ of finite subsets of $\mathbb{N}$ is *Ramsey* if for every finite partition

$$\mathcal{F} = \mathcal{F}_0 \cup \cdots \cup \mathcal{F}_k$$

there is an infinite set $M \subseteq \mathbb{N}$ such that at most one of the restrictions

$$\mathcal{F}_0 \restriction M, \ldots, \mathcal{F}_k \restriction M^3$$

is non-empty.

Exercise II.2.4. Show that the family $\mathbb{N}^{[k]}$ of all k-element subsets of $\mathbb{N}$ is Ramsey (and Sperner).

Lemma II.2.5. *For every Ramsey family $\mathcal{F}$ there is an infinite set M such that $\mathcal{F} \restriction M$ is Sperner.*

Proof. Let $\mathcal{F} = \mathcal{F}_0 \cup \mathcal{F}_1$, where $\mathcal{F}_0$ is the family of all $\subseteq$-minimal elements of $\mathcal{F}$ and apply the Ramsey property of $\mathcal{F}$. $\qquad\square$

Lemma II.2.6. *For every Ramsey family $\mathcal{F}$ there is an infinite set M such that $\mathcal{F} \restriction M$ is thin.*

Proof. Let $\mathcal{F} = \mathcal{F}_0 \cup \mathcal{F}_1$, where $\mathcal{F}_0$ is the family of all $\sqsubseteq$-minimal members of $\mathcal{F}$ and apply the Ramsey property of $\mathcal{F}$. $\qquad\square$

Lemma II.2.7 (Nash–Williams). *Every thin family is Ramsey.*

Proof. We present the original 1965 proof which involves the first instance of what can be called 'combinatorial forcing'. We let the variable $M, N, P, \ldots$ run over infinite subsets of $\mathbb{N}$, and $s, t, m, \ldots$ to run over finite subsets of $\mathbb{N}$.

A set $\mathcal{D}$ of infinite subsets on $\mathbb{N}$ is *open* if $N \subseteq M \in \mathcal{D}$ implies $N \in \mathcal{D}$. We say that such a $\mathcal{D}$ is *dense below M* if for every $N \subseteq M$ there is a $P \subseteq N$ in $\mathcal{D}$. A *dense-open-set assignment* on M is a family $\mathcal{D}_s$ ($s \in M^{[<\infty]}$) such that for all $s \in M^{[<\infty]}$,[4]

$$\mathcal{D}_s \text{ is dense-open below } M/s$$

Note the following property of this notion:

(1) For every dense-open-set assignment $\mathcal{D}_s$ ($s \in M^{[<\infty]}$) on M there is an $N \subseteq M$ such that $N/s \in \mathcal{D}_s$ for all $s \in N^{[<\infty]}$.

A function $E : \mathbb{N}^{[<\infty]} \longrightarrow \mathcal{P}(\mathbb{N}^{[<\infty]})$ is an *extensor* if $s \sqsubseteq t$ for all s and t in $E(s)$. Fixing an extensor E we give a definition:

---

[3]Recall that $\mathcal{F}_i \restriction M = \mathcal{F}_i \cap \mathcal{P}(M)$.
[4]$M/s = \{m \in M : m > \max(s)\}$.

Definition II.2.8 (Combinatorial forcing). We say that s is *inextensible* in M if $E(s) \cap P(M) = \emptyset$. We say that s is *strongly extensible* in M if $E(s) \cap P(N) \neq \emptyset$ for every $N \in [s, M]^5$.

Note the following property of these notions:

(2) For every M there is $N \subseteq M$ such that every finite subset of N is either inextensible in N or is strongly extensible in N.

Back to the proof of Nash–Williams' Lemma. It suffices to show that for every pair $\mathcal{F}_0 \subseteq \mathcal{F}$ of thin families of finite subsets of $\mathbb{N}$ and every M there is $N \subseteq M$ such that

$$\text{either } \mathcal{F}_0 \restriction N = \emptyset \text{ or } \mathcal{F} \restriction N \subseteq \mathcal{F}_0$$

Given such $\mathcal{F}_0 \subseteq \mathcal{F}$ define an extensor E by

$$t \in E(s) \text{ iff } t \sqsupseteq s \text{ and } t \text{ has an initial segment in } \mathcal{F}_0.$$

By (2) shrinking the given M we may assume that every finite subset of M is either inextensible in M, or is strongly extensible.

For $s \in M^{[<\infty]}$ let $\mathcal{D}_s$ be the collection of all $P \subseteq M$ such that either
(a) $s \cup \{n\}$ is inextensible in M for all $n \in P$, or
(b) $s \cup \{n\}$ is strongly extensible in M for all $n \in P$.

Clearly, $\mathcal{D}_s$ ($s \in M^{[<\infty]}$) is a dense-open-set assignment on M, so by (1), there is $N \subseteq M$ such that
(c) $N/s \in \mathcal{D}_s$ for all $s \in M^{[<\infty]}$.

Case 1. $\emptyset$ is inextensible in M (and therefore in N). In this case we have that $\mathcal{F}_0 \restriction N = \emptyset$.

Case 2. $\emptyset$ is strongly extensible in M (and therefore in N). Note that by the definition of extension E, an s is extensible in M iff there is $n \in M/s$ such that $s \cup \{n\}$ is extensible in M. So by the choice of $N \subseteq M$ an $s \subseteq N$ is extensible in N iff for some (all) $n \in N/s$, $s \cup \{n\}$ is extensible in N. It follows that:
(3) Every finite subset s of N is (strongly) extensible in N.

Using (3) and the fact that $\mathcal{F}$ is thin, one concludes that N cannot contain a member of $\mathcal{F} \setminus \mathcal{F}_0$ (see the definition of extension E). $\square$

Corollary II.2.9. *For every front $\mathcal{F}$ there is an infinite set $M \subseteq \bigcup \mathcal{F}$ such that the restriction $\mathcal{F} \restriction M$ is a barrier on M.*

Proof. Given a front $\mathcal{B}$, let $\mathcal{B}_0$ be the set of all $\subseteq$-minimal elements of $\mathcal{B}$. By Nash–Williams' lemma there is an infinite $M \subseteq \bigcup \mathcal{B}$ such that either $\mathcal{B}_0 \restriction M = \emptyset$ or $(\mathcal{B} \setminus \mathcal{B}_0) \restriction M = \emptyset$. Note that the second alternative holds. $\square$

It follows that in the definition of b.q.o., we may restrict ourselves to continuous maps of the form $\hat{f} \colon \mathbb{N}^{[\infty]} \longrightarrow Q$ where $f \colon \mathcal{B} \longrightarrow Q$ has its domain $\mathcal{B}$ to be

---

$^5[s, M] = \{P : s \sqsubseteq P \subseteq s \cup M\}$

a barrier rather than a front. This gives us a way to finitize the outcome of the Definition II.1.13 . For a finite set $\emptyset \neq s \subseteq \mathbb{N}$, let

$$_*s = s \setminus \{\min(s)\}.$$

For finite nonempty subsets s and t of $\mathbb{N}$, put

$$s \vartriangleleft t \text{ iff } {}_*s \sqsubset t$$

where $\sqsubset$ is the strict version of the initial segment ordering $\sqsubseteq$. Thus,

$$\{2\} \vartriangleleft \{3\}, \{2,3,4\} \vartriangleleft \{3,4,6\}, \dots$$

The relation $\vartriangleleft$ captures the shift-operator on $\mathbb{N}^{[\infty]}$ as shown in the following reformulation of Definition 3.3.

Lemma II.2.10. *A quasi-ordered set Q is b.q.o. iff for every $f \colon \mathcal{B} \longrightarrow Q$ defined on some barrier $\mathcal{B}$ there exists $s \vartriangleleft t$ in $\mathcal{B}$ such that $f(s) \leq f(t)$.* $\qquad\square$

Definition II.2.11. For an integer $k \geqslant 1$ and a barrier $\mathcal{B}$, let

$$\mathcal{B}[k] := \{s_0 \cup s_1 \cup \dots \cup s_{k-1} : s_0 \vartriangleleft s_1 \vartriangleleft \dots \vartriangleleft s_{k-1} \text{ in } \mathcal{B}\}$$

Lemma II.2.12. *If $\mathcal{B}$ is a barrier then so is any of its finite stretches $\mathcal{B}[k]$.*

Proof. Given an infinite set M, let $M = M_0, \dots, M_{i+1} = {}_*M_i, \dots, M_{k-1} = {}_*M_{k-2}$ be the sequence of first k shifts of M. For each $i < k$ there is unique $s_i \in \mathcal{B}$ such that $s_i \sqsubset M_i$. Then

$$s_0 \vartriangleleft s_1 \vartriangleleft \dots \vartriangleleft s_i \vartriangleleft \dots \vartriangleleft s_{k-1} \tag{II.1}$$

and $s = s_0 \cup s_1 \cup \dots \cup s_{k-1}$ is a member of $\mathcal{B}[k]$ such that $s \sqsubset M$. Note also since $\mathcal{B}$ satisfies II.2.1(1) every member s of $\mathcal{B}[k]$ has a unique decomposition $s = s_0 \cup \dots \cup s_{k-1}$ into members of $\mathcal{B}$. From this one easily concludes that $\mathcal{B}[k]$ also satisfies II.2.1(1'), and so $\mathcal{B}[k]$ is a barrier. $\qquad\square$

Lemma II.2.13. *If Q is b.q.o., then for every barrier $\mathcal{B}$ and $f \colon \mathcal{B} \longrightarrow Q$ there exists an infinite $M \subseteq \mathbb{N}$ such that $f(s) \leq f(t)$ for all $s, t \in \mathcal{B} \restriction M$ such that $s \vartriangleleft t$.*

Proof. Apply Nash–Williams' generalization of Ramsey's theorem (Lemma II.2.7) to the coloring $\mathcal{B}[2] = \mathcal{C} \cup \mathcal{D}$ defined by letting $s \in \mathcal{C}$ iff $f(s_0) \leq f(s_1)$ where $s_0, s_1 \in \mathcal{B}$ are unique such that $s_0 \vartriangleleft s_1$ and $s = s_0 \cup s_1$. $\qquad\square$

Corollary II.2.14. *The product of two b.q.o. is a b.q.o.*

Proof. Consider a map $f : \mathcal{B} \to Q_0 \times Q_1$ from a barrier $\mathcal{B}$ into the cartesian product of the quasi-orderings. Let $f_i : \mathcal{B} \to Q_i$ $(i < 2)$ be the composition of f with the corresponding projection. Applying Lemma II.2.13 successively to f and then to f_1 we get an infinite set $M \subseteq \mathbb{N}$ such that $f_i(s) \leq f_i(t)$ for all $s \vartriangleleft t$ from $\mathcal{B} \restriction M$ and $i < 2$. It follows that $f(s) \leq f(t)$ for all $s \vartriangleleft t$ from $\mathcal{B} \restriction M$, as required. $\qquad\square$

The following lemma gives us an ordinal index on which one can carry inductive arguments about fronts and barriers.

Lemma II.2.15. *Every front is lexicographically well-ordered.*

Proof. The lexicographical ordering on finite subsets of $\mathbb{N}$ to which one refers is defined as:

$$s <_{\text{lex}} t \text{ iff } \min(s \bigtriangleup t) \in s.$$

Suppose a given front $\mathcal{B}$ contains an infinite sequence $(s_i)_{i=0}^{\infty}$ that decreases relative to $<_{\text{lex}}$. Then in particular

$$\min(s_i \bigtriangleup s_j) \neq \min(s_j \bigtriangleup s_k) \tag{II.2}$$

for all $i < j < k$ in $\mathbb{N}$. Applying Ramsey's theorem we get an infinite set $X \subseteq \mathbb{N}$ such that either
(a) $\min(s_i \bigtriangleup s_j) < \min(s_j \bigtriangleup s_k)$ for all $i < j < k$ in X, or
(b) $\min(s_i \bigtriangleup s_j) > \min(s_j \bigtriangleup s_k)$ for all $i < j < k$ in X.

Clearly the case (b) is impossible so we are left with (a). Applying Ramsey's theorem once again we find an infinite subset Y of X such that for some $\rho \in \{\sqsubset, =, \sqsupset\}$,
(c) $s_i \cap \min(s_i \bigtriangleup s_j) \rho s_j \cap \min(s_j \bigtriangleup s_k)$ for all $i < j < k$ in Y.

Clearly the case when ρ is equal to $\sqsupset$ is impossible. The case when ρ is equal to $\sqsubset$ gives
(d) $s_i \cap \min(s_i \wedge s_j) \sqsubset s_j \cap \min(s_j \bigtriangleup s_k)$ for all $i < j < k$ in Y

and this contradicts property (2) of $\mathcal{B}$. So we are left with the case $\rho \,==$ which together with (a) gives us that $s_i <_{\text{lex}} s_j$ whenever $i < j$ belongs to Y contradicting our assumption that $(s_i)_{i=0}^{\infty}$ decreases relative to $<_{\text{lex}}$. $\square$

Definition II.2.16. For two families $\mathcal{B}$ and $\mathcal{C}$ of finite (nonempty) subsets of $\mathbb{N}$, let

$$\mathcal{C} \oplus \mathcal{B} = \{s \cup t : s \in \mathcal{B}, \, t \in \mathcal{C}, \text{ and } \max(s) < \min(t)\}.$$

Example II.2.17. $\mathbb{N}^{[k]} \oplus \mathbb{N}^{[1]} = \mathbb{N}^{[k+1]}$

The following fact about this operation is immediate.

Lemma II.2.18. *If $\mathcal{B}_0$ and $\mathcal{B}_1$ are fronts on the same domain then so is $\mathcal{B}_1 \oplus \mathcal{B}_0$.* $\square$

Exercise II.2.19. Show that the lexicographical rank of $\mathcal{B}_1 \oplus \mathcal{B}_0$ is equal to the sum of lexicographical ranks of $\mathcal{B}_0$ and $\mathcal{B}_1$.

Definition II.2.20. Given two families $\mathcal{B}, \mathcal{C} \subseteq \text{FIN}$, let

$$\mathcal{B} \otimes \mathcal{C} = \{s_1 \cup \cdots \cup s_n : s_1 < \cdots < s_n, \, s_i \in \mathcal{B}, \, \{\min s_i\}_{i=1}^{n} \in \mathcal{C}\}.$$

Exercise II.2.21. Prove that if $\mathcal{B}, \mathcal{C}$ are fronts on M, then so is $\mathcal{B} \otimes \mathcal{C}$.

The following immediate fact gives us a way to construct fronts of arbitrary high lexicographical ranks.

Lemma II.2.22. *For an infinite sequence $(\mathcal{B}_i)_{i \in \mathbb{N}}$ of fronts on $\mathbb{N}$, the family*

$$\mathcal{B} = \bigcup_{i=0}^{\infty} \mathcal{B}_0 \oplus \mathcal{B}_1 \oplus \cdots \oplus \mathcal{B}_i \oplus \{i\}.$$

is also a front on $\mathbb{N}$. $\square$

In particular, the construction of Lemma II.2.22 gives us the following example of a barrier of lexicographical rank ω^ω.

Example II.2.23. $\mathcal{S} = \bigcup_{i=0}^{\infty} \mathbb{N}^{[1]} \oplus \mathbb{N}^{[1]} \oplus \cdots \oplus \mathbb{N}^{[1]} \oplus \{i\}$ where at the level i we take $\mathbb{N}^{[1]} \oplus \cdots \oplus \mathbb{N}^{[1]}$ i times is an example of what one can call *Schreier barrier.*

In general we have the following fact.

Lemma II.2.24. *For every countable indecomposable[6] ordinal δ there is a barrier on $\mathbb{N}$ of lexicographical rank equal to δ.*

Proof. The lexicographical rank of $\mathbb{N}^{[1]}$ is clearly equal to ω. To see the inductive step, suppose

$$\omega^\alpha = \sum_{i=0}^{\infty} \omega^{\alpha_i}$$

where $\alpha_0 \le \alpha_1 \le \cdots < \alpha$. For each $i \in \omega$ pick a barrier $\mathcal{B}_i$ on $\mathbb{N}$ of rank ω^{α_i} and let

$$\mathcal{B} = \bigcup_{i=0}^{\infty} \mathcal{B}_0 \oplus \mathcal{B}_1 \oplus \cdots \oplus \mathcal{B}_i \oplus \{i\}.$$

Then $\mathcal{B}$ is a front of rank ω^α. Note that $\mathcal{B}$ is not necessarily a barrier. Note, however, that $\mathcal{B}$ is 'uniform' in the sense that it looks the same (and so in particular it has the same lexicographical rank) when restricted to an arbitrary infinite subset of $\mathbb{N}$. Applying II.2.9 and rescaling back to $\mathbb{N}$ we get the conclusion of the Lemma. $\square$

Exercise II.2.25. Compute the lexicographical rank of the Schreier barrier.

Banach space theory offers many applications of fronts and barriers (see for example [2], [3], [7], [28], [68]) though the experts in this area prefer to think about them as compact (under the topology of pointwise convergence) families of finite subsets of $\mathbb{N}$ by considering in fact not fronts $\mathcal{B}$ themselves but their downwards closures

$$\widehat{\mathcal{B}} = \{s \subseteq \mathbb{N} : (\exists t \in \mathcal{B})s \subseteq t\}.$$

For example, instead of Schreier's barrier introduced above in Example II.2.23 they consider the corresponding *Schreier family*

$$\widehat{\mathcal{S}} = \{s \subseteq \mathbb{N} : |s| \le \min(s) + 1\}.$$

---

[6]A limit ordinal δ is called *indecomposable* if there is no $\alpha, \beta < \delta$ such that $\alpha + \beta = \delta$

During the last ten years or so (see [3], [2] and [68]) Schreier family has been generalized into the transfinite as follows:

$$\widehat{S}_0 = \{\{n\} : n \in \mathbb{N}\} \cup \{\emptyset\},$$

$$\widehat{S}_{\alpha+1} = \{\bigcup_{k=0}^{n} E_k : n \leq E_0 < E_1 < \cdots < E_n, \ E_k \in \widehat{S}_\alpha\},$$

$$\widehat{S}_\alpha = \{E : E \in \widehat{S}_{\delta_n}, \ n \in \mathbb{N}, \ n \leq E\},$$

for δ a countable limit ordinal and $(\delta_n)_{n=0}^{\infty}$ an increasing sequence of smaller ordinals converging to δ.

The reader is invited to examine the corresponding sequence S_α ($\alpha < \omega_1$) of fronts and compute their lexicographical ranks. While the sequence S_α ($\alpha < \omega_1$) leaves out certain lexicographical ranks the main difficulty with S_α ($\alpha < \omega_1$) is however their non uniformity, i.e., the difficulty in taking restrictions to infinite subsets of $\mathbb{N}$ while preserving their initial structure, a feature that is very important in Ramsey theory. The purpose of the next section is to examine a uniform hierarchy of fronts and barriers that has been quite useful in Ramsey theory over the last three decades and which hopefully will find its uses in Banach space theory as well.

Exercise II.2.26. Show that for each countable ordinal α the Schreier family $\widehat{S}_\alpha$ is *spreading*, i.e., it has the property that a finite set $\{n_0, \ldots, n_k\}$ enumerated increasingly belongs to $\widehat{S}_\alpha$ if one can find a set $\{m_0, \ldots, m_k\} \in \widehat{S}_\alpha$ of the same size and enumerated increasingly such that $m_i \leq n_i$ for all $i \leq k$.

II.3 Uniform Fronts and Barriers

For a family $\mathcal{B}$ of infinite subsets of $\mathbb{N}$ and $n \in \mathbb{N}$ set

$$\mathcal{B}_n = \{s \in \mathcal{B} : n = \min(s)\}$$

$$\mathcal{B}_{\{n\}} = \{s \in \mathbb{N}^{[<\infty]} : \{n\} \cup s \in \mathcal{B}, \ n < \min(s)\}.$$

Definition II.3.1. Let α be a countable ordinal and let M be an infinite subset of $\mathbb{N}$. We say that a family $\mathcal{B}$ of finite subsets of $\mathbb{N}$ is α-*uniform on* M provided that:
(a) $\alpha = 0$ implies $\mathcal{B} = \{\emptyset\}$
(b) $\alpha = \beta + 1$ implies that $\emptyset \notin \mathcal{B}$ and $\mathcal{B}_{\{n\}}$ is β-uniform on M/n[7] for all $n \in M$.
(c) $\alpha > 0$ limit implies that there is an increasing sequence $\{\alpha_n\}_{n \in M}$ of ordinals converging to α such that $\mathcal{B}_{\{n\}}$ is α_n-uniform on M/n for all $n \in M$.

Note the following three properties of uniform families proved by straightforward induction on α.

Lemma II.3.2. *Every α-uniform family on M is a front on M.* □

---

[7] $M/n = \{m \in M : m > n\}$.

The second property is the main reason behind our choice of the word 'uniform' in the name for these families.

Lemma II.3.3. *If $\mathcal{B}$ is α-uniform on M and if N is an infinite subset of M, then $\mathcal{B} \restriction N$ is α-uniform on N.*   □

So in particular (see II.2.9), every α-uniform family $\mathcal{B}$ has a restriction $\mathcal{B} \restriction M$ that is not just an α-uniform family but also an α-uniform *barrier*.

Lemma II.3.4. *If $\mathcal{B}$ is uniform (i.e., α-uniform for some α) on M, then $\mathcal{B}$ is a maximal thin family on M.*   □

Exercise II.3.5. Show that for every positive integer k, the family $\mathbb{N}^{[k]}$ is the only k-uniform family on $\mathbb{N}$.

Exercise II.3.6. Find examples of ω-uniform and $(\omega + 1)$-uniform families on $\mathbb{N}$. Are they unique?

The next immediate property (proved again by a straightforward induction on α and β) shows that the notion of uniform families behaves well with respect the operations of taking sums and products between two families of finite subsets of $\mathbb{N}$.

Lemma II.3.7. (a) *Suppose that $\mathcal{B}$ and $\mathcal{C}$ are α and β-uniform on M, respectively. Then $\mathcal{B} \oplus \mathcal{C}$ is $\alpha + \beta$-uniform on M.*
(b) *If $\mathcal{B}$ is an α-uniform front on M, and $\mathcal{C}$ is a β-uniform front on M, then $\mathcal{B} \otimes \mathcal{C}$ is an $\alpha \cdot \beta$-uniform front on M.*   □

The following is one of the most important results connecting arbitrary families of finite subsets of $\mathbb{N}$ with fronts and barriers. It is also the key result in the development of the topological Ramsey theory of $\mathbb{N}^{[\infty]}$ (see Section II.6).

Lemma II.3.8 (Galvin). *For every family $\mathcal{B}$ of finite subsets of $\mathbb{N}$ and every infinite $M \subseteq \mathbb{N}$ there exists an infinite $N \subseteq M$ such that the restriction $\mathcal{B} \restriction N$ is either empty or it contains a barrier.*

Proof. We may assume that the given M is equal to $\mathbb{N}$ and use variables M, N, $P, \ldots$ for infinite subsets of $\mathbb{N}$ and $s, t, n, \ldots$ for finite subsets of $\mathbb{N}$.

Definition II.3.9 (Combinatorial forcing). We say that M *accepts* s if every $X \in [s, M]^8$ has an initial segment in $\mathcal{B}$. If there is no $N \subseteq M$ which accepts s we say that M *rejects* s. We say that M *decides* s if M either accepts or rejects s.

Note the following monotonicity properties of this forcing relation:
(i) If M accepts (rejects) s then every $N \subseteq M$ accepts (rejects) s.
(ii) For every M and s there exists $N \subseteq M$ which decides s.
(iii) M accepts s if and only if M accepts $s \cup \{n\}$ for all $n \in M/s^9$.

---

[8]$[s, M] = \{X \in \mathbb{N}^{[\infty]} : s \sqsubseteq X \subseteq s \cup M\}$.
[9]$M/s = \{m \in M : m > \max(s)\}$.

(iv) If M rejects s then M does not accept $s \cup \{n\}$ for all but finitely many $n \in M/s$.

To finish the proof a simple diagonalization will give us an M which decides all of its finite subsets. If M accepts $\emptyset$ we have reached one of the conclusions of the lemma. If M rejects $\emptyset$, let $m_0 \in M$ be minimal such that M rejects $\{n\}$ for all $n \geq m_0$ in M (use (iv) and that M decides all of its finite subsets). Having defined an increasing sequence $m_0 < m_1 < \cdots < m_k$ of members of M with the property that M rejects s for all $s \subseteq \{m_0, \ldots, m_k\}$ let $m_{k+1} > m_k$ be the minimal $m \in M$ such that M rejects $s \cup \{n\}$ for all $n \geq m$ in M (use (iv) and that M decides all of its finite subsets). Finally let $N = \{m_i : i \in \mathbb{N}\}$. Then $\mathcal{B} \upharpoonright N = \emptyset$, finishing the proof of Lemma II.3.8. $\qquad \square$

Corollary II.3.10. *For every family $\mathcal{B}$ of finite subsets of $\mathbb{N}$ and for every infinite $M \subseteq \mathbb{N}$ there exists an infinite $N \subseteq M$ such that $\mathcal{B} \upharpoonright N = \emptyset$ or $\mathcal{B} \upharpoonright N$ contains an uniform barrier.*

Proof. By lemma II.3.8 we may assume to have an infinite $M \subseteq \mathbb{N}$ such that $\mathcal{B} \upharpoonright M$ contains a barrier. Shrinking $\mathcal{B}$ we may assume $\mathcal{B}$ is a barrier on M. Now the proof that we can shrink M further to have $\mathcal{B} \upharpoonright M$ uniform is by induction on, say, lexicographical rank of $\mathcal{B} \upharpoonright M$. To see that this can be done note that

$$\text{rank}(\mathcal{B}_n \upharpoonright M) < \text{rank}(\mathcal{B} \upharpoonright M)$$

for every $n \in M$ $\qquad \square$

Theorem II.3.11. *The following are equivalent for a family $\mathcal{F}$ of finite subsets of $\mathbb{N}$:*

(a) *$\mathcal{F}$ is Ramsey.*
(b) *There is an infinite $M \subseteq \mathbb{N}$ such that $\mathcal{F} \upharpoonright M$ is Sperner.*
(c) *There is an infinite $M \subseteq \mathbb{N}$ such that $\mathcal{F} \upharpoonright M$ is either empty or uniform on M.*
(d) *There is an infinite $M \subseteq \mathbb{N}$ such that $\mathcal{F} \upharpoonright M$ is thin.*
(e) *There is an infinite $M \subseteq \mathbb{N}$ such that for every infinite $N \subseteq M$ the restriction $\mathcal{F} \upharpoonright N$ cannot contain two disjoint uniform families on N.*

Proof. (a) $\to$ (b) Apply the Ramsey property of $\mathcal{F}$ to the partition $\mathcal{F} = \mathcal{F}_0 \cup \mathcal{F}_1$ where $\mathcal{F}_0$ is the set of all $\subseteq$-minimal elements of $\mathcal{F}$.
(b) $\to$ (c) This follows from Corollary II.3.10.
(c) $\to$ (d) This follows from Lemma II.3.4.
(d) $\to$ (e) By Lemma II.3.4 the union of two disjoint uniform families on the same set cannot be thin.
(e) $\to$ (a) Let $\mathcal{F} = \mathcal{F}_0 \cup \cdots \cup \mathcal{F}_k$ be a given finite partition of $\mathcal{F}$. By Corollary II.3.10 and Lemma II.3.3 we can find an infinite M such that for all $i \leq k$ the restriction $\mathcal{F}_i \upharpoonright M$ is either empty or uniform. Going to a further subset of M we may assume that $\mathcal{F} \upharpoonright M$ satisfies the conclusion of (5). It follows that at most one $\mathcal{F}_i \upharpoonright M$ is nonempty. $\qquad \square$

The following general comparability property between different barriers is quite useful.

Lemma II.3.12. *Suppose $\mathcal{B}$ and $\mathcal{C}$ are two uniform barriers on the same infinite set M. Then there is an infinite subset N of M such that either for every $s \in \mathcal{B} \upharpoonright N$ there is $t \in \mathcal{C} \upharpoonright N$ such that $s \subseteq t$, or vice versa for every $t \in \mathcal{C} \upharpoonright N$ there is $s \in \mathcal{B} \upharpoonright N$ such that $t \subseteq s$.*

Proof. Let $\mathcal{B}_0$ be the collection of all $s \in \mathcal{B}$ which are not included in any member of $\mathcal{C}$. Similarly, let $\mathcal{C}_0$ be the collection of all $t \in \mathcal{C}$ that are not included in any member of $\mathcal{B}$. By Theorem II.3.11 we can find an infinite $N \subseteq M$ such that :

(a) $\mathcal{B}_0 \upharpoonright N = \emptyset$ or $\mathcal{B} \upharpoonright N \subseteq \mathcal{B}_0$, and

(b) $\mathcal{C}_0 \upharpoonright N = \emptyset$ or $\mathcal{C} \upharpoonright N \subseteq \mathcal{C}_0$.

Now note that since $\mathcal{B}$ and $\mathcal{C}$ are barriers it is not possible to have at the same time the first alternative in (a) and the first alternative in (b). $\qquad\square$

Given two barriers $\mathcal{B}$ and $\mathcal{C}$ we say that $\mathcal{C}$ *dominates* $\mathcal{B}$ if every member of $\mathcal{B}$ is included in some member of $\mathcal{C}$.

Recall the standard identification of the power set of $\mathbb{N}$ with the Cantor space $2^{\mathbb{N}}$ via characteristic functions. This will give us the natural topology on the power set of $\mathbb{N}$. Referring to this topology note the following property of uniform barriers:

Lemma II.3.13. *The topological closure $\overline{\mathcal{B}}$ of an uniform barrier $\mathcal{B}$ is equal to its downward closure relative to the inclusion, or in other words,*

$$\overline{\mathcal{B}} = \widehat{\mathcal{B}},$$

where $\widehat{\mathcal{B}} = \{s \subseteq \mathbb{N} : (\exists t \in \mathcal{B})s \subseteq t\}$. $\qquad\square$

Note also the following pleasant properties of barriers relative to the operation $\widehat{\mathcal{B}}$ of taking the downwards closure under inclusion.

Lemma II.3.14. (a) *If $\mathcal{B}$ is a barrier on M the $\widehat{\mathcal{F} \upharpoonright N} = \widehat{\mathcal{F}} \upharpoonright N$ for every $N \subseteq M$.*
(b) *If $\mathcal{F}$ is a barrier on M, then for every $N \subseteq M$ such that $M \setminus N$ is infinite, $\widehat{\mathcal{F} \upharpoonright N} = \mathcal{F}[N]$, and so in particular $\mathcal{F}[N]$ is downwards closed under inclusion.* $\qquad\square$

Exercise II.3.15. Show that the Cantor–Bendixson rank of the topological closure $\overline{\mathcal{B}}$ of an α-uniform barrier $\mathcal{B}$ is equal to $\alpha + 1$.

Corollary II.3.16. *Suppose that $\mathfrak{A}$ is an α uniform barrier on M and $\mathcal{B}$ is a β uniform barrier on M, with $\alpha < \beta$. Then there is some $N \in M^{[\infty]}$ such that $\mathfrak{A} \upharpoonright N \subseteq \mathcal{B}$*

Proof. By Lemma II.3.12, there is some infinite subset $N \subseteq M$ such that either $\mathfrak{A} \upharpoonright N \subseteq \overline{\mathcal{B}}$ or $\mathcal{B} \upharpoonright N \subseteq \mathfrak{A}$. We work to show that the second alternative is impossible. Suppose not; since $\mathcal{B} \upharpoonright N$ is also a β-uniform barrier on N, Cantor–Bendixon rank of $\overline{\mathcal{B} \upharpoonright N}$ is $\beta + 1$, hence the Cantor–Bendixon rank of $\overline{\mathfrak{A}}$ is at least $\beta + 1$, which is impossible since $\mathfrak{A}$ is α-uniform. $\qquad\square$

Lemma II.3.17. *The lexicographical rank of an α-uniform barrier is equal to ω^α.*

Proof. Induction on α. □

Note that Lemma II.3.13 relates uniform barriers on infinite subsets of $\mathbb{N}$ and compact *hereditary* [10] families of finite subsets of $\mathbb{N}$. In order to examine this relationship even more precisely, it is convenient to have the notation

$$\mathcal{F}[M] = \{s \cap M : s \in \mathcal{F}\}$$

for the *trace* of $\mathcal{F}$ on M, where $\mathcal{F}$ is an arbitrary family of subsets of $\mathbb{N}$ and $M \subseteq \mathbb{N}$. This must not to be confused with the *restriction*

$$\mathcal{F} \upharpoonright M = \mathcal{F} \cap \mathcal{P}(M)$$

already used several times above as they are seldom equal. There is however an exception as the following fact shows.

Lemma II.3.18. *If $\mathcal{F}$ is a hereditary family of sets, then $\mathcal{F}[M] = \mathcal{F} \upharpoonright M$ for every set M.* □

Definition II.3.19. A family $\mathcal{F}$ of finite sets is called *pre-compact* if its topological closure $\overline{\mathcal{F}}$ consists only of finite sets.

Note the following immediate property of this notion.

Lemma II.3.20. *Suppose that $\mathcal{F}$ is a hereditary family of finite subsets off $\mathbb{N}$. Then $\mathcal{F}$ is compact if and only if it is pre-compact.* □

Lemma II.3.21. *Suppose that $\mathcal{F}$ is a compact hereditary family of finite subsets of $\mathbb{N}$. Then there is an infinite set M such that the set $\mathcal{F}[M]^{\max}$ of $\subseteq$-maximal elements of $\mathcal{F}[M]$ is a uniform barrier on M. Moreover, $\mathcal{F}[M]^{\max}$ is also equal to the set $\mathcal{F}[M]^{\subseteq-\max}$ of $\subseteq$-maximal elements of $\mathcal{F}[M]$.*

Proof. The proof is by induction on the Cantor–Bendixon rank α of the family $\mathcal{F}$, the minimal countable ordinal α such that the αth-derivative of $\mathcal{F}$ is finite. If $\alpha = 0$, then $\mathcal{F}$ is finite and in this case the conclusion of the lemma is immediate. Let us consider the case $\alpha > 0$. Going to a restriction of the form $\mathcal{F}[\mathbb{N}/n_0] = \mathcal{F} \upharpoonright (\mathbb{N}/n_0)$, we may assume that $\mathcal{F}^{(\alpha)} = \{\emptyset\}$. For $n \in \mathbb{N}$, we set

$$\mathcal{F}_{\{n\}} = \{t \in \mathbb{N}^{[<\infty]} : t \cup \{n\} \in \mathcal{F}\} \subseteq \mathcal{F}. \tag{II.3}$$

Claim. For every n, the section $\mathcal{F}_{\{n\}}$ is a compact hereditary family of Cantor-Bendixon rank strictly smaller than α.

Proof of the claim. It is clear that $\mathcal{F}_{\{n\}}$ is pre-compact since it is a subset of $\mathcal{F}$. To show that it is hereditary, consider $t \subseteq u$ with $u \in \mathcal{F}_{\{n\}}$. Then $\{n\} \cup t \subseteq \{n\} \cup u \in \mathcal{F}$

---

[10] A family $\mathcal{F}$ of sets is *hereditary* if $A \subseteq B \in \mathcal{F}$ implies $A \in \mathcal{F}$.

and so $\{n\} \cup t \in \mathcal{F}$, since $\mathcal{F}$ is hereditary. This implies that $u \in \mathcal{F}_{\{n\}}$, as desired. It follows that $\mathcal{F}$ is compact. An easy induction on γ shows that

$$\{t \cup \{n\} : t \in (\mathcal{F}_{\{n\}})^{(\gamma)}\} \subseteq (\mathcal{F})^{(\gamma)}. \tag{II.4}$$

This gives that the Cantor–Bendixon rank of $\mathcal{F}_{\{n\}}$ is strictly small than α since otherwise $\{t \cup \{n\} : t \in (\mathcal{F}_{\{n\}})^{(\alpha)}\} \subseteq \{\emptyset\}$ which leads to a contradiction. $\square$

Applying the inductive hypothesis, we can build a sequence (M_i) of infinite subsets of $\mathbb{N}$ such that, setting $m_i = \min M_i$, then for every i, we have that $m_i < M_{i+1}$ and that $\mathcal{F}_{\{m_i\}}[M_{i+1}]^{\max} = \mathcal{B}_i$ is a uniform barrier on M_{i+1}. Let $M_\infty = \{m_i\}$, and let

$$\mathcal{B} = \bigcup_i \mathcal{B}_i \restriction M_\infty \oplus \{m_i\}. \tag{II.5}$$

It is clear that $\mathcal{B}$ is a uniform family on M_∞ but not necessarily a barrier on M_∞. Let $M = \{m_i\}_{i \in I} \subseteq M_\infty$ be such that $\mathcal{B} \restriction M$ is a uniform barrier. Note that

$$\mathcal{B} \restriction M = \bigcup_{i \in I} \mathcal{B}_i \restriction M \oplus \{m_i\}. \tag{II.6}$$

Let us show that $\mathcal{F}[M]^{\max} = \mathcal{B} \restriction M$. Consider first an $s \in \mathcal{F}[M]^{\max}$, and let $i \in I$ be such that $m_i = \min s$. Fix $t \in \mathcal{F}$ such that $s = t \cap M \subseteq t \cap M_{i+1}$. Since $m_i \in t$ and $\mathcal{F}$ is hereditary, we obtain that $_*s \in \mathcal{F}_{\{m_i\}}[M_{i+1}]$. Let $u \in \mathcal{F}_{\{m_i\}}[M_{i+1}]^{\max} = \mathcal{B}_i$ be such that $_*s \subseteq u$.

Case 1. $_*s = u$. Then $_*s \in \mathcal{B}_i$, and $_*s \subseteq M$, so $_*s \in \mathcal{B}_i \restriction M$, and therefore $s = \{m_i\} \cup _*s \in \mathcal{B} \restriction M$.

Case 2. $_*s \subsetneq u$. Consider the infinite set $_*s \cup (M/s) \subseteq M_{i+1}$. Since $\mathcal{B}_i$ is a barrier on M_{i+1} we can find $v \in \mathcal{B}_i$ such that $v \sqsubseteq _*s \cup (M/s)$. Note that in this case $v \subseteq M/m_i$.

Subcase 2.1. $_*s$ is a strict initial part of v. Then

$$s = \{m_i\} \cup _*s \subsetneq \{m_i\} \cup v = (\{m_i\} \cup v) \cap M. \tag{II.7}$$

Let $w \in \mathcal{F}_{\{m_i\}}$ be such that $v = w \cap M_{i+1}$. Then

$$s \subsetneq (\{m_i\} \cup v) \cap M = (\{m_i\} \cup (w \cap M_{i+1})) \cap M = (\{m_i\} \cup w) \cap M \in \mathcal{F}[M], \tag{II.8}$$

since $\{m_i\} \cup w \in \mathcal{F}$. But this contradicts the maximality of s.

Subcase 2.2 $v \sqsubseteq _*s$. Then $v \sqsubseteq _*s \subsetneq u$, and $u, v \in \mathcal{B}_i$, which contradicts the fact that $\mathcal{B}_i$ is Sperner.

Let us now show the other inclusion $\mathcal{B} \restriction M \subseteq \mathcal{F}[M]^{\max}$. So, consider an $s \in \mathcal{B} \restriction M$, and let $m_i = \min s$. Then $_*s \in \mathcal{B}_i \restriction M$, hence $_*s \in \mathcal{F}_{\{m_i\}}[M_{i+1}]^{\max}$. So $_*s = t \cap M_{i+1}$, for some $t \in \mathcal{F}_{\{m_i\}}$. Since $_*s \subseteq M/m_i$ we obtain

$$s = (\{m_i\} \cup _*s) \cap M = (\{m_i\} \cup (t \cap M_{i+1})) \cap M \subseteq (\{m_i\} \cup t) \cap M \in \mathcal{F}[M]. \tag{II.9}$$

Hence, there is some $u \in \mathcal{F}[M]^{\max} \subseteq \mathcal{B} \restriction M$ such that $s \subseteq u$. But $\mathcal{B} \restriction M$ is Sperner, so we must have $s = u$, as desired. $\square$

Theorem II.3.22. *Suppose that $\mathcal{F}$ is a pre-compact family. Then there is an infinite set M such that*

(a) $\mathcal{F}[M]^{\max} - \mathcal{F}[M]^{\subseteq -\max}$ *is a uniform barrier on M, and*

(b) $\mathcal{F}[M] = \widehat{\mathcal{F}[M]}^{\subseteq} = \widehat{\mathcal{F}[M]}^{\subseteq} = \overline{\mathcal{F}[M]}$, *and so in particular the trace $\mathcal{F}[M]$ is a hereditary family.*

Proof. By Lemma II.3.21 applied to the compact hereditary family $\widehat{\mathcal{F}}$ there is some M such that $\widehat{\mathcal{F}}[M]^{\max}$ is a uniform barrier on M. Note that $\overline{\mathcal{F}[M]} = \widehat{\mathcal{F}}[M]$, so $\mathcal{F}[M]^{\max} = \widehat{\mathcal{F}}[M]^{\max}$. Let $\mathcal{B} = \mathcal{F}[M]^{\max}$. Choose $N \subseteq M$ such that $M \setminus N$ is infinite. By Lemma II.3.14 (b), we know that

$$\mathcal{B}[N] = \widehat{\mathcal{B}} \restriction N. \tag{II.10}$$

Note also that

$$\mathcal{B} \restriction N = (\mathcal{F}[M]^{\max}) \restriction N \subseteq (\mathcal{F}[M]^{\max})[N] \subseteq \mathcal{F}[M][N] = \mathcal{F}[N]. \tag{II.11}$$

It follows that $\mathcal{F}[N]^{\max} = \mathcal{B} \restriction N$, and therefore

$$\mathcal{F}[N] \subseteq \widehat{\mathcal{B} \restriction N} = \mathcal{B}[N] \subseteq \mathcal{F}[N]. \tag{II.12}$$

$\square$

Corollary II.3.23. *Suppose that $\mathcal{F}_0$ and $\mathcal{F}_1$ are two pre-compact families. Then there is an infinite set M such that either $\mathcal{F}_0[M] \subseteq \mathcal{F}_1[M]$ or $\mathcal{F}_1[M] \subseteq \mathcal{F}_0[M]$.*

Proof. Let $M_1 \subseteq M_0$ be such that $\mathcal{F}_i[M_i]$ is the closure of a uniform barrier $\mathcal{B}_i$ on M_i ($i = 0, 1$). Let $M \subseteq M_1$ be such that $\mathcal{B}_i \restriction M \subseteq \widehat{\mathcal{B}_j \restriction M}$ for $i \neq j \in \{0, 1\}$. By Lemma II.3.14 (a) and Lemma II.3.18

$$\widehat{\mathcal{B}_i \restriction M} = \widehat{\mathcal{B}}_i \restriction M = \widehat{\mathcal{B}}_i[M] = \mathcal{F}_i[M_i][M] = \mathcal{F}_i[M], \tag{II.13}$$

so $\mathcal{F}_i[M] \subseteq \mathcal{F}_j[M]$. $\square$

Exercise II.3.24. Show that if (C_n) is an infinite sequence of clopen subsets of $\alpha + 1$ where α is an ordinal $< \omega^\omega$ then either
(a) (C_n) has an infinite subsequence (C_{n_k}) with nonempty intersection, or
(b) (C_n) has a subsequence (C_{n_k}) such that for some positive integer l, every intersection of at most l of C_{n_k} is nonempty but every intersection of at least $l + 1$ C_{n_k}'s is empty. (Hint: let $\mathcal{F} = \{\{n : \gamma \in C_n\} : \gamma \in \alpha + 1\}$, apply Theorem II.3.22 and argue that the resulting barrier $\mathcal{B}$ must be of some finite rank l).

There is another notion of heredity which is enjoyed by many uniform barriers. It rests on another ordering on $\mathbb{N}^{[<\infty]}$ reminiscent to the one appearing in the w.q.o. theory (see Theorem II.1.8 above): $s \preceq t$ iff the only strictly increasing map $\sigma : t \to s$ satisfies that $n \geq \sigma(n)$ for all $n \in t$. We say that a family $\mathcal{F}$ of finite subsets of $\mathbb{N}$ is *spreading* on M if $s \preceq t \subseteq M$ and $s \in \mathcal{F} \restriction M$ implies $t \in \mathcal{F}$.

This notion can be quite useful but unfortunately is not shared by closures of all uniform barriers as the following example shows.

Example II.3.25. For $n \in \mathbb{N}$, let $f_n : \mathbb{N} \to \mathbb{N}$ by

$$f_n(i) = \max\{1, i - n + 1\}.$$

For $n \in \mathbb{N}$, let

$$\mathcal{B} = \bigcup_{n=0}^{\infty} \mathcal{C}_n \oplus \{n\}.$$

Then $\overline{\mathcal{B}} \upharpoonright M$ is not a spreading family for any infinite set M: Fix $M = \{m_k\}_{k=0}^{\infty}$ increasing enumeration. Notice that $\mathcal{B} = \{s \in \mathrm{FIN} : |_*s| = f_{\min s}(\min(_*s))\}$. Let $s = \{m_0, m_2, \ldots, m_k\}$ and $t = \{m_1, m_2, \ldots, m_k\}$ where $k = f_{m_0}(m_2) = m_2 - m_0 + 1$. It is clear that $s \preceq t$ and $s \in \mathcal{B}$, but $t \notin \overline{\mathcal{B}}$ since if $u \sqsupseteq t$, then $|_*t| = k$, but $f_{m_1}(m_2) = m_2 - m_1 + 1 < k$.

To get spreading one needs to make a slight variation on Definition II.3.1.

Definition II.3.26. Let α be a countable ordinal and let M be an infinite subset of $\mathbb{N}$. We say that a family $\mathcal{B}$ of finite subsets of $\mathbb{N}$ is α-*extrauniform* on M provided that:
(a) $\alpha = 0$ implies $\mathcal{B} = \{0\}$,
(b) $\alpha = \beta + 1$ implies $\mathcal{B} = \mathcal{C} \oplus M^{[1]}$ for some β-extrauniform family $\mathcal{C}$ on M.
(c) $\alpha > 0$ limit implies that there is a strictly increasing sequence $(\alpha_m)_{m \in M}$ of ordinals converging to α such that $\mathcal{B}_{\{m\}}$ is α_m-extrauniform on M/m for all $m \in M$.

Proposition II.3.27. *Suppose that $\mathcal{B}$ is extrauniform on M. Then there is some infinite set $N \subseteq M$ such that $\overline{\mathcal{B}} \upharpoonright N$ is spreading in N.*

Proof. Suppose that $\mathcal{B}$ is α-extrauniform on M. The proof goes by induction on α. Suppose $\alpha = \beta + 1$. Then $\mathcal{B} = \mathcal{C} \oplus M^{[1]}$ for some β-extrauniform family $\mathcal{C}$ on M, so there is some $N \in M^{[\infty]}$ such that $\overline{\mathcal{C}} \upharpoonright N$ is spreading. Is it easy to see that $\overline{\mathcal{B}} \upharpoonright M$ is also spreading. Now suppose that α is limit. Then fix an increasing sequence of ordinals $(\alpha_m)_{m \in M}$ with limit α such that for every $m \in M$, $\mathcal{B}_{\{m\}}$ is α_m-extrauniform on M/m.

We find a strictly decreasing sequence $(M_k)_{k=0}^{\infty}$, of infinite subsets of M such that:
(a) The sequence $n_k = \min M_k$ is strictly increasing in k,
(b) $\overline{\mathcal{B}_{\{n_0\}}} \upharpoonright M_0$ is spreading,
(c) $\overline{\mathcal{B}_{\{n_{k+1}\}}} \upharpoonright M_{k+1}$ is spreading and $\mathcal{B}_{\{n_k\}} \upharpoonright M_{k+1} \subseteq \overline{\mathcal{B}_{\{n_{k+1}\}}} \upharpoonright M_{k+1}$.

Let $N = \{n_k\}_{k=0}^{\infty}$. We claim that $\overline{\mathcal{B}} \upharpoonright N$ is spreading on N: For suppose that $s \preceq t \subseteq N$, $s \in \mathcal{B} \upharpoonright N$, and set $n_k = \min s$, $n_l = \min t$. Since $\overline{\mathcal{B}_{\{n_k\}}} \upharpoonright N_k$ is spreading on N_k, and since $_*s \preceq _*t$ and $_*t \subseteq N_k$ we have that $_*t \in \overline{\mathcal{B}}_{n_k}$. If $k = l$, then $t = \{n_k\} \cup {}_*t \in \overline{\mathcal{B}}$; if $k < l$, then since $_*t \subseteq N_{k+1}$ and $\mathcal{B}_{\{n_k\}} \upharpoonright N_{k+1} \subseteq \overline{\mathcal{B}_{\{n_{k+1}\}}} \upharpoonright N_{k+1}$ we obtain that $_*t \in \overline{\mathcal{B}_{\{n_{k+1}\}}}$ and so $t \in \overline{\mathcal{B}}$. This finishes the proof. $\square$

Given a family $\mathcal{B}$ of finite sets of $\mathbb{N}$, let

$$_*\mathcal{B} = \{_*s \,:\, s \in \mathcal{B}\},$$

Note that if $\mathcal{B}$ is an uniform family on some set M and of some rank > 0, then $_*\mathcal{B}$ is also an uniform family on $_*M$.

Corollary II.3.28. *Let $\mathcal{B}$ be an α-uniform family on M for some ordinal $\alpha > 0$. Then there is an infinite subset $N \subseteq M$ such that the family $_*(\overline{\mathcal{B}} \upharpoonright N)$ is spreading on $_*N$.*

Proof. Choose an α-extrauniform family $\mathcal{C}$ on M. By Proposition II.3.27, Lemma II.3.12 and Corollary II.3.16 there is some infinite subset $N \subseteq M$ such that $\overline{\mathcal{C}} \upharpoonright N$ is spreading in N and either
(a) $\mathcal{B} \upharpoonright N \subseteq \overline{\mathcal{C}} \upharpoonright N \subseteq \overline{\mathcal{B}} \upharpoonright N \oplus N^{[\leq 1]}$ or
(b) $\mathcal{C} \upharpoonright N \subseteq \overline{\mathcal{B}} \upharpoonright N \subseteq \overline{\mathcal{C}} \upharpoonright N \oplus N^{[\leq 1]} \subseteq \overline{\mathcal{B}} \upharpoonright N \oplus N^{[\leq 1]}$.
Using this one easily proves that $_*(\overline{\mathcal{B}} \upharpoonright N)$ is spreading on $_*N$. $\qquad\square$

Corollary II.3.29. *For every pre-compact family $\mathcal{F}$ of finite subsets of $\mathbb{N}$ there is an infinite set M such that $_*\mathcal{F}[M]$ is spreading.*

Proof. Use the previous result and Theorem II.3.22. $\qquad\square$

We finish this section by mentioning two applications. The first requires the following definition.

Definition II.3.30. (a) For a compact hereditary family $\mathcal{B}$ of finite subsets of $\mathbb{N}$ and a real $\theta \in (0,1)$, let $T[\mathcal{B},\theta]$ be the corresponding Tsirelson-like space defined to be the completion of $c_{00}(\mathbb{N})$ under the norm defined implicitly by

$$\|x\| = \max\left\{ \|x\|_\infty, \sup_{(E_i)_{i=1}^n} \theta \sum_{i=1}^n \|E_i x\| \right\}$$

where in the last sup $(E_i)_{i=1}^n$ is running over all finite sets $E_1 < \cdots < E_n$ such that there is some $\{k_1 < \cdots < k_n\} \in \mathcal{B}$ such that $k_1 \leq E_1 < k_2 \leq E_2 < \cdots < k_n \leq E_n$.
(b) Given two infinite dimensional Banach spaces X and Y we write $X \hookrightarrow Y$ iff there is some closed subspace of Y isomorphic to X.

Given a compact hereditary family $\mathcal{B}$ and $\theta \in (0,1)$ one is interested in knowing whether the space $T[\mathcal{B},\theta]$ shares some of the properties of the Tsirelson space $T = T[\overline{\mathcal{S}},1/2]$ and in determining when two spaces of the form $T[\mathcal{B},\theta]$ contain isomorphic copies of a common infinite-dimensional Banach space. It turns out that answer to these questions requires a fine analysis of uniform fronts and barriers that naturally follows the one exposed above. Here is an example of a result whose proof uses this kind of analysis in an essential way.

Theorem II.3.31 ([51]). *Suppose that $\mathcal{B}, \mathcal{B}'$ are two compact, hereditary, and spreading families of finite subsets of $\mathbb{N}$ with Cantor–Bendixon ranks*

$$i(\mathcal{B}) = \omega^{\alpha_0} n_0 + \cdots + \omega^{\alpha_k} n_k \quad and \quad i(\mathcal{B}') = \omega^{\alpha'_0} n'_0 + \cdots + \omega^{\alpha'_k} n'_k,$$

respectively, written in their Cantor normal forms. Suppose that $\theta, \theta' \in (0, 1)$. Then the following conditions are equivalent:
(a) *There is $X \hookrightarrow T[\mathcal{B}, \theta], T[\mathcal{B}', \theta']$.*
(b) *For every $X \hookrightarrow T[\mathcal{B}, \theta]$ and $Y \hookrightarrow T[\mathcal{B}', \theta']$ there is $Z \hookrightarrow X, Y$.*
(c) *There is some integer l such that either*
    (i) *$\theta^l = \theta'$ and $\max\{\omega^{\alpha_0}, n_0\}^l = \max\{\omega^{\alpha'_0}, n'_0\}$, or*
    (ii) *$\theta'^l = \theta$ and $\max\{\omega^{\alpha'_0}, n'_0\}^l = \max\{\omega^{\alpha_0}, n_0\}$.*

To state the second application, let $\mathfrak{M}_{\text{FIN}}$ be the collection of all *finite convex means* on $\mathbb{N}$, i.e., finitely supported maps $\mu \colon \mathbb{N} \to [0, 1]$ such that

$$\sum_{n=0}^{\infty} \mu(n) = 1.$$

For $F \subseteq \mathbb{N}$ and $\mu \in \mathfrak{M}_{\text{FIN}}$, let $\mu(F) = \sum_{u \in F} \mu(F)$.

The following is a generalization of a result in [4] and [70] where a similar conclusion is reached for the Schreier hierarchy.

Lemma II.3.32. *Let α and β be countable ordinals such that β is bigger or equal to the minimal indecomposable ordinal $> \alpha$. Let $\mathcal{A}$ be an α-uniform barrier on some infinite set M and let $\mathcal{B}$ be a β-uniform barrier on some set $N \supseteq M$. Then for every $\varepsilon > 0$ there is $\mu \in \mathfrak{M}_{\text{FIN}}$ such that $\operatorname{supp}(\mu) \in \overline{\mathcal{B}}$ and $\mu(s) < \varepsilon$ for all $s \in \mathcal{A}$.*

Proof. The proof is by induction on α. Notice that the requirement on β is equivalent to saying that $\beta \geq \omega^{\alpha_0 + 1}$, where $\alpha = \omega^{\alpha_0} n_0 + \cdots + \omega^{\alpha_k} n_k$ is the Cantor's normal form of α. By Lemma II.3.12 it suffices to consider only the case of $\beta = \omega^{\alpha_0 + 1}$. There are two cases:
Case 1. Suppose first that $\alpha = \omega^{\alpha_0}$, $\beta = \omega^{\alpha_0 + 1}$. The proof is by induction on α_0. If $\alpha_0 = 0$, then $\mathcal{B}$ is an ω-uniform barrier on M, while $\mathcal{A}$ is 1-uniform on M, hence $\mathcal{A} = M^{[1]}$. Fix $s \in \mathcal{B} \upharpoonright M$ such that $\varepsilon \cdot |s| < 1$, and let $\mu = (1/|s|)\chi_s$, where χ_s is the characteristic function of s.

Suppose that $\alpha_0 > 0$. Fix a sequence $(\gamma_m)_{m \in M}$ such that $\gamma_m \uparrow_m \omega^{\alpha_0}$ and such that for every $m \in M$, $\mathcal{A}_{\{m\}}$ is γ_m-uniform on M/m. Fix also k such that $1/k < \varepsilon/2$. Let $N \in M^{[\infty]}$ be such that

$$\mathcal{A} \upharpoonright N \otimes N^{[k]} \subseteq \overline{\mathcal{B} \upharpoonright N}, \tag{II.14}$$

(this is possible since $\mathcal{A} \otimes M^{[k]}$ is $\omega^{\alpha_0} \cdot k$-uniform on M, while $\mathcal{B}$ is $\omega^{\alpha_0 + 1}$-uniform on M). Now let $t_1 \in \mathcal{A} \upharpoonright N$ and $\mu \in \mathfrak{M}_{\text{FIN}}$ be such that $\operatorname{supp} \mu \subseteq t_1$. Since for every $n \in N \cap [1, \max s_1]$, $(\mathcal{A} \upharpoonright N)_{\{n\}}$ is γ_n-uniform, and $\gamma_n < \omega^{\alpha_0}$, by inductive hypothesis (for the pair $\gamma_n < \omega^{\alpha_0}$) we can find $\mu_2 \in \mathfrak{M}_{\text{FIN}}$ such that $\operatorname{supp} \mu_2 \subseteq t_2 \in \mathcal{A} \upharpoonright (N/n_1)$, and such that for every $s \in \mathcal{A} \upharpoonright N$ with $\min s \leq \max t_1$, we obtain that $\mu_2(s) < 1/2$.

In general we can find $\mu_1, \ldots, \mu_k \in \mathfrak{M}_{\text{FIN}}$ in such a way that:
(a) For every $i = 2, \ldots, k$, $\mu_i \subseteq t_i \in \mathcal{A} \upharpoonright (N/(\max t_{i-1}))$, and
(b) for every $i = 2, \ldots, k$, and every $s \in \mathcal{A} \upharpoonright N$ such that $\min s \leq \max t_{i-1}$, $\mu_i(s) \leq 1/2^{i-1}$.

Let $\mu = (1/k) \sum_{i=1}^{k} \mu_i$. It is clear that $\mu \in \mathfrak{M}_{\text{FIN}}$, and that

$$\operatorname{supp} \mu = \bigcup_{i=1}^{k} \operatorname{supp} \mu_i \subseteq \bigcup_{i=1}^{k} t_i \in \mathcal{A} \upharpoonright N \otimes N^{[k]} \subseteq \overline{\mathcal{B}}. \tag{II.15}$$

Now fix $s \in \mathcal{A}$. Let $\bar{s} \in \mathcal{A} \upharpoonright N$ be such that $s \cap N \subseteq \bar{s}$. (This is possible: Since $\mathcal{A} \upharpoonright N$ is a barrier on N, we may find $\bar{s} \in \mathcal{A} \upharpoonright N$ such that either $\bar{s} \subsetneq s \cap N$ or $s \cap N \subseteq \bar{s}$; the first alternative is not possible since it implies that $\bar{s} \subsetneq s$ and both are elements of the barrier $\mathcal{A}$.) So we have that

$$\mu(s) = \mu(s \cap N) \leq \mu(\bar{s}) = \frac{1}{k} \left(\mu_{i_0}(\bar{s}) + \cdots + \mu_k(\bar{s}) \right) \leq$$

$$\leq \frac{1}{k} \left(1 + \frac{1}{2^{i_0}} + \cdots + \frac{1}{2^{k-1}} \right) \leq \frac{2}{k} < \varepsilon. \tag{II.16}$$

Case 2. General case $\alpha = \omega^{\alpha_0} n_0 + \cdots + \omega^{\alpha_k} n_k$. Let $\mathcal{C}$ be an ω^{α_0}-uniform barrier on M. Then $\mathcal{C} \otimes M^{[n_0+1]}$ is $\omega^{\alpha_0}(n_0+1)$-uniform on M. Since $\omega^{\alpha_0}(n_0+1) > \alpha$, by Proposition II.3.16 there is some $N \in M^{[\infty]}$ such that $\mathcal{A} \upharpoonright N \subseteq \overline{\mathcal{C} \upharpoonright N \otimes N^{[n_0+1]}}$. Now fix $\varepsilon > 0$. Then there is some $\mu \in \mathfrak{M}_{\text{FIN}}$ with support in $\overline{\mathcal{B} \upharpoonright N}$ and such that for every $s \in \mathcal{C} \upharpoonright N$, $\mu(s) < \varepsilon/(n_0+1)$. We work to show that $\mu(s) < \varepsilon$ for every $s \in \mathcal{A}$. Fix $s \in \mathcal{A}$. Since $\operatorname{supp} \mu \subseteq N$, we have that $\mu(s) = \mu(s \cap N)$. Again we can find $\bar{s} \in \mathcal{A} \upharpoonright N$ such that $s \cap N \subseteq \bar{s}$, and since $\mu(s) = \mu(s \cap N) \leq \mu(\bar{s})$, we may assume that $s \in \mathcal{A} \upharpoonright N$. Now using that $\mathcal{A} \upharpoonright N \subseteq \overline{\mathcal{C} \upharpoonright N \otimes N^{[n_0+1]}}$, we can find $s_1 < \cdots < s_{n_0+1}$, $s_i \in \mathcal{C} \upharpoonright N$, such that $s \subseteq s_1 \cup \cdots \cup s_{n_0+1}$. Hence,

$$\mu(s) \leq \sum_{i=1}^{n_0+1} \mu(s_i) < \frac{\varepsilon}{n_0+1}(n_0+1) = \varepsilon, \tag{II.17}$$

as desired. $\square$

Remark II.3.33. One might wonder if the convex mean μ satisfying conclusion of Lemma II.3.32 can have a simpler form in comparison to the one obtained in the previous proof. For example, can one produce such mean μ of the form $\mu_t = (1/|t|)\chi_t$? It turns out that such a simple form of μ cannot be found if, for example, $\mathcal{A}$ is equal to the Schreier barrier $\{s \in \mathbb{N}^{[<\infty]} : |s| = \min s + 1\}$. Note that in this case for every $t \in \mathbb{N}^{[<\infty]}$ there is some $s \in \mathcal{A}$ such that $|s \cap t| \geq (1/2)|t|$ or in other words $\mu_t(s) \geq 1/2$.

Corollary II.3.34 (Pták). *For every pre-compact family $\mathcal{F}$ of finite subsets of $\mathbb{N}$ and every $\varepsilon > 0$ there is $\mu \in \mathfrak{M}_{\mathrm{FIN}}$ such that $\mu(s) < \varepsilon$ for all $s \in \mathcal{F}$.*

Proof. This follows from Theorem II.3.22 and Lemma II.3.32. □

Corollary II.3.35 (Mazur). *For every bounded weakly null sequence (x_n) in some Banach space X and every $\varepsilon > 0$ there is a finite convex combination $y = \sum_{i=1}^{k} \lambda_i x_{n_i}$ such that $\|y\| \leq \varepsilon$.*

Proof. We may assume that $\|x_n\| \leq 1$ for all n. For $x^* \in B_{X^*}$, let

$$F_{x^*} = \{n \in \mathbb{N} : |x^*(x_n)| \geq \varepsilon/2\}.$$

Let $\mathcal{F} = \{F_{x^*} : x^* \in B_{X^*}\}$. Note that $\mathcal{F}$ is a pre-compact family of finite subsets of $\mathbb{N}$. By Corollary II.3.34 there is $\mu \in \mathfrak{M}_{\mathrm{FIN}}$ such that $\mu(F) < \varepsilon/2$ for all $F \in \mathcal{F}$. Let $G = \mathrm{supp}(\mu)$ and

$$y = \sum_{n \in G} \mu(n) x_n.$$

Then y is the required convex combination. To see this consider an $x^* \in B_{X^*}$. Then

$$|x^*(y)| = \Big| \sum_{n \in F_{x^*} \cap G} \mu(n) x^*(x_n) + \sum_{n \in G \setminus F_{x^*}} \mu(n) x^*(x_n) \Big|$$
$$\leq \sum_{n \in F_{x^*} \cap G} \mu(n) |x^*(x_n)| + \sum_{n \in G \setminus F_{x^*}} \mu(n) |x^*(x_n)|$$
$$< \mu(F_x) + \mu(G \setminus F_{x^*}) \dot{\varepsilon}/2 < \varepsilon/2 + \varepsilon/2 = \varepsilon.$$

It follows that $\|y\| \leq \varepsilon$. □

Remark II.3.36. (a) There is also a direct way from Mazur's theorem to Ptak's lemma. To see this, fix a pre-compact family $\mathcal{F}$ of finite subsets of $\mathbb{N}$ and $\varepsilon > 0$. Define an infinite sequence $(f_n)_n \subseteq C(\overline{\mathcal{F}})$ as follows:

$$f_n(s) = \begin{cases} 0 & \text{if } n \notin s \\ 1 & \text{if } n \in s. \end{cases}$$

Then (f_n) is a weakly null sequence in the Banach space $(C(\overline{\mathcal{F}}), \|\cdot\|_\infty)$. By Mazur's theorem there exist a finite sequence $n_1 < \cdots < n_k$ of integers and a sequence $(\lambda_i)_{i=1}^{k} \subseteq [0,1]$ such that

$$\|\sum_{i=1}^{k} \lambda_i f_{n_i}\|_\infty < \varepsilon. \tag{II.18}$$

Define $\mu = \sum_{i=1}^{k} \lambda_i \chi_{\{n_i\}} \in \mathfrak{M}_{\mathrm{FIN}}$. Then $\mu(s) < \varepsilon$ for all $s \in \mathcal{F}$ as required for the conclusion of Ptak's Lemma.

(b) Note that Schreier space $\mathcal{S}$ introduced in Example I.2.13 has the property that every average $(1/|s| \sum_{i \in s} e_i)$ of its natural basis (e_n) has norm at least $1/4$. For Tsirelson space T (see Example I.2.14) something stronger is true: Every weakly null sequence (x_n) of T has a subsequence (x_{n_k}) such that all averages $(1/|s|\sum_{i\in s} x_{n_i})$ have norm at least $1/4$.

(c) If a normalized basic sequence (x_n) has ℓ_1 as spreading model, then every average on (x_n) has norm far away from zero. Note that the natural basis of $\mathcal{S}$, and every normalized block sequence (x_n) of T have ℓ_1 as a spreading model. Rosenthal's Theorem II.9.8 below clarifies the situation.

II.4 Canonical Equivalence Relations on Uniform Fronts and Barriers

The purpose of this section is to examine what are the essential equivalence relation that one can put on an uniform barrier $\mathcal{B}$ modulo the possibility of going to an arbitrary restriction of the form $\mathcal{B} \restriction M$ for M an infinite subset of $\mathbb{N}$. It will be instructive to first understand the canonical equivalence relations on barriers $M^{[k]}$ of finite rank. The following equivalence relations on $\mathbb{N}^{[k]}$ suggest themselves.

Example II.4.1. For a subset $I \subseteq \{0, \ldots, k-1\}$ define the following equivalence relation E_I on the uniform barrier $\mathbb{N}^{[k]}$:

$$\{m_0, \ldots, m_{k-1}\}_< E_I \{n_0, \ldots, n_{k-1}\}_< \text{ iff } m_i = n_i \text{ for all } i \in I.$$

The following classical result shows that E_I for $I \subset \{0, \ldots, k-1\}$ exhaust the list of all canonical equivalence relations on $\mathbb{N}^{[k]}$.

Theorem II.4.2 (Erdös–Rado). *For every integer $k \geq 1$ and every equivalence relation E on $\mathbb{N}^{[k]}$ there is an infinite $M \subseteq \mathbb{N}$ and $I \subseteq \{0, \ldots, k-1\}$ such that*

$$E \restriction M^{[k]} = E_I \restriction M^{[k]}$$

Proof. The proof is by induction on k. The case $k = 1$ is clear. So assume $k > 1$ and we have an equivalence relation E on $\mathbb{N}^{[k]}$. Let $\mathcal{E}$ be the collection of all equivalence relations on $\{0, 1, \ldots, 2d+1\}^{[k]}$. Define $c\colon \mathbb{N}^{[2k+2]} \longrightarrow \mathcal{E}$ by letting $c(S)$ the copy of the equivalence relation $E \restriction (S^{[k]})$ under the order preserving map from s onto $\{0, 1, \ldots, 2d+1\}$. By Ramsey's theorem there is an infinite $M \subseteq \mathbb{N}$ such that $c \restriction M^{[2k+2]}$ is constantly equal to some $E_0 \in \mathcal{E}$. If $E \restriction M^{[k]}$ is the equality relation then we are done. So assume that there exist $s \neq t$ in $M^{[k]}$ such that $s \, E \, t$. Let $s = \{m_0, m_1, \ldots, m_{k-1}\}$, $t = \{n_0, n_1, \ldots, n_{k-1}\}$ and let $j < k$ be minimal i such that $m_i \neq n_i$. It follows that the equation $x \, E \, y$ for $x, y \in M^{[k]}$ does not depend on the jth member of x and the jth member of y, so we are done by the inductive hypothesis. $\square$

Corollary II.4.3. *For every regressive[11] $f\colon \mathbb{N}^{[k]} \longrightarrow \mathbb{N}$ there is an infinite $M \subseteq \mathbb{N}$ such that $f(s) = f(t)$ for every pair $s, t \in M^{[k]}$ with $\min(s) = \min(t)$.*

---

[11]i.e., $f(s) < \min(s)$ for all $s \in \mathbb{N}^{[k]}$ with $\min(s) \neq 0$.

Remark II.4.4. The finite form of this result (which follows from it via a simple compactness argument) reads: For every k and m there is n such that for every regressive $f \colon n^{[k]} \longrightarrow n$ there is a min-homogeneous $X \subseteq n$ of size m. It turns out that this result is not provable in Peano arithmetic.

Corollary II.4.5 (Ramsey). *For every integer $k \geq 1$ and every equivalence relation E with finitely many classes there exists an infinite $M \subseteq \mathbb{N}$ such that $s \mathbin{E} t$ for all $s, t \in M^{[k]}$.* □

Exercise II.4.6. Prove the finite version of the Erdös–Rado canonization theorem by working just in PA.

Definition II.4.7. An equivalence relation E defined on some uniform barrier $\mathcal{B}$ is *canonical* if there is an uniform barrier $\mathcal{C}$ (not necessarily on the same domain as $\mathcal{B}$) and a mapping $f \colon \mathcal{B} \longrightarrow \mathcal{C}$ such that
(a) $f(s) \subseteq s$ for all $s \in \mathcal{B}$
(b) for $s, t \in \mathcal{B}$, $s \mathbin{E} t$ iff $f(s) = f(t)$.

Theorem II.4.8 (Pudlak–Rödl). *For every equivalence relation E defined on some (uniform) barrier $\mathcal{B}$ there is an infinite set $M \subseteq \mathbb{N}$ such that $E \restriction (\mathcal{B} \restriction M)$ is canonical.*

Typical equivalence relations are defined using a map $h \colon \mathcal{B} \to \mathbb{N}$ by $s \mathbin{E_h} t$ iff $h(s) = h(t)$. So let us examine behaviors of such maps $h \colon \mathcal{B} \longrightarrow \mathbb{N}$.

Lemma II.4.9. *Suppose $\mathcal{B}$ is an uniform barrier on some infinite set $M \subseteq \mathbb{N}$ and that $h \colon \mathcal{B} \longrightarrow \mathbb{N}$ is such that $h(s) \notin s$ for all $s \in \mathcal{B}$. Then there is infinite $N \subseteq M$ such that $(h[\mathcal{B} \restriction N]) \cap N = \emptyset$.*

Proof. Let $\mathcal{B}$ be α-uniform for some α. The proof is by induction on α. □

Lemma II.4.10. *Suppose $h_0 \colon \mathcal{B}_0 \longrightarrow \mathbb{N}$ and $h_1 \colon \mathcal{B}_1 \longrightarrow \mathbb{N}$ are 1-1 mappings defined on two uniform barriers $\mathcal{B}_0$ and $\mathcal{B}_1$ on the same domain $M \subseteq \mathbb{N}$. Then there is an infinite $N \subseteq M$ such that either*
(1) $\mathcal{B}_0 \restriction N = \mathcal{B}_1 \restriction N$ *and* $h_0 \restriction (\mathcal{B}_0 \restriction N) = h_1 \restriction (\mathcal{B}_1 \restriction N)$, *or*
(2) $h_0[\mathcal{B}_0 \restriction N] \cap h_1[\mathcal{B}_1 \restriction N] = \emptyset$.

Proof. Define a partition $\mathcal{B}_0 = \mathcal{B}_0^{(1)} \cup \mathcal{B}_0^{(2)}$ by letting

$$s \in \mathcal{B}_0^{(1)} \text{ iff } s \in \mathcal{B}_1 \text{ and } h_0(s) = h_1(s).$$

By the Ramsey property of $\mathcal{B}_0$ we can find an infinite $N \subseteq M$ such that $\mathcal{B}_0 \restriction N \subseteq \mathcal{B}_0^{(1)}$ or $(\mathcal{B}_0 \restriction N) \cap \mathcal{B}_0^{(1)} = \emptyset$. In the first case we have $\mathcal{B}_0 \restriction N \subseteq \mathcal{B}_1 \restriction N$ which by maximality of $\mathcal{B}_0 \restriction N$ means $\mathcal{B}_0 \restriction N = \mathcal{B}_1 \restriction N$ and so we have the conclusion (1).

So suppose that for every $s \in \mathcal{B}_0 \restriction N$ we have that either $s \notin \mathcal{B}_1$ or $s \in \mathcal{B}_1$ but $h_0(s) \neq h_1(s)$. Define now

$$g_0 \colon \mathcal{B}_0 \restriction N \longrightarrow \mathbb{N} \text{ and } g_1 \colon \mathcal{B}_0 \restriction N \longrightarrow \mathbb{N}$$

as follows: given $s \in \mathcal{B}_0 \upharpoonright N$, if there is $t \in \mathcal{B}_1$ such that $h(s) = h_1(t)$ and if the unique t has the property that $t \setminus s \neq \emptyset$, let $g_0(s) = \min(t \setminus s)$; otherwise $g_0(s) = 0$. The function g_1 is defined symmetrically, i.e., $g_1(s) \neq 0$ implies that $g_1(s) \in s \setminus t$ for the unique $t \in \mathcal{B}_1$ such that $h_0(s) = h_1(t)$. By the previous Lemma there exists an infinite $P \subseteq N$ such that

$$g_i [\mathcal{B}_i \upharpoonright P] \cap P = \emptyset \text{ for } i = 0, 1.$$

Then

$$h_0 [\mathcal{B}_0 \upharpoonright P] \cap h_1 [\mathcal{B}_1 \upharpoonright P] = \emptyset. \qquad \square$$

Corollary II.4.11. *Suppose $h_0 \colon \mathcal{B}_0 \longrightarrow \mathbb{N}$ and $h_1 \colon \mathcal{B}_1 \longrightarrow \mathbb{N}$ are 1-1 mappings defined on uniform barriers acting on the same set M. Suppose $\mathcal{B}_0$ is α_0-uniform, $\mathcal{B}_1$ is α_1-uniform, and $\alpha_0 \neq \alpha_1$. Then there is an infinite $N \subseteq M$ such that*

$$h_0 [\mathcal{B}_0 \upharpoonright N] \cap h_1 [\mathcal{B}_1 \upharpoonright N] = \emptyset.$$

Proof of Theorem II.4.8. Suppose $\mathcal{B}$ is α-uniform on $\mathbb{N}$. The proof is by induction on this α. If $\alpha = 0$, the conclusion is clear, so assume $\alpha > 0$. For $n \in \mathbb{N}$, define an equivalence relation E_n on $\mathcal{B}_{\{n\}}$ by letting

$$s E_n t \text{ iff } (\{n\} \cup s) E (\{n\} \cup t).$$

Using the inductive hypothesis and diagonalizing we get uniform barriers $\mathcal{C}_n$ $(n \in \mathbb{N})$ and mappings

$$f_n \colon \mathcal{B}_{\{n\}} \xrightarrow{1-1} \mathcal{C}_n$$

such that
$(a)_n$ $f_n(s) \subseteq s$ for all $s \in \mathcal{B}_{\{n\}}$
$(b)_n$ for $s, t \in \mathcal{B}_{\{n\}}$, $s E_n t$ iff $f_n(s) = f_n(t)$.
    For each $n \in \mathbb{N}$ let

$$\psi_n \colon \mathcal{C}_n \longrightarrow \mathcal{B}_{\{n\}}/E_n$$

be the mapping such that for $v \in \mathcal{C}_n$,

$$f_n(s) = v \text{ for all } s \in \psi_n(v).$$

A simple diagonalization argument using Lemma II.4.10 will give us an infinite set $M \subseteq \mathbb{N}$ such that for all $m < n$ in M, either
(1) $\mathcal{C}_m \upharpoonright (M/n) = \mathcal{C}_n \upharpoonright (M/n)$ and $\psi_m \upharpoonright (\mathcal{C}_m \upharpoonright (M/n)) = \psi_n \upharpoonright (\mathcal{C}_n \upharpoonright (M/n))$.
(2) $\psi_m [\mathcal{C}_m \upharpoonright (M/n)] \cap \psi_n [\mathcal{C}_m \upharpoonright (M/n)] = \emptyset$.
    Applying Ramsey's theorem we have the following two cases to consider:
Case 1: For all $n < m$ in M,

$$\mathcal{C}_m \upharpoonright (M/n) = \mathcal{C}_n \upharpoonright (M/n)$$

and

$$\psi_m \upharpoonright (\mathcal{C}_m \upharpoonright (M/n)) = \psi_n \upharpoonright (\mathcal{C}_n \upharpoonright (M/n)).$$

Let $\mathcal{C} = \mathcal{C}_{\min(M)}$, $N = M \setminus \{\min(M)\}$, and let

$$f: \mathcal{B} \restriction N \longrightarrow \mathcal{C}$$

be defined by

$$f(s) = f_{\min(s)}(s \setminus \{\min(s)\}).$$

Then f canonizes $E \restriction (\mathcal{B} \restriction N)$.

Case 2: $\psi_m [\mathcal{C}_m \restriction (M/n)] \cap \psi_n [\mathcal{C}_m \restriction (M/n)] = \emptyset$.
For $m < n$ in M and $u \subseteq M \cap (m, n]$, let

$$\mathcal{C}_{m(u)n} = \{v : n < \min(v) \text{ and } u \cup v \in \mathcal{C}_m\}$$

and define

$$\psi_{m(u)n}: \mathcal{C}_{m(u)n} \longrightarrow \mathcal{B}_{\{n\}}/E_n$$

by

$$\psi_{m(u)n}(v) = \psi_n(u \cup v).$$

Shrinking M, we may assume that

$$\mathcal{C}_{m(u)n} \text{ is uniform on } M/n$$

for all $m < n$ and $u \subseteq (m, n] \cap M$. Performing another diagonalization procedure using Lemma II.4.10 we get an infinite set $N \subseteq M$ such that for all $m < n$ in N and $u \subseteq N \cap (m, n]$, either

(3) $\mathcal{C}_{m(u)n} \restriction (N/n) = \mathcal{C}_n \restriction (N/n)$ & $\psi_{m(u)n} \restriction (\mathcal{C}_{m(u)n} \restriction (N/n)) = \psi_n \restriction (\mathcal{C}_n \restriction (N/n))$, or

(4) $\psi_{m(u)n} [\mathcal{C}_{m(u)n} \restriction (N/n)] \cap \psi_n [\mathcal{C}_n \restriction (N/n)] = \emptyset$.
　　Let

$$\mathcal{C} = \bigcup_{n \in N} \{\{n\} \cup t : t \in \mathcal{C}_n \restriction (N/n)\}$$

and define $f: \mathcal{B} \restriction N \longrightarrow \mathcal{C}$ by

$$f(s) = \{\min(s)\} \cup f_{\min(s)}(s \setminus \{\min(s)\}).$$

It is now routine to check that f canonizes $\mathcal{B} \restriction N$.　　　　　□

　　We finish this section with the following fact which shows that there is essentially only one solution to the canonization problem for equivalence relations defined on uniform barriers.

Theorem II.4.12. *Suppose E is a canonical equivalence relation defined on some uniform barrier $\mathcal{B}$ and witnessing that this is so by two pair $(f_0, \mathcal{C}_0)$ and $(f_1, \mathcal{C}_0)$ as in Definition II.4.7. Then there is an infinite $M \subseteq \mathbb{N}$ such that*

$$\mathcal{C}_0 \restriction M = \mathcal{C}_1 \restriction M \text{ and } f_0 \restriction (\mathcal{B} \restriction M) = f_1 \restriction (\mathcal{B} \restriction M).$$

Proof. For $i < 2$ let $\varphi_i \colon \mathcal{C}_i \to \mathcal{B}/E$ be such that for all $s \in \mathcal{C}_i$,

$$f_i(t) = s \text{ for all } t \in \varphi_i(s).$$

Apply now Lemma II.4.10 and note that the alternative (2) is impossible since

$$\varphi_0[\mathcal{C}_0 \upharpoonright M] = (\mathcal{B} \upharpoonright M)/E = \varphi_1[\mathcal{C}_1 \upharpoonright M].$$

The alternative II.4.10 (1) gives us the required conclusion. $\qquad\square$

Exercise II.4.13. Deduce Theorem II.4.2 from Theorem II.4.8.

II.5 Unconditional Subsequences of Weakly Null Sequences

Recall the notion of an *unconditional basic sequence* (x_n) in some Banach space X, a basic sequence for which we one can find a constant C such that

$$\left\| \sum_{n=0}^{k} \varepsilon_n a_n x_n \right\| \leq C \left\| \sum_{n=0}^{k} a_n x_n \right\|$$

for every sequence $(a_n)_{n=0}^{k}$ of scalars and every sequence $(e_n)_{n=0}^{k}$ of signs. Note that every separable Banach space is spanned by an unconditional basic sequence. In fact there are separable Banach spaces that are not even embeddable into one with unconditional basic sequences. Thus as in the case of Schauder basic sequences one is left to analyzing the possibilities of having unconditional basic subsequences. It turns out that weakly null sequences (x_n) in Banach spaces do sometimes have basic unconditional subsequences (x_{n_k}). First of all, note that they do always have Schauder basic subsequences:

Lemma II.5.1. *Every normalized weakly null sequence in some Banach space X has a subsequence (x_{n_k}) that is a Schauder basis of its closed linear span.*

Proof. We may assume that X itself has a Schauder basis (e_i). Renorming X we may assume that the basis constant of (e_i) is 1. Choose an $\varepsilon > 0$. Let P_j be the sequence of projections on finite initial segments relative to the basis (e_i). Put $n_0 = 0$ and choose j_0 such that $\|x_0 - P_{j_0}(x_0)\| < \varepsilon/2$. Since $\lim_{n \to \infty} e_i^*(x_n) = 0$ for all i, we can find n_1 and $j_1 > j_0$ such that

$$\|P_{j_0}(x_{n_1})\| + \|x_{n_1} - P_{j_1}(x_{n_1})\| < \frac{\varepsilon}{4},$$

and so on, at stage k we can find $n_k > n_{k-1}$ and $j_k > j_{k-1}$ such that

$$\|P_{j_{k-1}}(x_{n_k})\| + \|x_{n_k} - P_{j_k}(x_{n_k})\| < \frac{\varepsilon}{2^{k+1}}.$$

Proceeding in this way one obtains a subsequence (x_{n_k}) that is a Schauder basis with constant at most $1 + 2\varepsilon$. $\qquad\square$

We shall now see that the problem whether a given weakly null sequence (x_n) has an unconditional basic sequence may depend on the nature of the vectors x_n as well as the Banach space X in which we are working.

Lemma II.5.2 (Rosenthal). *Suppose (x_n) is a sequence of characteristic functions of subsets of some set Γ and that (x_n) is weakly null inside the Banach space $\ell_\infty(\Gamma)$. Then (x_n) has an unconditional basic subsequence.*

Proof. For $\gamma \in \Gamma$, let

$$F_\gamma = \{n \in \mathbb{N} : x_n(\gamma) = 1\}.$$

Let $\mathcal{F} = \{F_\gamma : \gamma \in \Gamma\}$. Then $\mathcal{F}$ is a pre-compact family of finite subsets on $\mathbb{N}$, i.e., the pointwise closure $\overline{\mathcal{F}}$ of $\mathcal{F}$ in $2^{\mathbb{N}}$ consists only of finite sets. By Theorem II.3.22 there is an infinite set $M \subseteq \mathbb{N}$ such that $F[M] = \overline{\mathcal{B}}$ for some uniform barrier $\mathcal{B}$ on M. We claim that the subsequence $(x_m)_{n \in M}$ is 2-unconditional. To see this let F be a finite initial segment of M and let a_n $(n \in F)$ be a given sequence of scalars from $[-1, 1]$ and let ε_n $(n \in F)$ be a given sequence of signs. Fix $\gamma \in \Gamma$. Then

$$\left| \sum_{n \in F} \varepsilon_n a_n x_n(\gamma) \right| = \left| \sum_{n \in F \cap F_\gamma} \varepsilon_n a_n \right|.$$

Let

$$G = \{n \in F \cap F_\gamma : \varepsilon_n = 1\} \text{ and } H = \{n \in F \cap F_\gamma : \varepsilon_n = -1\}.$$

Then we can find α and $\beta \in \Gamma$ such that

$$G = F \cap F_\alpha \text{ and } H = F \cap F_\beta.$$

Then

$$\left| \sum_{n \in F \cap F_\gamma} \varepsilon_n a_n \right| \leq \left| \sum_{n \in F \cap F_\alpha} a_n \right| + \left| \sum_{n \in F \cap F_\beta} a_n \right|$$

$$= \left| \sum_{n \in F} a_n x_n(\alpha) \right| + \left| \sum_{n \in F} a_n x_n(\beta) \right|$$

$$\leq 2 \left\| \sum_{n \in F} a_n x_n \right\|_\infty$$

It follows that $\left\| \sum_{n \in F} \varepsilon_n a_n x_n \right\|_\infty \leq 2 \left\| \sum_{n \in F} a_n x_n \right\|_\infty$, as required. $\qquad\square$

Note that the proof establishes the following slightly more general fact.

Lemma II.5.3. *Suppose that (x_n) is a semi-normalized weakly null sequence in some Banach space of the form $\ell_\infty(\Gamma)$ such that there is some finite set V of positive real numbers such that the range of x_n is contained in V, for every n. Then (x_n) has an unconditional basic sequence.*

Proof. For every $\gamma \in \Gamma$, let $F_\gamma = \{n \in \mathbb{N} : x_n(\gamma) \neq 0\}$. Let $\mathcal{F} = \{F_\gamma : \gamma \in \Gamma\}$, which is a pre-compact set. Then there is some infinite set $M \subseteq \mathbb{N}$ such that $\mathcal{F}[M] = \overline{\mathcal{B}}$ for some uniform barrier $\mathcal{B}$ on M. We claim that the subsequence

$(x_m)_{n \in M}$ is $4(\max V / \min V)$-unconditional. To see this we are going to show that for every subset $N \subseteq M$ and a sequence of scalars $(a_m)_{m \in M}$

$$\| \sum_{n \in N} a_n x_n \|_\infty \le 2 \frac{\max V}{\min V} \| \sum_{m \in M} a_m x_m \|_\infty. \tag{II.19}$$

Fix $s \subseteq I \sqsubset M$, and $\gamma \in \Gamma$. Set $s^+ = \{n \in s \cap F_\gamma : a_n > 0\}$, and $s^- = \{n \in s \cap F_\gamma : a_n < 0\}$. Then

$$|\sum_{n \in s} a_n x_n(\gamma)| = \sum_{n \in s^+} a_n x_n(\gamma) - \sum_{n \in s^-} |a_n| x_n(\gamma). \tag{II.20}$$

We can find now $\alpha, \beta \in \Gamma$ such that $s^+ = F_\alpha \cap I$ and $s^- = F_\beta \cap I$. Then $\sum_{n \in I} a_n x_n(\alpha) = \sum_{n \in s^+} a_n x_n(\alpha)$ and $\sum_{n \in I} a_n x_n(\beta) = \sum_{n \in s^+} a_n x_n(\beta)$. So,

$$|\sum_{n \in s} a_n x_n(\gamma)| = \sum_{n \in s^+} a_n x_n(\gamma) - \sum_{n \in s^-} |a_n| x_n(\gamma) \le$$

$$\le \frac{\max V}{\min V} \sum_{n \in I} a_n x_n(\alpha) - \frac{\min V}{\max V} \sum_{n \in I} |a_n| x_n(\beta) =$$

$$= \frac{\max V}{\min V} \sum_{n \in I} a_n x_n(\alpha) + \frac{\min V}{\max V} \sum_{n \in I} a_n x_n(\beta) \le$$

$$\le \frac{\max V}{\min V} \left| \sum_{n \in I} a_n x_n(\alpha) \right| + \frac{\min V}{\max V} \left| \sum_{n \in I} a_n x_n(\beta) \right| \le \tag{II.21}$$

$$\le 2 \frac{\max V}{\min V} \| \sum_{n \in I} a_n x_n \|_\infty \tag{II.22}$$

$\square$

Let us now extend Lemma II.5.3 to the case when the finite set V has both positive and negative values. To state the corresponding combinatorial result we need a piece of notation. For a given set V, let

$$\mathrm{Fn}(\mathbb{N}, V)$$

be the set of functions f with domain a finite subset of $\mathbb{N}$ and range included in V, i.e. $f : \mathrm{dom} f \to V$, where $\mathrm{dom} f$ is a finite subset of $\mathbb{N}$. From now on we fix a finite set $V \subseteq \mathbb{R}$. Every map

$$\Phi : \mathcal{B} \to \mathrm{Fn}(\mathbb{N}, V)$$

considered below will satisfy that $\mathrm{dom} \Phi(s) = s$ for every $s \in \mathcal{B}$.

Lemma II.5.4. *Suppose that $\mathcal{B}$ is a uniform barrier on M and that $\Phi : \mathcal{B} \to \mathrm{Fn}(\mathbb{N}, V)$. Then there is some infinite set $N \subseteq M$ such that $\Phi \upharpoonright (\mathcal{B} \upharpoonright N)$ is 1-Lipschitz, i.e., for every $s, t \in \mathcal{B} \upharpoonright N$ if $s \cap t \sqsubseteq s, t$ then $\Phi(s) \upharpoonright (s \cap t) = \Phi(t) \upharpoonright (s \cap t)$.*

Proof. We find a decreasing chain of infinite sets (M_i) such that, if we set $m_i = \min M_i$, then $m_i < M_{i+1}$ and for every i and every $u \subseteq \{n_0, ..., n_i\}$, if $s, t \in \mathcal{B}_{\{u\}}$ then $\Phi(u \cup s) \restriction u = \Phi(u \cup t) \restriction u$. Then it is clear that the desired result will hold for $N = \{m_i\}$. So, suppose we have defined $(M_j)_{j \leq i}$. For a fixed $u \subseteq \{n_0, ..., n_i\}$ consider the finite coloring $c_u : \mathcal{B}_u \to {}^uV$ defined for $s \in \mathcal{B}_u$ by $c_u(s) = \Phi(u \cup s) \restriction u$. Then there will always be some $P \subseteq M_i/m_i$ such that c_u is monochromatic on $\mathcal{B}_u \restriction P$. Since there are only finitely many $u \subseteq \{n_0, ..., n_i\}$ we can find a single infinite set $M_{i+1} \subseteq M_i/m_i$ such that for every $u \subseteq \{n_0, ..., n_i\}$, the coloring c_u is constant on $\mathcal{B}_u \restriction M_{i+1}$. $\qquad\square$

Remark II.5.5. (a) Whenever $\Phi : \mathcal{B} \to \mathrm{Fn}(\mathbb{N}, V)$ is 1-Lipschitz then Φ has a natural extension $\Phi : \widehat{\mathcal{B}} \to \mathrm{Fn}(\mathbb{N}, V)$ defined for $t \in \widehat{\mathcal{B}}$ we set $\Phi(t) = \Phi(s) \restriction t$, where $s \in \mathcal{B}$ is arbitrary member of $\mathcal{B}$ such that $t \sqsubseteq s$.
(b) In this context the following notation is going to be useful. Fix $\Phi : \mathcal{B} \to \mathrm{Fn}(\mathbb{N}, V)$, and $m \in M$. The corresponding section-mapping $\Phi_m : \mathcal{B}_{\{m\}} \to \mathrm{Fn}(\mathbb{N}, V)$ is defined by

$$\Phi_m(s) = \Phi(\{m\} \cup s) \restriction s,$$

where $s \in \mathcal{B}_{\{m\}}$. Now for $v \in V$, we define $\varphi^v : \mathcal{B} \to \mathbb{N}^{[<\infty]}$ by

$$\varphi^v(s) = \Phi(s)^{-1}\{v\} = \{n \in s : \Phi(s)(n) = v\}. \tag{II.23}$$

The corresponding section-mappings, denoted by φ_m^v, are defined by

$$\varphi_m^v(s) = \{n \in s : \Phi(\{m\} \cup s)(n) = v\},$$

where $s \in \mathcal{B}_{\{m\}}$.
(c) Suppose that $\mathcal{B}$ is a barrier on M, and that $\varphi : \mathcal{B} \to \mathrm{Fn}(\mathbb{N}, V)$ is 1-Lipschitz. Suppose that N is an infinite subset of M such that $M \setminus N$ is also infinite. Then for every $s \in \overline{\mathcal{B} \restriction N}$ there is some $t \in \mathcal{B} \restriction N$ such that $\Phi(t) \restriction N = \Phi(s)$.

Lemma II.5.6. *Suppose that $\mathcal{B}$ is a barrier on M, and suppose that $\Phi : \mathcal{B} \to \mathrm{Fn}(\mathbb{N}, V)$. There is an infinite subset $N \subseteq M$ such that for every $s \in \widehat{\mathcal{B} \restriction N}$ and every $v \in V$ there is some $t = t(s, v) \in \mathcal{B}$ such that*
(a) $\Phi(t)"(t \cap N) \subseteq \{v\}$[12], *and*
(b) $\varphi^v(s) = \varphi^v(t) \cap N$.

Proof. Fix $v \in V$. We may assume that $\mathcal{B}$ is an α-uniform barrier on M, and that Φ is 1-Lipschitz. We may also assume that the mapping $s \mapsto \Phi(s)(\min s) \in V$ is constant, with value v_0. The proof is by induction on the ordinal α. We first note the following
Claim. There is an infinite subset N of M such that for every $n \in N$ and every $s \in \widehat{\mathcal{B}_{\{n\}} \restriction N}$ there is some $t \in \mathcal{B}_{\{n\}}$ such that $\Phi_{\{n\}}(t)"(t \cap N) \subseteq \{v\}$, $\varphi_n^v(s) = \varphi_n^v(t) \cap N$ and $\min \varphi_n^v(t) \geq \min \varphi_n^v(s)$. Moreover every infinite subset P of N has the same properties. $\qquad\square$

---

[12]Notation: For a mapping ϕ and a set X, we let $\phi"X = \{\phi(x) : x \in X\}$.

From now on, we fix $N \subseteq M$ with the properties stated in the claim. There are two cases to consider:

Case 1: $v_0 = v$. Then N satisfies the conclusion of the Lemma.

Case 2: $v_0 \neq v$. Then using Corollary II.3.23 we can find a decreasing sequence (N_i) of infinite subsets of N such that if we set $n_i^0 = \min N_i$ and $n_i^1 = \min {}_*N_i$ then $N_{i+1} \subseteq N_i/n_i^1$ and there are $j(i) \neq k(i) \in \{0,1\}$ such that

$$(\varphi_{n_i^{j(i)}}^{v}{}''(\mathcal{B}_{\{n_i^{j(i)}\}} \upharpoonright (N_i/n_i^1)))[P] \subseteq (\varphi_{n_i^{k(i)}}^{v}{}''(\mathcal{B}_{\{n_i^{k(i)}\}} \upharpoonright (N_i/n_i^1)))[P] \qquad \text{(II.24)}$$

with both traces hereditary and with $P = \{n_i^{j(i)}\}$. Then P satisfies the conclusion of the Lemma. $\qquad\qquad\square$

Theorem II.5.7. *Suppose that (x_n) is a normalized weakly-null sequence of $\ell_\infty(\Gamma)$ with the property that*

$$\inf\{|x_n(\gamma)| : n \in \mathbb{N}, \gamma \in \Gamma\} = \delta > 0. \qquad \text{(II.25)}$$

Then there is an infinite set N and the closure $\mathcal{C}$ of a α-uniform barrier on $\mathbb{N}$ for some ordinal $\alpha > 0$ such that $(x_n)_{n \in N}$ is $\delta/4$-equivalent to the natural basis (t_i) of the corresponding Schreier space $\mathcal{S}_\mathcal{C}$ and so is, in particular, unconditional.

Proof. Consider the corresponding family $\mathcal{F} - \{F_\gamma : \gamma \in \Gamma\}$, where $F_\gamma = \{n \in \mathbb{N} : x_n(\gamma) \neq 0\}$. Notice that condition (II.25) implies that $\mathcal{F}$ is a pre-compact family. Use Theorem II.3.22 to find an infinite M and a uniform barrier on M such that $\mathcal{F}[M]^{\max} = \mathcal{B}$ and $\mathcal{F}[M] - \widehat{\mathcal{B}}$. Note that since each x_n is normalized, for every M and every $m \in M$ there is some $\gamma \in \Gamma$ such that $m \in F(\gamma)$, so $\mathcal{F}[M]$ contains the singletons $\{m\}$ ($m \in M$). Hence $\mathcal{B}$ is α-uniform with $\alpha > 0$.

For every $s \in \mathcal{B}$ we choose $\gamma(s) \in \Gamma$ such that $s = F(\gamma(s)) \cap M$. Define $\Phi : \mathcal{B} \to \text{Fn}(\mathbb{N}, \{-1, 1\})$ for $s \in \mathcal{B}$ by

$$\Phi(s)(n) = \begin{cases} x_n(\gamma(s))/|x_n(\gamma(s))| & \text{if } n \in s \\ 0 & \text{otherwise.} \end{cases} \qquad \text{(II.26)}$$

Let $N \subseteq M$ be the result given by the use of Theorem II.5.6 to Φ and $\mathcal{B}$. Let $\Theta : N \to \mathbb{N}$ be the order-preserving bijection between N and $\mathbb{N}$, and let $\mathcal{C} = \Theta''\widehat{\mathcal{B} \upharpoonright N} = \{\Theta''(s) : s \subseteq t \in \mathcal{B} \upharpoonright N\}$. It is clear that $\mathcal{C}$ is the closure of a uniform barrier on $\mathbb{N}$. We claim that the subsequence $(x_n)_{n \in N}$ is $\delta/4$-equivalent to the natural basis of the corresponding Schreier space $\mathcal{S}_\mathcal{C}$. Fix a sequence of scalars $(a_n)_{n \in N}$. Let $\gamma \in \Gamma$ be such that $|\sum_{n \in N} a_n x_n(\gamma)| \approx \|\sum_{n \in N} a_n x_n\|$. Note that

$$|\sum_{n \in N} a_n x_n(\gamma)| = |\sum_{n \in F_\gamma \cap N} a_n x_n(\gamma)| \leq \sum_{n \in F_\gamma \cap N} |a_n| \leq \|\sum_{n \in N} a_n t_{\Theta(n)}\|_{\mathcal{S}_\mathcal{C}}, \qquad \text{(II.27)}$$

since $F_\gamma \cap N \in \mathcal{F}[N] = \mathcal{F}[M][N] = \widehat{\mathcal{B}}[N] = \widehat{\mathcal{B} \upharpoonright N}$. Now let $s \in \mathcal{B} \upharpoonright N$ be such that

$$\|\sum_{n \in N} a_n t_{\Theta(n)}\|_{\mathcal{S}_\mathcal{C}} \approx \sum_{n \in s} |a_n|. \qquad \text{(II.28)}$$

For $i = 0, 1$, let

$$s_i = \{n \in s : (-1)^i a_n \geq 0\}, \tag{II.29}$$

Now we can find $\gamma(i, j) \in \Gamma$ $((i, j) \in \{0, 1\}^2)$ such that

$$F(\gamma(i, j)) \cap N = \{n \in F(\gamma(i, j)) \cap N : |x_n(\gamma(i, j))| = (-1)^j x_n(\gamma(i, j))\}$$

$$= \varphi^{(-1)^j}(s_i) = s(i, j). \tag{II.30}$$

Note that

$$s = \bigcup_{(i,j) \in \{0,1\}^2} s(i, j). \tag{II.31}$$

It can be shown that

$$4 \left\| \sum_{n \in N} a_n x_n \right\| \geq \sum_{(i,j) \in \{0,1\}^2} \left| \sum_{n \in N} a_n x_n(\gamma(i, j)) \right|$$

$$\geq \delta \sum_{(i,j) \in \{0,1\}^2} \sum_{n \in s(i,j)} |a_n| = \delta \sum_{n \in s} |a_n|, \tag{II.32}$$

which gives the desired result. $\square$

More details about the above proof can be found in [52]. Similar results are proved in [8] and [33] and the reader is also referred to these sources for a more complete information about the problem of finding unconditional subsequences of weakly null sequences.

Remark II.5.8. Besides the separation from 0 of the set of values of a given weakly null sequence (x_n), one can have various other simplifying conditions such as for example conditions on the corresponding pre-compact family of finite sets

$$\mathcal{F} = \{\{n : x_n(\gamma) \neq 0\} : \gamma \in \Gamma\}.$$

If we assume for example that there is an infinite set M such that $\mathcal{F}[M] = \overline{\mathcal{B}}$ for some uniform barrier $\mathcal{B}$ on M of some *finite* rank, then the sequence (x_n) has an unconditional basic subsequence. This is essentially the content of parts (a) and (b) of the next theorem; the part (c) says that some restriction on the uniform rank of $\mathcal{B}$ is necessary (and that some restrictions altogether are necessary!).

Theorem II.5.9 (Maurey–Rosenthal).   (a) *For every $\varepsilon > 0$, every weakly null normalized sequence $(x_n) \subseteq C(\alpha + 1)$ for some $\alpha < \omega^\omega$ has a $(2 + \varepsilon)$-unconditional subsequence.*

(b) *For every $\varepsilon > 0$, every weakly null normalized sequence $(x_n) \subseteq C(\omega^\omega + 1)$ has a $(4 + \varepsilon)$-unconditional subsequence.*

(c) *There is a weakly null normalized sequence $(x_n) \subseteq C(\omega^{\omega^2} + 1)$ with no unconditional subsequence.*

Proof. One can view (a) and (b) as natural application of the theory of fronts and barriers developed above. Given a weakly null sequence $(x_n) \subseteq C(\alpha + 1)$ for some $\alpha < \omega^\omega$, since $\alpha + 1$ is a countable index-set we can find a sequence (C_k) of clopen subsets of $\alpha + 1$ and a subsequence (x_{n_k}) of (x_n) such that

(1) $\chi_{C_k} \to 0$ pointwise on $\alpha + 1$

(2) $\sum_{k=0}^{\infty} \|x_{n_k} - x_{n_k} \chi_{C_k}\| < \infty$.

Thus, we may replace our sequence (x_n) with the sequence (y_k) where

$$y_k = x_{n_k} \chi_{C_k}$$

which now has the property that $D_k = \{\gamma : y_k(\gamma) \neq 0\}$ is a sequence of clopen subsets of $\alpha + 1$ with no $\gamma \in \alpha + 1$ belonging to infinitely many of the D_k's. It follows in particular that

(3) $\gamma \mapsto F_\gamma = \{k : y_k(\gamma) \neq 0\}$ is a continuous map from α into $\mathbb{N}^{[<\infty]}$.

Hence $\mathcal{F} = \{F_\gamma : \gamma \in \alpha + 1\}$ is a compact family of finite subsets of $\mathbb{N}$, so applying Theorem II.3.22 we can find a barrier $\mathcal{B}$ on M such that $\mathcal{F}[M] = \overline{\mathcal{B}}$. Since $\mathcal{F}[M]$ is a continuous image of $\alpha + 1$, its Cantor–Bendixon rank must be finite. It follows that $\mathcal{B} = M^{[l]}$ for some positive integer l.

Pick an $\varepsilon > 0$ and choose a finite set V of real numbers and for each k continuous

$$z_k : \alpha + 1 \to V$$

such that $\|y_k - z_k\|_\infty < \varepsilon / l$. It follows that for every positive integer m, every sequence $(a_k)_{k=0}^m \subseteq [-1, 1]$, and every $\gamma < \alpha + 1$,

$$|\sum_{k=0}^m a_k(y_k - z_k)(\gamma)| = |\sum_{k \in F_\gamma} a_k(y_k - z_k)(\gamma)| \leq \sup_k |a_k| |F_\gamma| \frac{\varepsilon}{l} \leq \varepsilon.$$

So if (z_{k_i}) is a $(2 + \varepsilon)$-unconditional subsequence of (z_k) then the corresponding (y_{k_i}) will be $(2 + \delta)$-unconditional subsequence of (y_k) with the possibility that δ, depending on ε, be arbitrarily small. Hence, we have reduced the part (a) of Theorem II.5.9 to Theorem II.5.7.

To prove the part (b) i.e., treat the case $\alpha = \omega^\omega$ one proceeds as above to the subsequence (x_{n_k}) for which one has a perturbation (y_k) supported by a sequence (C_k) of clopen subsets of $\omega^\omega + 1$ converging pointwise to 0 and then diagonalizes applying the part (a) to the restriction of the tail sequence to various compact sets C_{k_i} that have finite Cantor–Bendixon rank.

The idea behind the example witnessing Theorem II.5.9 (c) is also quite interesting so let us examine it. Pick an $0 < \varepsilon < 1$ and choose a strictly increasing sequence (m_i) such that

$$\sum_{i=0}^{\infty} \sum_{j \neq i} \min((\frac{m_i}{m_j})^{1/2}, (\frac{m_j}{m_i})^{1/2}) \leq \varepsilon / 2.$$

Choose a 1-1 function [13]

$$\sigma : \text{FIN}^{[<\infty]} \to M = \{m_i\}$$

such that $\varphi(E_0, \ldots, E_k) > E_k$ for all $(E_0, \ldots, E_k) \in \text{FIN}^{[<\infty]}$. A finite or infinite sequence (E_i) of elements of FIN is *special* if

(1) $|E_0| = 1$,

(2) $E_0 < E_1 < \cdots < E_i < \cdots$,

(3) $|E_{i+1}| = \sigma(E_0, \ldots, E_i)$ whenever $i + 1$ is in the domain of the sequence.

Let $\mathcal{S}_\sigma$ be the collection of all special sequences of finite nonempty subsets of $\mathbb{N}$. For each $s = (E_i) \in \mathcal{S}_\sigma$ we associate a *special functional*

$$f^s = \sum_{i \in \text{dom}(s)} \frac{\chi_{E_i}}{|E_i|^{1/2}}$$

and a *special sequence* (x_i^s) *of vectors* from c_{00} defined by

$$x_i^s = \frac{\chi_{E_i}}{|E_i|^{1/2}}.$$

Let $\mathcal{F}_\sigma$ be the collection of all special functions associated to finite or infinite special block sequences of finite nonempty subsets of $\mathbb{N}$. For $s = (E_i) \in \mathcal{S}_\sigma$ and $I \subseteq \mathbb{N}$ let

$$f^s I = \sum_{i \in \text{dom}(s)} \frac{\chi_{E_i \cap I}}{|E_i|^{1/2}},$$

the *restriction* of f^s to I. Let $\overline{\mathcal{F}}_\sigma$ be the collection of functionals of the form

$$f^s I \quad (s \in \mathcal{S}_\sigma, \ I \subseteq \mathbb{N} \text{ interval }).$$

Using $\overline{\mathcal{F}}_\sigma$ one defines the *Maurey–Rosenthal space* $\mathfrak{X}_{MR}$ as the completion of c_{00} under the norm

$$\|x\| = \sup\{\langle f, x \rangle : f \in \overline{\mathcal{F}}_\sigma\}.$$

Take an arbitrary infinite $s = (E_i)_{i=0}^{\infty}$ in $\mathcal{S}_\sigma$. The corresponding sequence of vectors (x_i^s) is semi-normalized and weakly null, and

$$\max_{k \leq n} \left| \sum_{i=0}^{k} a_i \right| \leq \left\| \sum_{i=0}^{n} a_i x_i^s \right\| \leq (1 + \varepsilon) \max_{k \leq n} \left| \sum_{i=0}^{k} a_i \right| \tag{II.33}$$

---

[13]$\text{FIN}^{[<\infty]}$ is the collection of all finite *block* sequences $E_0 < E_1 < \cdots < E_k$ of nonempty finite subsets of $\mathbb{N}$; see Section III.3.

for every n and every choice $(a_i)_{i=0}^n$ of scalars from $[-1, 1]$. These two inequalities result from the observation that up to a small error the norm of the vector of the term

$$y = \sum_{i=0}^k a_i x_i^s$$

is essentially equal to the absolute value of the evaluation of this vector at the corresponding special functional

$$f^{s \restriction k+1} = \sum_{i=0}^k \frac{\chi_{E_i}}{|E_i|^{1/2}}$$

(which happens to be equal to $|\sum_{i=0}^k a_i|$). To see this note that for any other (infinite) special sequence $t = (F_i) \in \mathcal{S}_\sigma$, there is $j = j(s, t)$ such that:

(4) $E_i = F_i$ for $i < j$,

(5) $|E_i| \neq |F_i|$ for $i > j$.

So our assumption about the growth of the sequence (m_i) will give us that, up to a small error, the evaluation of f^t on a vector of the form $y = \sum_{i=0}^k a_i x_i^s$ will be equal to the evaluation of $f^{s \restriction j}$ on y.

Note that the inequalities (II.33) show that the summing basis of c_0,

$$s_0 = (1, 0, 0, \dots), s_1 = (1, 1, 0, 0, \dots), s_2 = (1, 1, 1, 0, 0, \dots), \dots$$

is $(1 + \varepsilon)$-equivalent to any special sequence $(x_i^s)_{i=0}^\infty$ of vectors of $\mathfrak{X}_{MR}$. Note that every infinite subset N of $\mathbb{N}$ contains an infinite special block sequence $s = (E_i)_{i=0}^\infty$. This shows that every subsequence $(e_i)_{i \in N}$ of the standard basis $(e_i)_{i=0}^\infty$ of $\mathfrak{X}_{MR}$ contains a block subsequence $(x_i^s)_{i=0}^\infty$ which is $(1 + \varepsilon)$-equivalent to the summing basis of c_0. So in particular, no subsequence of $(e_i)_{i=0}^\infty$ is unconditional.

Let $\mathfrak{X}_{MR}$ be the bounded version of the Maurey–Rosenthal space, where one takes only the special functionals over special sequences $s = (E_i) \in \mathcal{S}_\sigma$ of finite lengths j determined by their starting sets E_0 in the sense that $E_0 = \{j\}$. It is not difficult to show that the bounded version $\mathfrak{X}_{MR}^b$ of the Maurey–Rosenthal space is isomorphic to $\mathcal{C}(\omega^{\omega^2} + 1)$ and that every subsequence of its unit vector basis finitely block represents the summing basis of c_0. $\qquad \square$

II.6 Topological Ramsey Theory

Definition II.6.1. A set $\mathfrak{X} \subseteq \mathbb{N}^{[\infty]}$ is *Ramsey* if for every s and M there is $N \in [s, M]$ such that

$$[s, N] \subseteq \mathfrak{X} \text{ or } [s, N] \cap \mathfrak{X} = \emptyset.$$

Example II.6.2 (Baumgartner, Mathias). Choose a nonprincipal ultrafilter $\mathcal{U}$ on $\mathbb{N}$, and let

$$\mathfrak{X} = \{M = (m_i)_{i \in \mathbb{N}} \in \mathbb{N}^{[\infty]} : \bigcup_{i \in \mathbb{N}} [m_{2i}, m_{2i+1}) \in \mathcal{U}\}.$$

Then $\mathfrak{X}$ is not Ramsey, since $M \in \mathfrak{X}$ iff $M \setminus \{\min(M)\} \notin \mathfrak{X}$.

Exercise II.6.3. Show that there is a non-Ramsey $\mathfrak{X} \subseteq \mathbb{N}^{[\infty]}$ which has the property of Baire relative to the product topology of $\mathbb{N}^{[\infty]}$. (Hint: Intersect $\mathfrak{X}$ of Example II.6.2 with $\{M \in \mathbb{N}^{[\infty]} : \forall m, n \in M \; m \equiv n \,(\text{mod } 2)\}$.)

Lemma II.6.4 (Galvin). *Metrically open subsets of $\mathbb{N}^{[\infty]}$ are Ramsey.*

Proof. Fix an open set $\mathcal{O} \subseteq \mathbb{N}^{[\infty]}$, and consider

$$\mathcal{F} = \{s \in \mathbb{N}^{[<\infty]} : [s, \mathbb{N}] \subseteq \mathcal{O}\}.$$

Applying Lemma II.3.8 we obtain $N \subseteq M$ such that $\mathcal{F} \upharpoonright N$ is either empty or it contains a barrier. In the first case $[\emptyset, N] \cap \mathcal{O} = \emptyset$, and in the second $[\emptyset, N] \subseteq \mathcal{O}$. Relativizing to an arbitrary $[s, M]$ in place of $[\emptyset, N]$ we get the conclusion of the lemma. $\qquad\square$

Corollary II.6.5 (Nash–Williams). *For every pair $\mathcal{O}_0$ and $\mathcal{O}_1$ of disjoint metrically open subsets of $\mathbb{N}^{[\infty]}$ there exists $M \subseteq \mathbb{N}$ such that $M^{[\infty]} \cap \mathcal{O}_i = \emptyset$ for some $i = 0, 1$.*

Theorem II.6.6 (Galvin–Prikry). *Metrically Borel subsets of $\mathbb{N}^{[\infty]}$ are Ramsey.*

Proof. Use induction on Borel rank to show that for every $[s, M]$ and every Borel $\mathfrak{X} \subseteq \mathbb{N}^{[\infty]}$ there is $N \in [s, M]$ such that

$$\mathfrak{X} \cap [s, N] \text{ is relatively clopen in } [s, N].$$

Suppose $\mathfrak{X} = \bigcup_{k \in \mathbb{N}} \mathfrak{X}_k$ is a Borel set written as union of a sequence of Borel sets of smaller rank for which the conclusion holds. Build a decreasing sequence $M_0 \supseteq M_1 \supseteq \cdots \supseteq M_k \supseteq \cdots$ of infinite sets such that $m_k = \min(M_k) < m_{k+1} = \min(M_{k+1})$ and such for all k and $s \subseteq \{m_0, \ldots, m_{k-1}\}$, $[s, M_k] \subseteq \mathfrak{X}_k$ or $[s, M_k] \cap \mathfrak{X}_k = \emptyset$. Let $M = \{m_k : k \in \mathbb{N}\}$. Then $M \cap [\emptyset, M]$ is a relatively open set, so we finish by Lemma II.6.4. $\qquad\square$

We finish this section with Farahat's proof of Rosenthal's ℓ_1-theorem that uses Galvin–Prikry theorem.

Theorem II.6.7 (Rosenthal). *Suppose that (x_n) is an infinite normalized sequence of elements of some Banach space X. Then (x_n) has an infinite subsequence (x_{n_k}) which is either weakly Cauchy or is equivalent to the natural basis of ℓ_1.*

It suffices to show that for every set Γ and an uniformly bounded sequence (f_n) of real-valued functions on Γ there is a subsequence (f_{n_k}) which is either pointwise convergent on Γ or is equivalent (relative to the supremum norm on Γ) to the unit vector basis of ℓ_1.

Recall the notion of *independence* for a sequence $((A_n, B_n))_{n=0}^{\infty}$ of pairs of disjoint subsets of Γ:

$$\left(\bigcap_{i \in I} A_i\right) \cap \left(\bigcap_{j \in J} B_j\right) \neq \emptyset$$

for all finite nonempty $I, J \subseteq \mathbb{N}$ with $I \cap J = \emptyset$. The relevance of this notion to Rosenthal's theorem is explained by the following lemma.

Lemma II.6.8. *Suppose that for some uniformly bounded $(f_n)_{n=0}^{\infty} \subseteq \ell_{\infty}(\Gamma)$ and some $\varepsilon < \delta$, the sequence*

$$(\{\xi \in \Gamma : f_n(\xi) < \varepsilon\}, \{\xi \in \Gamma : f_n(\xi) > \delta\})_{n=0}^{\infty}$$

is independent. The (f_n) is equivalent to the natural basis of ℓ_1.

Proof. Set $A_n = \{\xi \in \Gamma : f_n(\xi) < \varepsilon\}$, $B_n = \{\xi \in \Gamma : f_n(\xi) > \delta\}$ $(n = 0, 1, \dots)$. Choose a finite sequence $(a_i)_{i=0}^{k} \subseteq [-1, 1]$. It suffices to show that

$$\left\| \sum_{i=0}^{k} a_i f_i \right\|_{\infty} \geq \left(\frac{\delta - \varepsilon}{2}\right) \sum_{i=0}^{k} |a_i|. \tag{II.34}$$

Let $I^+ = \{i \leq k : a_i \geq 0\}$, $I^- = \{i \leq k : a_i < 0\}$. By independence we can choose

$$\xi \in (\bigcap_{i \in I} A_i) \cap (\bigcap_{j \in J} B_j) \tag{II.35}$$

and

$$\eta \in (\bigcap_{i \in I} B_i) \cap (\bigcap_{j \in J} A_j). \tag{II.36}$$

Then

$$\sum_{i=0}^{k} a_i f_i(\xi) \geq \sum_{j \in J} |a_j| \delta - \sum_{i \in I} |a_i| \varepsilon \tag{II.37}$$

and

$$\sum_{i=0}^{k} a_i f_i(\eta) \leq -\sum_{i \in I} |a_j| \delta + \sum_{j \in J} |a_j| \varepsilon. \tag{II.38}$$

From (II.38)–(II.37) we get

$$\sum_{i=0}^{k} a_i (f_i(\xi) - f_i(\eta)) \geq (\delta - \varepsilon) \sum_{i=0}^{k} |a_i| \tag{II.39}$$

since the left-hand side is bounded by $2\|\sum_{i=0}^{k} a_i f_i\|_{\infty}$, we are done. $\square$

Call a sequence $((A_n, B_n))_{n=0}^\infty$ of pairs of disjoint subsets of Γ *convergent* if

$$\left(\bigcap_m \bigcup_{n \geq m} A_n\right) \cap \left(\bigcap_m \bigcup_{n \geq m} B_n\right) = \emptyset.$$

Lemma II.6.9. *Any sequence $((A_n, B_n))_{n=0}^\infty$ of pairs of disjoint subsets of Γ contains an infinite independent subsequence or an infinite convergent sequence.*

Proof. Let $\mathfrak{X}$ be the collection of all infinite $M = (m_i)_{i=0}^\infty \subseteq \mathbb{N}$ such that for all k,

$$(\bigcap_{i=0}^{k} A_{m_{2i}}) \cap (\bigcap_{i=0}^{k} B_{m_{2i+1}}) = \emptyset. \tag{II.40}$$

Clearly, $\mathfrak{X}$ is a closed subset of $\mathbb{N}^{[\infty]}$. Applying the Galvin–Prikry theorem we get an $M \in \mathbb{N}^{[\infty]}$ such that either $M^{[\infty]} \subseteq \mathfrak{X}$ or else $M^{[\infty]} \cap \mathfrak{X} = \emptyset$. Note that if the second alternative holds, the sequence $((A_m, B_m))_{m \in M}$ is convergent. So assume that $M^{[\infty]} \subseteq \mathfrak{X}$. Let $M = (m_i)_{i=0}^\infty$ be the increasing enumeration of M. Then it is easily checked that the sequence $((A_{m_{2i+1}}, B_{m_{2i+1}}))_{i=0}^\infty$ is independent. $\square$

We are now ready to finish the proof of Rosenthal's ℓ_1-theorem: Let $(\varepsilon_k, \delta_k)$ $(k \in \mathbb{N})$ be an enumeration of all pairs (ε, δ) of rationals such that $\varepsilon < \delta$. By a successive application of Lemma II.6.9 we build a decreasing sequence $(M_k)_{k=0}^\infty$ of infinite subsets of $\mathbb{N}$ such that $(m_k = \min(M_k))_{k=0}^\infty$ is strictly increasing and such that for each k the sequence

$$(\{\xi \in \Gamma : f_n(x) < \varepsilon_k\}, \{\xi \in \Gamma : f_n(x) \geq \delta_k\})_{n \in M_k} \tag{II.41}$$

is either independent or convergent. If for some k the corresponding sequence is independent, Lemma II.6.8 gives us that $(f_n)_{n \in M_k}$ is equivalent to the natural basis of ℓ_1. On the other hand if this never happens the diagonal sequence $(f_{m_k})_{k=0}^\infty$ is pointwise convergent on Γ. $\square$

II.7   The Theory of Better-Quasi-Orderings

Recall the notion of better-quasi-ordering introduced above in Section II.1. We shall now see how topological Ramsey theory can be effectively used in proofs of some of the deepest results of b.q.o. theory.

Lemma II.7.1. *A quasi-ordering Q is b.q.o. iff for every Borel function $f \colon \mathbb{N}^{[\infty]} \longrightarrow Q$ there is an infinite $X \subseteq \mathbb{N}$ such that $f(X) \leq f(X \setminus \{\min(X)\})$.*

Proof. Find first $M \in \mathbb{N}^{[\infty]}$ such that the image $f\left[M^{[\infty]}\right]$ is countable (see Theorem II.8.7 below) and then apply the Galvin–Prikry theorem to shrink even further to make the restriction $f \restriction M^{[\infty]}$ continuous. $\square$

Exercise II.7.2. Prove that if Q is b.q.o., then so is Q^ω quasi-ordered by : $(x_i) \leq (y_k)$ iff there is a strictly increasing sequence $k_0 < k_1 < \cdots < k_i < \cdots$ of integers such that $x_i \leq y_{k_i}$ for all i.

(Hint: Assume there is a bad function $f\colon \mathbb{N}^{[\infty]} \longrightarrow Q^\omega$, i.e., $f(M) \nleq f(M \setminus \{\min(M)\})$ for all $M \in \mathbb{N}^{[\infty]}$ and find $M \in \mathbb{N}^{[\infty]}$ and $g\colon M^{[\infty]} \longrightarrow Q$ such that for every $N \in M^{[\infty]}$, $g(N)$ is a term of $f(N)$ and $g(N) \nleq g(N \setminus \{\min(N)\})$.)

Exercise II.7.3. Let $\mathfrak{X}_i (i \in I)$ be a given uncountable family of Borel subsets of $\mathbb{N}^{[\infty]}$. Show that there is infinite $M \subseteq \mathbb{N}$ such that, either
(a) $M^{[\infty]} \subseteq \mathfrak{X}_i$ for infinitely many $i \in I$, or
(b) $M^{[\infty]} \cap \mathfrak{X}_i = \emptyset$ for infinitely many $i \in I$.

A *partial ranking* of a quasi-ordered set Q is a well-founded partial ordering $\leq'$ on Q such that $x \leq' y$ implies $x \leq y$. Given a set q.o. Q, a *bad Q-array* is a Borel map of the form $f\colon M^{[\infty]} \longrightarrow Q$ where M is an infinite subset of $\mathbb{N}$ and

$$f(X) \nleq f(X \setminus \{\min(X)\}) \text{ for all } X \text{ in } M^{[\infty]}.$$

Given a pair $f\colon M^{[\infty]} \longrightarrow Q$ and $g\colon N^{[\infty]} \longrightarrow Q$ of bad Q-arrays, let $f \leq^* g$ mean that $M \subseteq N$ and that

$$f(X) \leq' g(X) \text{ for all } X \text{ in } M^{[\infty]},$$

and let $f <^* g$ mean that $M \subseteq N$ and

$$f(X) <' g(X) \text{ for all } X \text{ in } M^{[\infty]}.$$

A Q-array g is *minimal bad* (relative to the partial ranking $\leq'$) if there is no bad Q-array $f <^* g$. The following is one of the key results of b.q.o. theory.

Theorem II.7.4 (Nash–Williams). *For every bad Q-array $g\colon M^{[\infty]} \longrightarrow Q$ (relative to the fixed partial ranking of Q) there is a minimal Q-array $f \leq^* g$.*

Proof. Starting with $g_0 = g$ we build for as long as it is possible a $\leq^*$-decreasing transfinite sequence

$$g_\alpha\colon M_\alpha^{[\infty]} \longrightarrow Q(\alpha \leq \gamma)$$

of bad Q-array as follows.

Suppose we have constructed $g_\alpha\colon M_\alpha^{[\infty]} \longrightarrow Q$. If g_α is a minimal bad Q-array we stop the construction. Otherwise, we choose a bad Q-array $f_\alpha\colon N_\alpha^{[\infty]} \longrightarrow Q$ such that $f_\alpha <^* g_\alpha$. By the Galvin–Prikry theorem we may assume that f_α is actually continuous on its domain $N_\alpha^{[\infty]}$. We may assume that $M_\alpha \setminus N_\alpha$ is infinite and we can pick an integer $n_\alpha \geq 0$ such that f_α is constant on the basic open neighborhood $[n_\alpha, N_\alpha]^{14}$. Let

$$M_{\alpha+1} = N_\alpha \cup (\{0, 1, \ldots, n_\alpha\} \cap M_\alpha)$$

---

[14] $[k, P] = \{X \in P^{[\infty]} : X \cap \{0, 1, \ldots, k-1\} = P \cap \{0, 1, \ldots, k-1\}\}$.

and define $g_{\alpha+1} : M_{\alpha+1}^{[\infty]} \longrightarrow Q$ by letting

$$g_{\alpha+1} \upharpoonright N_\alpha^{[\infty]} = f_\alpha$$
$$g_{\alpha+1} \upharpoonright M_{\alpha+1}^{[\infty]} \setminus N_\alpha^{[\infty]} = g_\alpha \upharpoonright M_{\alpha+1}^{[\infty]} \setminus N_\alpha^{[\infty]}.$$

Then $g_{\alpha+1}$ is a bad Q-array and $g_{\alpha+1} \leq^* g_\alpha$.

If β is a countable limit ordinal such that g_α is defined for all $\alpha < \beta$, let

$$M_\beta = \bigcap_{\alpha < \beta} M_\alpha$$

Claim. M_β is infinite.

Proof. Suppose not and pick m such that

$$M_\beta \subseteq \{0, 1, \ldots, m - 1\}.$$

For $\alpha < \beta$, let $m_\alpha = \min(M_\alpha \setminus m)$. Clearly $m_\alpha < m_{\alpha'}$ whenever $\alpha \leq \alpha' < \beta$. By the choice of m there must be infinitely many $\alpha < \beta$ such that $m_\alpha < m_{\alpha+1}$, or in other words, $m_\alpha \notin M_{\alpha+1}$. Going back to the definition of $f_\alpha, N_\alpha, M_{\alpha+1}$ and $g_{\alpha+1}$ from $g_\alpha : M_\alpha^{[\infty]} \longrightarrow Q$ we see that

$$m_\alpha < m_{\alpha+1} \text{ implies } n_\alpha \leq m$$

So we can find an infinite set $A \subseteq \beta$, an integer $0 \leq n \leq m$ and $s \subseteq \{0, 1, \ldots, n-1\}$ such that for all $\alpha \in A$,

$$n_\alpha = n \text{ and } N_\alpha \cap \{0, 1, \ldots, n\} = s$$

Consider a pair $\alpha < \overline{\alpha}$ of elements of A. Note that $N_{\overline{\alpha}} \in [n_\alpha, N_\alpha]$, and therefore

$$g_{\overline{\alpha}}(N_{\overline{\alpha}}) \leq' g_{\alpha+1}(N_{\overline{\alpha}}) = f_\alpha(N_{\overline{\alpha}}) = f_\alpha(N_\alpha) <' g_\alpha(N_\alpha).$$

This gives us an infinite $<'$-decreasing sequence contradicting the well-foundedness of $\leq'$. $\qquad \square$

Given that M_β is infinite define $g_\beta : M_\beta^{[\infty]} \longrightarrow Q$ by

$$g_\beta(X) = \lim_{\alpha \to \beta} g_\alpha(X)$$

Since $g_\alpha(X)$ is $\leq'$-decreasing in α, the limit exists. Note also that g_β is a Borel function.

Since M_α's are strictly decreasing, the process must stop at some countable ordinal γ in which case we have reached a minimal bad Q-array. $\qquad \square$

For a given quasi-ordered set Q let

$$Q^{Ord} = \bigcup_{\alpha \in Ord} Q^{\alpha}$$

quasi-ordered by: $(x_\xi)_{\xi<\alpha} \leq_1 (y_\eta)_{\eta<\beta}$ iff there is a strictly increasing sequence η_ξ ($\xi < \alpha$) of ordinals $< \beta$ such that

$$x_\xi \leq y_{\eta_\xi} \text{ for all } \xi < \alpha.$$

Theorem II.7.5 (Nash–Williams). *If Q is b.q.o., then so is Q^{Ord}.*

Proof. For $(x_\xi)_{\xi<\alpha}, (y_\eta)_{\eta<\beta} \in Q^{Ord}$, set $(x_\xi)_{\xi<\alpha} \leq' (y_\eta)_{\eta<\beta}$ iff $\alpha \leq \beta$ and $x_\xi = y_\xi$ for all $\xi < \alpha$. Suppose Q^{Ord} is not b.q.o. By Theorem II.7.4 we can consider a minimal bad Q^{Ord}-array

$$g: M^{[\infty]} \longrightarrow Q^{Ord}$$

relative to this particular partial ranking. For a given $X \in M^{[\infty]}$ we know that

$$g(X) \nleq_1 g(X \setminus \{\min(X)\}).$$

So there must be γ_X such that

$$g(X) \upharpoonright \gamma_X \leq_1 g(X \setminus \{\min(X)\})$$

but

$$g(X) \upharpoonright (\gamma_X + 1) \nleq_1 g(X \setminus \{\min(X)\}).$$

Let $f: M^{[\infty]} \longrightarrow Q^{Ord}$ defined by

$$f(X) = g(X) \upharpoonright \gamma_X$$

Note that f is a Borel function and that $f(X) <' g(X)$ for all $X \in M^{[\infty]}$. Applying the Galvin–Prikry theorem to the Borel set

$$\mathfrak{X} = \{X \in M^{[\infty]} : f(X) \nleq_1 f(X \setminus \{\min(X)\})\},$$

we get $N \in M^{[\infty]}$ such that $N^{[\infty]} \subseteq \mathfrak{X}$ or $N^{[\infty]} \cap \mathfrak{X} = \emptyset$. Note that the second alternative is not possible since this would make $f \upharpoonright N^{[\infty]}$ a bad Q^{Ord}-array contradicting the minimality of g. To simplify the notation, for $X \subseteq \mathbb{N}$, let $_*X$ denote $X \setminus \{\min(X)\}$. Then the inclusion $N^{[\infty]} \subseteq \mathfrak{X}$ means

$$g(X) \upharpoonright \gamma_X \leq_1 g(_*X) \upharpoonright \gamma_{_*X} \quad (X \in \mathbb{N}^{[\infty]}).$$

Combining this with the choice of $X \longmapsto \gamma_X$, we conclude that for all $X \in \mathbb{N}^{[\infty]}$,

$$g(X) \upharpoonright \gamma_X \leq_1 g(_*X) \upharpoonright \gamma_{_*X}$$

but

$$g(X) \upharpoonright (\gamma_X + 1) \nleq_1 g(_*X) \upharpoonright (\gamma_{_*X} + 1).$$

It follows that

$$g(X)_{\gamma_X} \not\leq g(_*X)_{\gamma_{*}x}$$

for all $X \in \mathbb{N}^{[\infty]}$. So starting from a minimal bad Q^{Ord}-array $g \colon M^{[\infty]} \longrightarrow Q^{Ord}$ we have found a bad Q-array $X \longmapsto g(X)_{\gamma_X}$ $(X \in \mathbb{N}^{[\infty]})$ which for every X picks a term out of the transfinite sequence $g(X)$. This finishes the proof.   □

The following result provides a converse for Theorem II.7.5.

Theorem II.7.6 (Pouzet). *The following are equivalent for a quasi-ordered set Q:*

1. *Q is b.q.o.*

2. *$Q^{<\omega_1}$ is w.q.o.*   □

Recall that a linearly ordered set L is *scattered* if it has no subset isomorphic to the rationals. The class $\mathcal{S}$ of scattered orders can be generated by the following recursive procedure:

Let $\mathcal{S}_0$ be the class of one-point linear orders. Given an ordinal $\xi > 0$ let $\mathcal{S}_\xi$ be the class of linear orderings isomorphic to either well-ordered or conversely well-ordered sums of linear orders from $\bigcup_{\eta < \xi} \mathcal{S}_\eta$. The well-known result of Hausdorff shows that $\mathcal{S} = \bigcup_{\xi \in Ord} \mathcal{S}_\xi$.

Exercise II.7.7. Prove Hausdorff's Theorem, that $\mathcal{S} = \bigcup_{\xi \in Ord} \mathcal{S}_\xi$.

Theorem II.7.8 (Laver). *The class of scattered linear orderings is b.q.o. under embeddability.*

Proof. For $L \in \mathcal{S} = \bigcup_{\xi \in Ord} \mathcal{S}_\xi$ let $\xi(L)$ be the minimal ξ such that $L \in \mathcal{S}_\xi$. Define $L <' M$ iff L is embeddable into M and $\xi(L) < \xi(M)$. Let $L \leq' M$ iff $L <' M$ or $L = M$. Note that $\leq'$ is a partial ranking on $\mathcal{S}$. If $\mathcal{S}$ is not b.q.o., by Theorem II.7.4 there is a minimal bad array

$$g \colon M^{[\infty]} \to \mathcal{S},$$

relative to this partial ranking. For each $X \in M^{[\infty]}$ we write

$$g(X) = L_0^X + L_1^X + \cdots + L_\alpha^X + \ldots \quad (\alpha < \beta_X)$$

or

$$g(X) = \cdots + L_\alpha^X + \cdots + L_1^X + L_0^X,$$

such that $\xi(L_\alpha^X) < \xi(g(X))$ for all $\alpha < \beta_X$. Shrinking M we may assume that $g(X)$ has, say, the first form for all $X \in M^{[\infty]}$. Then as in the proof of Theorem II.7.5 we find $N \in M^{[\infty]}$ and for each $X \in N^{[\infty]}$ and ordinal $\gamma_X < \beta_X$ such that

$$\sum_{\alpha < \gamma_X} L_\alpha^X \leq \sum_{\alpha < \gamma_X} L_\alpha^{X \setminus \{\min(X)\}}$$

but $L_{\gamma_X}^X \not\leq L_{\gamma_X}^{X \setminus \{\min(X)\}}$. It follows that

$$f(X) = L_{\gamma_X}^X \quad (X \in \mathbb{N}^{[\infty]})$$

defines a bad array such that $f <^* g$, a contradiction.   □

The following corollary was conjectured by R. Fraïssé in the 1950s and it represents one of the finest applications of Nash–Williams b.q.o. theory.

Corollary II.7.9 (Laver). *The class of countable linearly ordered sets is w.q.o.*

Remark II.7.10. The reader interested in a more complete picture of this interesting subject is referred to the excellent introductions to this area given by Laver [47], Milner [62] and Simpson [85].

II.8 Ellentuck's Theorem

Consider the refinement of the usual metrizable topology on $\mathbb{N}^{[\infty]}$ by letting the sets of the form

$$[s, M] \quad (s \in \mathbb{N}^{[<\infty]}, M \in \mathbb{N}^{[\infty]})$$

be open. This is the *Ellentuck topology* on $\mathbb{N}^{[\infty]}$.

Lemma II.8.1. *Every Ellentuck-open subset of $\mathbb{N}^{[\infty]}$ is Ramsey.*

Proof. Let $\mathfrak{X}$ be a given Ellentuck-open set, and let $[s, M]$ be a given basic-open set. We simplify notation by letting $s = \emptyset$. Shrinking M, we assume that M has the following property.

(1) If for some finite $s \subseteq M$ one can find an infinite $N \subseteq M$ such that $[s, N] \subseteq \mathfrak{X}$ then already $[s, M] \subseteq \mathfrak{X}$.

Let

$$\mathcal{F} = \{s \in M^{[<\infty]} : [s, M] \subseteq \mathfrak{X}\}.$$

Applying Galvin's Lemma to $\mathcal{F}$ we find $N \subseteq M$ such that either
(a) $\mathcal{F} \cap \mathcal{P}(N) = \emptyset$, or
(b) Every infinite $X \subseteq N$ contains an initial segment in $\mathcal{F}$.

If (b) holds, then clearly $[\emptyset, N] \subseteq \mathfrak{X}$. If on the other hand (a) holds then by (1) and the fact that $\mathfrak{X}$ is Ellentuck-open one gets that $[\emptyset, N] \cap \mathfrak{X} = \emptyset$ □

Definition II.8.2. A subset $\mathfrak{X}$ of $\mathbb{N}^{[\infty]}$ is *Ramsey-null* if for every basic-open set $[s, M]$ there is $N \in [s, M]$ such that $[s, N] \cap \mathfrak{X} = \emptyset$

Lemma II.8.3. *Every Ellentuck-nowhere-dense subset of $\mathbb{N}^{[\infty]}$ is Ramsey-null.*

Proof. Let $\mathfrak{X}$ be a given subset of $\mathbb{N}^{[\infty]}$ which is nowhere dense relative to the Ellentuck topology. Clearly we may assume $\mathfrak{X}$ is Ellentuck-closed. Let $[s, M]$ be a given basic-open set. By Lemma II.8.1, there is $N \in [s, M]$ such that

$$[s, N] \subseteq \mathfrak{X} \text{ or } [s, N] \cap \mathfrak{X} = \emptyset$$

Note that the first alternative is impossible □

Lemma II.8.4. *Every Ellentuck-meager subset of $\mathbb{N}^{[\infty]}$ is Ramsey-null.*

Proof. Let $\mathfrak{X} = \bigcup_{i \in \mathbb{N}} \mathfrak{X}_i$ be a given Ellentuck-meager set written as increasing union of Ellentuck-nowhere-dense sets. Let $[s, M]$ be a given basic set. We simplify the notation by assuming $s = \emptyset$. Starting with $M_0 = M$ we build an infinite decreasing sequence

$$M_0 \supseteq M_1 \supseteq \cdots \supseteq M_k \supseteq \cdots$$

of infinite subsets of $\mathbb{N}$ such that the sequence $\{m_i = \min(M_i)\}_{i \in \mathbb{N}}$ is strictly increasing and such that (applying Lemma II.8.3)

$$[s, M_{k+1}] \cap \mathfrak{X}_k = \emptyset$$

for all k and $s \subseteq \{m_0, \ldots, m_k\}$. Let $N = \{m_i : i \in \mathbb{N}\}$. Then $N \subseteq M$ and $[\emptyset, N] \cap \mathfrak{X} = \emptyset$. $\square$

Theorem II.8.5 (Ellentuck). *The following conditions are equivalent for a subset $\mathfrak{X}$ of $\mathbb{N}^{[\infty]}$:*

(a) $\mathfrak{X}$ *has the Ramsey property*

(b) $\mathfrak{X}$ *has the Baire property relative to the Ellentuck topology.*

    Moreover, $\mathfrak{X}$ is Ramsey-null if and only if $\mathfrak{X}$ is Ellentuck-meager.

Proof. Suppose an $\mathfrak{X} = \mathcal{O} \triangle \mathfrak{M}$ is a given subset of $\mathbb{N}^{[\infty]}$ written as the symmetric difference of an Ellentuck-open set $\mathcal{O}$ and an Ellentuck-meager set $\mathfrak{M}$. Let $[s, M]$ be a given basic set. By Lemma II.8.3. there is an $N \in [s, M]$ such that

$$[s, N] \cap \mathfrak{M} = \emptyset$$

It follows that $\mathfrak{X} \cap [s, N] = \mathcal{O} \cap [s, N]$ is an Ellentuck-open subset of $\mathbb{N}^{[\infty]}$ so by Lemma II.8.1 there is $P \in [s, N]$ such that $[s, P] \subseteq \mathfrak{X}$ or $[s, P] \cap \mathfrak{X} = \emptyset$. $\square$

Corollary II.8.6 (Silver). *The field of Ramsey subsets of $\mathbb{N}^{[\infty]}$ is closed under Souslin operation, and so in particular every metrically analytic subset of $\mathbb{N}^{[\infty]}$ is Ramsey.*

Theorem II.8.7 (Louveau–Simpson). *Suppose f is a Baire-measurable map from the Ellentuck space $\mathbb{N}^{[\infty]}$ into some metric space Z. Then there is $M \in \mathbb{N}^{[\infty]}$ such that $f \upharpoonright M^{[\infty]}$ is continuous and has separable range in Z.*

Proof. Otherwise we can assume to have a discrete family $\mathcal{U}$ of size continuum of open subsets Z such that for every infinite set $M \subseteq \mathbb{N}$,

$$\mathcal{U}_M = \{U \in \mathcal{U} : U \cap f[M^{[\infty]}] \neq \emptyset\}$$

is uncountable. Choose a Borel map $g : Z \longrightarrow \mathbb{R}$ such that

(a) $g \upharpoonright (Z \setminus \bigcup \mathcal{U}) \equiv 0$

(b) for every $U \in \mathcal{U}$, $g \upharpoonright U$ takes only one value $x_U \in \mathbb{R}$ such that $x_U \neq x_V$ whenever $U \neq V$.

(c) $\{x_U : u \in U\}$ is a set of reals which contains no perfect subset.

Consider the composition $h = f \circ g \colon \mathbb{N}^{[\infty]} \longrightarrow \mathbb{R}$. Then h is Baire-measurable relative to the Ellentuck topology on $\mathbb{N}^{[\infty]}$. Using Ellentuck's theorem, a simple diagonalization argument over a countable basis of $\mathbb{R}$ will give us an infinite $M \subseteq \mathbb{N}$ such that $h \upharpoonright M^{[\infty]}$ is continuous even relative to the metrizable topology of $M^{[\infty]}$. It follows that the range of $h \upharpoonright M^{[\infty]}$ is an uncountable analytic set of reals without a perfect subset, a contradiction. □

Lemma II.8.8. *Every analytic equivalence relation E on $\mathbb{N}^{[\infty]}$ either has a perfect set of pairwise inequivalent elements, or there is an infinite set $M \subseteq \mathbb{N}$ such that E on $M^{[\infty]}$ has at most countably many classes.*

Proof. We shall supply the proof working under the assumption that the Baire property of the Ellentuck topology on $\mathbb{N}^{[\infty]}$ is $\aleph_1$-additive. A simple absoluteness argument (which we leave to the reader) shows that the conclusion of II.8.8 cannot depend on this sort of assumptions.

So, suppose E is an analytic equivalence relation on $\mathbb{N}^{[\infty]}$ that has no perfect set of pairwise non-equivalent elements of $\mathbb{N}^{[\infty]}$. By Burges' theorem (see Kechris [42]) we conclude that the quotient $\mathbb{N}^{[\infty]}/E$ has size at most $\aleph_1$. Taking $\mathbb{N}^{[\infty]}/E$ with the discrete metric our assumption yields that the quotient map

$$\pi \colon \mathbb{N}^{[\infty]} \to \mathbb{N}^{[\infty]}/E$$

is Baire measurable relative to the Ellentuck topology on $\mathbb{N}^{[\infty]}$. Then Theorem II.8.7 gives us $M \subset \mathbb{N}^{[\infty]}$ such that $\pi[M^{[\infty]}]$ is countable, as required. □

Exercise II.8.9. Show how the assumption in the proof of Lemma II.8.8 can be avoided.

Lemma II.8.10. *Let (e_n) be a normalized basic sequence in some Banach space X. Then either there exist continuum many pairwise non isomorphic closed subspaces of X of the form $\overline{(e_{n_k})}$, or else there is a subsequence (e_{n_k}) of (e_n) such that there exists only countable many isomorphism types among spaces of the form $\overline{(e_{n_{k_l}})}$ where $(e_{n_{k_l}})$ is a subsequence of (e_{n_k}).*

Proof. For $M, N \in \mathbb{N}^{[\infty]}$ put $M \, E \, N$ iff the closed subspaces $\overline{(e_m)_{m \in M}}$ and $\overline{(e_n)_{n \in N}}$ are isomorphic. Note that E is an analytic relation on $\mathbb{N}^{[\infty]}$. Now, apply Lemma II.8.8. □

Going in this direction one has the following more refined result.

Theorem II.8.11 (Rosendal). *Let (e_n) be a basic sequence in some Banach space X. Then either there exist continuum many pairwise non isomorphic subspaces among spaces spanned by subsequences of (e_n) or there is a subsequence (e_{n_k}) of (e_n) such that all further subsequences of (e_{n_k}) span isomorphic spaces.* □

We finish this section by mentioning yet another result along these lines.

Theorem II.8.12 (Ferenczi-Rosendal). *Let X be a Banach space with an unconditional basis (e_n). Then either:*

(I) *There is continuum many isomorphism types among subspaces of X of the form $\overline{(e_{n_k})}$ where (e_{n_k}) is an infinite subsequence of (e_n), or*

(II) *There is a subspace Y spanned by a subsequence of (e_n) such that Y is isomorphic to its hyperplanes, to its square, and it admits a Schauder decomposition into countably many uniformly isomorphic copies of itself.*   □

II.9   Summability in Banach Spaces

Recall that a sequence (x_n) of elements of some Banach space X is *Cesàro-summable* if the corresponding sequence

$$\frac{x_1 + \cdots + x_n}{n}$$

of arithmetic means converges in norm. A classical result reads that every convergent sequence $(x_n) \subseteq \mathbb{R}$ is Cesàro-summable in $\mathbb{R}$ to the same limit and in fact the same is true for any other Banach space. A Banach space X is said to have the *Banach–Saks property* if every bounded sequence (x_n) in X has a Cesàro-summable subsequence. Example of such spaces are the spaces $L_p[0,1]$, $(1 < p < \infty)$ and this is what Banach and Saks have originally established.

Example II.9.1 (Schreier). The basis (e_n) of the Schreier space S (see Example I.2.13 above) is weakly null but has no Cesàro-summable subsequence since for every finite set $s \in \mathrm{FIN}$, we have that

$$\frac{1}{|s|} \|\sum_{i \in s} e_i\|_S \geq \frac{1}{4}.$$

The Cesàro-summability is just one of the methods of summability in Banach spaces. Thus, an $\omega \times \omega$-matrix (a_{ij}) is called a *regular method of summability* if given a sequence (x_n) of elements of some Banach space X converging in norm to some $x \in X$, then the sequence

$$y_i = \sum_{j=0}^{\infty} a_{ij} x_j$$

also converges in norm to x. It is well known that (a_{ij}) is a regular method of summability if and only if:
(a) The sequence $(\sum_{j=0}^{\infty} a_{ij})_i$ is uniformly bounded in $\mathbb{R}$.
(b) $\lim_{i \to \infty} \sum_{j=0}^{\infty} a_{ij} = 1$.
(c) $\lim_{i \to \infty} a_{ij} = 0$ for every j.

Theorem II.9.2 (Erdös–Magidor). *Let (x_n) be a bounded sequence in some Banach space X and let (a_{ij}) be a given method of summability. Then there is a subsequence (x_{n_k}) of (x_n) such that either:*

(1) *Every subsequence of* (x_{n_k}) *is* (a_{ij})-*summable and all of them have the same limit.*

(2) *No subsequence of* (x_{n_k}) *is* (a_{ij})-*summable.*

Proof. Let
$$\mathfrak{X} = \{ M \in \mathbb{N}^{[\infty]} : (x_n)_{n \in M} \text{ is } (a_{ij})\text{-summable}\}.$$

Note that $\mathfrak{X}$ is a Borel subset of $\mathbb{N}^{[\infty]}$, so by the Galvin–Prikry theorem there is $M \in \mathbb{N}^{[\infty]}$ such that $M^{[\infty]} \subseteq \mathfrak{X}$ or $M^{[\infty]} \cap \mathfrak{X} = \emptyset$. The second alternative gives us the alternative (2) of the theorem. So let us assume $M^{[\infty]} \subseteq \mathfrak{X}$ and refine M further and obtain a subsequence satisfying the alternative (a).

Let Y be the closed subspace of X spanned by the subsequence $(x_n)_{n \in M}$. Note that Y is separable, so we can fix a sequence (B_k) of open balls of Y forming a basis of its norm topology. For each k let
$$\mathfrak{X}_k = \{ P \in M^{[\infty]} : (x_n)_{n \in P} \text{ is } (a_{ij})\text{-summable to a point of } B_k\}.$$

Then each $\mathfrak{X}_k$ is a Borel subset of $M^{[\infty]}$ so by a sequence of successive applications of the Galvin–Prikry Theorem we can find a decreasing sequence
$$M \supseteq M_0 \supseteq M_1 \supseteq \cdots \supseteq M_k \supseteq \ldots$$

of infinite subsets of M such that
$$M_k^{[\omega]} \subseteq \mathfrak{X}_k \text{ or } M_k^{[\infty]} \cap \mathfrak{X}_k = \emptyset$$

for all k. We may assume that the sequence $(m_k = \min(M_k))$ is strictly increasing, and this will give us an infinite set $N = \{m_k : k \in \mathbb{N}\}$ which diagonalizes the sequence (M_k). Note that for an infinite set $P = (p_i) \subseteq N$ the limit of the sequence
$$y_i^P = \sum_{j=0}^{\infty} a_{ij} x_{p_j}$$

does not change if we make a finite change to the set P (see conditions (a), (b), (c) on (a_{ij})). It follows that for every k and every infinite $P, Q \subseteq N$ the sequence (y_i^P) and (y_i^Q) either both converge to a point in B_k or both converge to a point outside B_k. Hence we must have
$$\lim_{i \to \infty} \sum_{j=0}^{\infty} a_{ij} x_{p_j} = \lim_{i \to \infty} \sum_{j=0}^{\infty} a_{ij} x_{q_j}$$

for every pair $P = (p_i)$ and $Q = (q_i)$ of infinite subsets of N. It follows that the subsequence $(x_n)_{n \in N}$ satisfies the alternative (2). $\qquad\square$

Corollary II.9.3. *If a Banach space X has the Banach–Saks property, then every bounded sequence (x_n) in X has a subsequence (x_{n_k}) such that every subsequence of (x_{n_k}) is Cesàro-summable.* $\qquad\square$

Exercise II.9.4. Prove that the set $\mathfrak{X} \subseteq \mathbb{N}^{[\infty]}$ from the proof of Erdös–Magidor Theorem is $G_{\delta\sigma\delta}$. What is the complexity of sets $\mathfrak{X}_k$ from the same proof?

Exercise II.9.5. Prove that in a finite-dimensional Banach space X, a bounded sequence each subsequence of which is Cesàro-summable must itself be convergent in X.

Exercise II.9.6 (Nishiura–Waterman). Show that every Banach space with the Banach–Saks property is reflexive.

Lemma II.9.7. *Suppose (x_n) is a weakly null basic subsequence in some Banach space X such that no subsequence of (x_n) is Cesàro summable. Then (x_n) has a subsequence (x_{n_k}) with spreading model isomorphic to ℓ_1.*

Proof. We think of (x_n) as a sequence of continuous functions on the dual ball B_{X^*} with the weak* topology. Our assumption about (x_n) in this setting states that (x_n) converges pointwise to the constant 0 function and that for no subsequence (x_{n_k}) the sequence

$$\frac{x_{n_1} + \cdots + x_{n_k}}{k}$$

converges uniformly on B_{X^*}. For $M \in \mathbb{N}^{[\infty]}$, $q \in \mathbb{R}$ and $x^* \in B_{X^*}$, let

$$M(x^* \geq \varepsilon) = \{m \in M : x^*(x_m) \geq \varepsilon\}.$$

Similarly one defines $M(x^* \leq \varepsilon)$. Let

$$\mathcal{F}_{\geq\varepsilon}(M) = \{M(x^* \geq \varepsilon) : x^* \in B_{X^*}\}.$$

Similarly one defines $\mathcal{F}_{\leq\varepsilon}(M)$. Let

$$\mathfrak{X}_{\geq\varepsilon} = \{M \in \mathbb{N}^{[\infty]} : \mathcal{F}_{\geq\varepsilon}(M) \text{ contains sets of arbitrary large size}\}.$$

Similarly one defines $\mathfrak{X}_{\leq\varepsilon}$. Note that assuming as we may that the Banach space X inside which we are working is separable, all these sets $\mathfrak{X}_{\geq\varepsilon}$ and $\mathfrak{X}_{\leq\varepsilon}$ are analytic. We claim that there must be $\varepsilon > 0$ such that either $\mathfrak{X}_{\geq\varepsilon}$ or $\mathfrak{X}_{\leq-\varepsilon}$ contains a set of the form $M^{[\infty]}$ for some $M \subseteq \mathbb{N}$. Otherwise, by Silver's theorem we can choose a decreasing sequence $M_1 \supseteq \cdots \supseteq M_k \supseteq \ldots$ of infinite subsets of $\mathbb{N}$ such that

$$M_k^{[\infty]} \cap (\mathfrak{X}_{\geq 1/k} \cup \mathfrak{X}_{\leq -1/k}) = \emptyset$$

for all k. Pick infinite M such that $k \in M_\ell$ for all $\ell, k \in M$ with $\ell < k$. Then for all $\varepsilon > 0$ there is a positive integer $n(\varepsilon)$ such that the set

$$M(x^*, \varepsilon) = \{m \in M : |x^*(x_m)| \geq \varepsilon\}$$

has size at most $n(\varepsilon)$. From this one easily concludes that if (m_k) is the increasing enumeration of M, then the sequence

$$\frac{x_{m_1} + \cdots + x_{m_k}}{k}$$

converges uniformly on B_{X^*} contradicting our assumption.

So, fix an $\varepsilon > 0$ such that $\mathfrak{X}_{\geq \varepsilon}$ or $\mathfrak{X}_{\leq -\varepsilon}$ contains a set of the form $M^{[\infty]}$ for some infinite $M \subseteq \mathbb{N}$. By symmetry we may consider only the first case and to save on notation, we assume that for every infinite $M \subseteq \mathbb{N}$ the family $\mathcal{F}_{\geq \varepsilon}(M)$ contains sets of arbitrarily large cardinality. Note that this means that for every $k > 0$ and $M \in \mathbb{N}^{[\infty]}$ there is $N \in M^{[\infty]}$ such that $N^{[k]} \subseteq \mathcal{F}_{\geq \varepsilon}(N)$. So diagonalizing we obtain an infinite set $M \subseteq \mathbb{N}$ such that

$$\mathcal{F}_{\geq \varepsilon}(M) = \mathcal{S}(M) = \{E \subseteq M : |E| \leq |\{m \in M : m < \min(E)\}|\}.$$

Pick an $E \in \mathcal{S}(M)$ and a sequence a_n ($n \in E$) of non negative scalars. Choose an $x^* \in B_{X^*}$ such that $E \subseteq M(x^* \geq \varepsilon)$. Then

$$\| \sum_{n \in E} a_n x_n \| \geq |x^*(\sum_{n \in E} a_n x_n)|$$

$$\geq \varepsilon | \sum_{n \in E} a_n | = \varepsilon \sum_{n \in E} a_n.$$

Let (m_k) be the sequence which increasingly enumerates M. Applying the Brunel–Sucheston Theorem I.2.12 we conclude that some subsequence $(x_{m_{k_i}})$ of (x_{m_k}) has a spreading model (e_i) that is unconditional. It follows that

$$\| \sum_{i=0}^{k-1} a_i e_i \| \geq (\varepsilon/2) \sum_{i=0}^{k-1} |a_i|,$$

for every k and every choice a_i ($i < k$) of positive scalars. Since (e_i) is unconditional this inequality (with $\varepsilon/2$ replaced by some other constant $C > 0$) remains true for every choice of scalars a_i ($i < k$). Hence (e_i) equivalent to the unit vector basis of ℓ_1. This finishes the proof. $\square$

Combining Corollary II.9.3 and Lemma II.9.7 we obtain the following.

Theorem II.9.8 (Rosenthal). *Let (x_n) be a weakly null sequence in some Banach space X. Then either:*

(1) *(x_n) has a subsequence all of whose subsequences are Cesàro-summable, or*

(2) *(x_n) has a subsequence with spreading model isomorphic to ℓ_1.* $\square$

Exercise II.9.9. Show that for every uniformly bounded sequence in some space of the form $\ell_\infty(\Gamma)$ the following are equivalent:

(1) There is $f \in \ell_\infty(\Gamma)$ such that

$$\lim_{n \to \infty} [\sum_{k_1 < \cdots < k_n} \|(\frac{1}{n} \sum_{i=1}^{n} f_{k_i})\|_\infty] = 0.$$

(2) Every subsequence of (f_n) is Cesàro-summable in $\ell_\infty(\Gamma)$ and each to the same limit.

Remark II.9.10. For further results in this area we refer the reader to the paper of Argyros, Mercourakis and Tsarpalias [7] which generalizes Cesàro-summability into a transfinite sequence of summability methods and prove the corresponding analogue of Rosenthal's theorem.

II.10   Summability in Topological Abelian Groups

In this section we examine the summability problem for series not only in Banach spaces but more generally in (Hausdorff) topological abelian groups. Given such a group G, we say that a series $\sum_{n=0}^\infty x_n$ of elements of G is *unconditionally convergent* if $\sum_{k=0}^\infty x_{n_k}$ is convergent in G for every subsequence (x_{n_k}) of (x_n).

Lemma II.10.1. *Let $\sum_{n=0}^\infty x_n$ be an unconditionally convergent series in a topological abelian group G. Suppose that $(A_n)_{n=0}^\infty$ and $(B_n)_{n=0}^\infty$ are sequences of Borel subsets of G such that $A_n \cup (B_n + x_n) = G$ for all n. Then there is a subsequence (x_{n_k}) of (x_n) such that $y = \sum_{k=0}^\infty x_{n_k}$ belongs to infinitely many A_n or infinitely many B_n.*

Proof. For $M \in \mathbb{N}^{[\infty]}$, let

$$x_M = \sum_{m \in M} x_m.$$

For $k \in \mathbb{N}$ let

$$\mathcal{N}_k = \{M \in \mathbb{N}^{[\infty]} : x_m \notin A_n \cup B_n \text{ for all } n \geq k\}.$$

Note that all $\mathcal{N}_k$'s are Borel subsets of $\mathbb{N}^{[\infty]}$, so by the Galvin–Prikry or Ellentuck theorem it suffices to show that all of them are Ramsey null. For suppose that $[l, M] \subseteq \mathcal{N}_k$ for some l and k. Pick $n > \max\{k, l\}$ in M. Let $N = M \setminus \{n\}$. Then $N \in [l, M] \subseteq \mathcal{N}_k$ so in particular neither x_M nor x_N belongs to $A_n \cup B_n$. However, note that by the assumption of the lemma, $x_M \notin A_n$ implies that $x_M \in B_n + x_n$, or equivalently $x_M - x_n = x_N \in B_n$, a contradiction. $\square$

Exercise II.10.2. (a) Show that Lemma II.10.1 is true if the sets A_n and B_n are assumed to be sequentially open in G.
(b) Show that if the sets A_n and B_n in Lemma II.10.1 are assumed only to have the property of Baire in G, then $\bigcap_m \bigcup_{n \geq m} (A_n \cup B_n)$ is comeager in G.
(c) Show that if G is locally compact and if the sets A_n and B_n of Lemma II.10.1 are assumed to be Haar measurable, then $\bigcap_m \bigcup_{n \geq m} (A_n \cup B_n)$ has the full measure.

Corollary II.10.3 (Uniform boundedness principle). *Suppose that $\mathcal{F}$ is a pointwise bounded family of continuous linear operators from a Banach space X into a normed space Y. Then $\mathcal{F}$ is uniformly bounded.*

Proof. If $\sup_{T \in \mathcal{F}} \|T\| = \infty$, then we can find two infinite sequences $(x_n) \subseteq X$ and $(T_n) \subseteq \mathcal{F}$ such that $\|x_n\| \leq 2^{-n}$ and $T_n(x_n) \geq n$ for all n. For $n \in \mathbb{N}$, let

$$A_n = B_n = \{x \in X : \|T_n(x)\| \geq \frac{n}{2}\}.$$

Then A_n (and B_n) is a closed subset of X and $A_n \cup (B_n + x_n) = X$ for all n. The series $\sum_{n=0}^{\infty} x_n$ is clearly unconditionally convergent, so by Lemma II.10.1 there is $x \in X$ such that $T_n(x) \geq n/2$ for infinitely many n, so the family $\mathcal{F}$ is not pointwise bounded. $\qquad\square$

Corollary II.10.4 (Automatic continuity principle). *Suppose that $T : X \to Y$ is a Borel linear map from a Banach space X into a normed space Y. Then T is continuous.*

Proof. Otherwise we can find an infinite sequence $(x_n) \subseteq X$ such that $\|x_n\| \leq 2^{-n}$ and $\|T(x_n)\| \geq n$ for all n. For $n \in \mathbb{N}$, let

$$A_n = B_n = \{x \in X : \|T(x)\| \geq n/2\}. \tag{II.42}$$

Then A_n and B_n are Borel sets such that $A_n \cup (B_n + x_n) = X$ for all n. By Lemma II.10.1 there is $x \in X$ such that $\|T(x)\| \geq n/2$ for infinitely many n, a contradiction. $\qquad\square$

Corollary II.10.5 (Schur's ℓ_1-theorem). *In the space ℓ_1 every weakly convergent sequence is norm convergent.*

Proof. Suppose $(x_n) \subseteq \ell_1$ is weakly null but for some $\varepsilon > 0$, $\|x_n\| \geq \varepsilon$ for all n. Since (x_n) is weakly null, there is an infinite subsequence (x_{n_k}) of (x_n) and a corresponding infinite sequence of disjointly and finitely supported functionals $(f_k) \subseteq S_{\ell_\infty}$ such that $|f_k(x_{n_k})| > \varepsilon/2$ for all k. For $k \in \mathbb{N}$, set

$$A_k = B_k = \{f \in \ell_\infty : |f(x_{n_k})| > \varepsilon/4\}.$$

The sets A_k and B_k are w^*-open and $\ell_\infty = A_k \cup (B_k + f_k)$ for all k. Since (f_k) is normalized and disjointly supported the series $\sum_{k=0}^{\infty} f_k$ is unconditionally convergent in (ℓ_∞, w^*). By Lemma II.10.1 there is $f \in \ell_\infty$ such that $|f(x_{n_k})| > \varepsilon/4$ for infinitely many k. This contradicts the fact that (x_n) is weakly null. $\qquad\square$

Exercise II.10.6 (Von Neumann). Find a weakly null sequence in ℓ_2 which is not norm null.

Corollary II.10.7 (Orlicz–Pettis Theorem). *A series $\sum_{n=0}^{\infty} x_n$ of some normed space X is unconditionally convergent relative to the norm topology iff it is unconditionally convergent relative to the weak topology.*

Proof. First of all note that we may assume that X is a separable Banach space and that it suffices to show that $\|x_n\| \to_n 0$ under the assumption that $\sum_{n=0}^{\infty} x_n$ is unconditionally convergent relative to the weak topology of X. Otherwise, there is an $\varepsilon > 0$ and an infinite subsequence (x_{n_k}) of (x_n) such that $\|x_{n_k}\| \geq \varepsilon$ for all k. By Lemma II.5.1, going to a subsequence we may assume that (x_{n_k}) is a semi-normalized basis of its closed linear span Y. The corresponding sequence $(f_k) \subseteq Y^*$ of the evaluation functionals relative to the basis (x_{n_k}) is also semi-normalized, is w^*-null and has the property that

$$|f_k(x_{n_k})| > \varepsilon \text{ for all } k.$$

For $k \in \mathbb{N}$, let

$$A_k = B_k = \{x \in Y : |f_k(x)| > \delta/2\}.$$

Then A_k and B_k are weakly open in Y and $Y = A_k \cup (B_k + x_{n_k})$ for all k. By Lemma II.10.1 there is $x \in Y$ such that $|f_k(x)| > \varepsilon/2$ for infinitely many k, contradicting the fact that (f_k) is w^*-null.   □

Corollary II.10.8 (Nikodym boundedness principle). *Let $\mathcal{F}$ be a family of countable additive set functionals defined on a measurable space $(X, \mathcal{B})$ such that*

$$b_E = \sup_{\mu \in \mathcal{F}} |\mu(E)| < \infty$$

for every $E \in \mathcal{B}$. Then

$$\sup_{E \in \mathcal{B}} b_E < \infty.$$

Proof. Suppose that $\sup_{E \in \mathcal{B}} b_E = \infty$. Pick $E_0 \in \mathcal{B}$ and $\mu_0 \in \mathcal{F}$ such that $|\mu_0(E_0)| > 1 + b_X$. Then $|\mu_0(E_0)| > 1$ and

$$|\mu_0(X \setminus E_0)| \geq |\mu_0(E_0)| - |\mu_0(X)| > 1.$$

At least one of $\sup_{E \subseteq E_0} b_E$ or $\sup_{E \subseteq X \setminus E_0} b_E$ must be infinite. Renaming $X \setminus E_0$ as E_0 if needed, we may assume that $\sup_{E \subseteq X \setminus E_0} b_E = \infty$. Then we can repeat the same argument inside $X \setminus E_0$ getting $E_1 \subseteq X \setminus E_0$ such that $\mu(E_1) > 2$ and $\sup_{E \subseteq X \setminus (E_0 \cup E_1)} b_E = \infty$, and so on. This gives us an infinite sequence $(E_n)_{n=0}^{\infty}$ of disjoint elements of $\mathcal{B}$ and a sequence $(\mu_n)_{n=0}^{\infty}$ such that $|\mu_n(E_n)| > 1 + n$ for all n.

Let G be the group of all bounded scalar-valued $\mathcal{B}$-measurable functions on X with the topology of pointwise convergence on the family of all countably additive set-functionals on $\mathcal{B}$. Then the series $\sum_{n=0}^{\infty} \chi_{E_n}$ is unconditionally convergent in G and the limits are characteristic functions of members of $\mathcal{B}$. Let

$$A_n = B_n = \{f \in G : |\int f d\mu_n| > (n+1)/2\}.$$

Then A_n and B_n are open in G, and $A_n \cup (B_n + \chi_n) = G$ for all n. By Lemma II.10.1 there exists an $E \in \mathcal{B}$ such that $|\mu_n(E)| > n + 1/2$ for infinitely many n, contradicting the assumption that $b_E < \infty$.   □

Corollary II.10.9 (Vitali–Hahn–Saks Theorem). *Suppose that $(X, \mathcal{B}, \mu)$ is a measure space (i.e., μ is non-negative on $\mathcal{B}$) and let $(\nu_n)_{n=0}^{\infty}$ be a sequence of μ-continuous additive measures defined on $\mathcal{B}$. Suppose that $\lim_{n \to \infty} \nu_n(E)$ exists for all $E \in \mathcal{B}$. Then the sequence $(\nu_n)_{n=0}^{\infty}$ is uniformly μ-continuous.*

Proof. Otherwise, one can find an infinite subsequence $(\nu_{n_k})_{k=0}^{\infty}$ of (ν_n) and a sequence $(E_k)_{k=0}^{\infty} \subseteq \mathcal{B}$ such that $\mu(E_k) \to 0$ but for some $\varepsilon > 0$, $|\nu_{n_k}(E_k)| > \varepsilon$ for all k. Using the μ-continuity of the ν_n's, and going to a subsequence of (ν_{n_k}) if needed, we may assume that for every k

$$|\nu_{n_k}(E)| < \varepsilon/3 \text{ for all } E \subseteq \bigcup_{l > k} E_k \text{ in } \mathcal{B}.$$

Let $F_k = E_k \setminus \bigcup_{l > k} E_k$ for $k \in \mathbb{N}$. Then (F_k) is an infinite sequence of disjoint sets from $\mathcal{B}$ such that $|\nu_{n_k}(F_k)| > (2/3)\varepsilon$ and $|\nu_{n_k}(F_l)| < \varepsilon/3$ for all $k < l$. In particular,

$$|\nu_{n_{k+1}}(F_{k+1}) - \nu_{n_k}(F_k)| > \varepsilon/3$$

for all k. As in the previous proof applying Lemma II.10.1 we get a set $E \in \mathcal{B}$ such that $|\nu_{n_{k+1}}(E) - \nu_{n_k}(E)| > \varepsilon/3$ for infinitely many k's. This means that $(\nu_n(E))_{n=0}^{\infty}$ is not a convergent sequence, a contradiction. $\square$

Remark II.10.10. We have taken Lemma II.10.1 (but not its proof) from a paper of Matheron [53] We refer the reader to papers [15] and [19] for some further uses of these ideas.

Definition II.10.11. A topological group G is *Mazur–Orlicz complete* if every null sequence (x_n) in G contains a subsequence (x_{n_k}) such that the series $\sum_{k=0}^{\infty} x_{n_k}$ is convergent in G. If in addition $\sum_{k=0}^{\infty} x_{n_k}$ is unconditionally convergent then we say that G is *unconditionally Mazur–Orlicz complete* .

Clearly every completely metrizable abelian group is unconditionally Mazur–Orlicz complete. The following result is a partial converse to this implication.

Theorem II.10.12 (Burzyk–Klis–Lipecki). *Every metrizable Mazur–Orlicz complete abelian group G is a Baire space.*

Proof. Consider a decreasing sequence $(U_n)_{n=0}^{\infty}$ of dense open subsets of G. We shall show that 0 is in the closure of the intersection from which it follows easily that $\bigcap_{n=0}^{\infty} U_n$ is dense in G. Let d be a fixed translation invariant metric on G giving us its topology and let $\varepsilon > 0$. A simple recursive construction will give us a sequence $(x_n)_{n=0}^{\infty} \subseteq G$ and a sequence $(V_n)_{n=0}^{\infty}$ of open subsets of G such that for all n:
(1) $d(0, x_n) \leq \varepsilon \cdot 2^{-n-1}$,
(2) $\overline{V_n} \subseteq U_n$, and
(3) $\sum_{k=0}^{n} \sigma_k x_k \in \bigcap_{k=0}^{n} V_k^{\sigma_k}$ for all $(\sigma_k)_{k=0}^{n} \subseteq \{0, 1\}$,
where for a subset S of G we let $S^0 = X$ and $S^1 = S$. Applying the Mazur–Orlicz completeness of G to (x_n) we get a subsequence (x_{n_i}) such that the series $\sum_{i=0}^{\infty} x_{n_i}$

converges in G. It follows that $\sum_{i=0}^{\infty} x_{n_i} \in \bigcap_{n=0}^{\infty} U_n$ and $d(0, \sum_{i=0}^{\infty} x_{n_i}) \leq \delta$, as required. $\qquad\square$

Corollary II.10.13. *Suppose that G is a metrizable abelian group which has the property of Baire in its completion. Then G is complete if and only if it is Mazur–Orlicz complete.* $\qquad\square$

Corollary II.10.14. *Let Y be a subspace of a Banach space X. If Y is an analytic subset of X and if it is Mazur–Orlicz complete then it must in fact be closed.* $\qquad\square$

Remark II.10.15. This corollary has been first observed by A.R.D. Mathias who was using Silver's theorem to prove it (see [39] and [88]).

Note that there are interesting groups that are unconditionally Mazur–Orlicz complete but are not Baire spaces. One such example is the group $G = (\ell_1, w)$. They of course cannot be metrizable. It turns out however that non-metrizable group do share some of the properties of the metrizable ones as the following result shows.

Theorem II.10.16. *Let G be an unconditionally Mazur–Orlicz complete abelian group. Then every Borel subgroup H of G which is Mazur–Orlicz complete must in fact be closed in G.*

Proof. Suppose H is not closed in G and pick a sequence $(x_n) \subseteq H$ converging to a point of $G \setminus H$. Applying the unconditional Mazur–Orlicz completeness of G and going to a subsequence of (x_n) we may assume that the series $\sum_{n=0}^{\infty} x_n$ is unconditionally convergent in G. For $n \in \mathbb{N}$ let $y_n = x_{n+1} - x_n$. Note that $\sum_{n=0}^{\infty} y_n \notin H$. For $M \in \mathbb{N}^{[\infty]}$, let

$$y_M = \sum_{k=0}^{\infty} \sum_{n=M(2k)}^{M(2k+1)-1} y_n,$$

where $(M(k))_{k=0}^{\infty}$ is the increasing enumeration of M. Let $\mathfrak{X}_M$ be the collection of all $M \in \mathbb{N}^{[\infty]}$ such that $y_M \in H$. Then $\mathfrak{X}_H$ is a Borel subset of $\mathbb{N}^{[\infty]}$, so Galvin–Prikry theorem applies to $\mathfrak{X}_H$. Pick an $M \in \mathbb{N}^{[\infty]}$ such that $M^{[\infty]} \subseteq \mathfrak{X}_H$ or $M^{[\infty]} \cap \mathfrak{X}_H = \emptyset$. Note that for $N \in \mathbb{N}^{[\infty]}$ and $n \in \mathbb{N}$ at most one of the elements y_N and $y_{N \setminus \{n\}}$ can belong to $\mathfrak{X}_H$ since $\sum_{n=0}^{\infty} y_n \notin H$. It follows that $M^{[\infty]} \cap \mathfrak{X}_H = \emptyset$. For $k \in \mathbb{N}$, set

$$z_k = \sum_{n=M(2k)}^{M(2k+1)-1} y_k.$$

The fact that the series $y_M = \sum_{k=0}^{\infty} z_k$ is convergent in G means in particular that the sequence (z_k) of elements of H converges to 0. Applying the Mazur–Orlicz completeness of H to (z_k) we get that a subsequence (z_{k_i}) such that $\sum_{i=0}^{\infty} z_{k_i}$ is convergent in H. Note however that $\sum_{i=0}^{\infty} z_{k_i}$ is equal to y_N for some $N \in M^{[\infty]}$, so $M^{[\infty]} \cap \mathfrak{X}_H \neq \emptyset$, a contradiction. $\qquad\square$

Chapter III

Ramsey Theory of Finite and Infinite Block Sequences

III.1 Hindman's Theorem

Hindman's theorem in its finite unions form (rather than non-repeating sums form) can be considered as the first non-trivial result of so-called block Ramsey theory. To state this theorem, let FIN denote the collection of all finite nonempty subsets of $\mathbb{N}$. A finite or infinite sequence $X = (x_i)$ of elements of FIN is a *block sequence* if

$$x_i < x_j{}^1 \text{ whenever } i < j.$$

For a block sequence $X = (x_i)$ set

$$[X] = \{x_{i_0} \cup \cdots \cup x_{i_k} : k \in \mathbb{N},\ i_0 < \cdots < i_k < |X|\}.$$

We call $[X]$ a partial *subsemigroup* of $(\text{FIN}, \cup)$ generated by $X = (x_i)$. Note that when $X = (x_i)_{i=0}^{\infty}$ is infinite then $([X], \cup)$ is isomorphic to $(\text{FIN}, \cup)$ via isomorphism which sends $\{i\}$ to x_i. So the following well-known result gives us a basic pigeon-hole principle for the semigroup FIN.

Theorem III.1.1 (Hindman). *For every finite coloring of* FIN *there is an infinite block sequence* X *in* FIN *such that* $[X]$ *is monochromatic.*

Proof. Let γFIN be the collection of all ultrafilters $\mathcal{U}$ on FIN such that for all $n \in \mathbb{N}$,

$$\{x \in \text{FIN} : n < x\} \in \mathcal{U}.$$

For $\mathcal{U}, \mathcal{V} \in \gamma$FIN let

$$\mathcal{U} \cup \mathcal{V} = \{A \subseteq \text{FIN} : \{x : \{y : x \cup y \in A\} \in \mathcal{V}\} \in \mathcal{U}\}.$$

---

[1]For $x, y \in$ FIN by $x < y$ we denote the fact that $\max(x) < \min(y)$.

Note that $\mathcal{U} \cup \mathcal{V}$ is an ultrafilter and that it belongs to γFIN. So we have defined an operation $\cup$ on γFIN which is easily seen to be associative, i.e.,

$$(\mathcal{U} \cup \mathcal{V}) \cup \mathcal{W} = \mathcal{U} \cup (\mathcal{V} \cup \mathcal{W}).$$

One can view γFIN as a closed subset of the Stone–Čech compactification βFIN of FIN taken with its discrete topology. Thus a basic open set of γFIN is determined by a subset A of FIN as below:

$$A^* = \{\mathcal{U} \in \gamma\text{FIN} : A \in \mathcal{U}\}. \qquad \square$$

Lemma III.1.2. *For each* $\mathcal{V} \in \gamma$FIN *the function* $\mathcal{U} \mapsto \mathcal{U} \cup \mathcal{V}$ *is a continuous function from* γFIN *into* γFIN.

Proof. It suffices to show that for every $A \subseteq$ FIN, the set

$$T = \{\mathcal{U} \in \gamma\text{FIN} : A \in \mathcal{U} \cup \mathcal{V}\}$$

is open in γFIN. Let $B = \{x : \{y : x \cup y \in A\} \in \mathcal{V}\}$. Then $T = B^*$, and so T is open in γFIN. $\qquad \square$

Lemma III.1.3 (Ellis). *Suppose* $(S, *)$ *is a compact semigroup such that* $x \mapsto x * y$ *is a continuous map for each* $y \in S$. *Then* $(S, *)$ *has an idempotent.*

Proof. By compactness, S contains a minimal closed nonempty subsemigroup T. Pick $t \in T$. By continuity of $x \mapsto x * t$ the subsemigroup $T * t$ is compact and therefore closed subsemigroup of T. It follows that $T * t = T$ so

$$V = \{x \in T : x * t = t\}.$$

is nonempty. It is also closed being a preimage of the point t under $x \mapsto x * t$. It follows that $V = T$ and therefore $t * t = t$, as required. $\qquad \square$

The following lemma is the last step in the proof of Theorem III.1.1.

Lemma III.1.4. *Suppose* $\mathcal{U}$ *is an idempotent of* $(\gamma$FIN$, \cup)$. *Then every* $A \in \mathcal{U}$ *contains a subsemigroup of* (FIN$, \cup)$ *generated by an infinite block sequence.*

Proof. Since $A \in \mathcal{U} = \mathcal{U} \cup \mathcal{U}$ we can find $x_0 \in A$ such that

$$A_1 = \{y \in A : x_0 < y \text{ and } x_0 \cup y \in A\} \in \mathcal{U}.$$

Then again since $A_1 \in \mathcal{U} \cup \mathcal{U}$ we can find $x_1 \in A_1$ such that

$$A_2 = \{y \in A_1 : x_1 < y \text{ and } x_1 \cup y \in A_1\} \in \mathcal{U},$$

and so on. This procedure gives us a decreasing sequence $A = A_0 \supseteq A_1 \supseteq \cdots \supseteq A_n \supseteq \cdots$ of elements of $\mathcal{U}$ and an infinite block sequence $X = (x_n)$ such that $x_n \in A_n$ and

$$A_{n+1} = \{y \in A_n : x_n < y \text{ and } x_n \cup y \in A_n\}.$$

Inductively on $k \in \mathbb{N}$ one shows that for every sequence $n_0 < n_1 < \cdots < n_k$ of elements of $\mathbb{N}$,

$$x_{n_0} \cup x_{n_1} \cup \cdots \cup x_{n_k} \in A_{n_0}.$$

This is clearly so for $k = 0$. To see the recursive step, let $y = x_{n_1} \cup \cdots \cup x_{n_k}$. Then by the inductive hypothesis, $y \in A_{n_1}$. Since $A_{n_1} \subseteq A_{n_0+1}$ we get that $y \in A_{n_0+1}$ and therefore $x_{n_0} \cup y \in A_{n_0}$ as required. $\qquad\square$

Exercise III.1.5. Show that $\mathcal{V} \mapsto \mathcal{U} \cup \mathcal{V}$ is not necessarily a continuous function on γFIN, when $\mathcal{V}$ is taken to be an arbitrary member of γFIN.

For an integer $k \geq 1$, let $\mathrm{FIN}^{[k]}$ denote the collection of all block sequences of elements of FIN of length k. Similarly, let $\mathrm{FIN}^{[\infty]}$ denote the set of all infinite block sequences of elements of FIN. The following is an analogue of the Ramsey theorem in the context of FIN.

Theorem III.1.6 (Taylor). *For every integer $k \geq 1$ and every finite coloring of $\mathrm{FIN}^{[k]}$ there exists an infinite block sequence $X = (x_i)$ such that $[X]^{[k]}$ is monochromatic.*

Before proving the theorem, we give some definitions: For two block sequences $X = (x_i)$ and $Y = (y_i)$ of finite subsets of $\mathbb{N}$ we put

$$X \leq Y \text{ iff } x_i \subset [Y] \text{ for all } i.$$

For a block sequence $X = (x_i) \subseteq \mathrm{FIN}$ and $s \in \mathrm{FIN}$, let

$$X/s = (x_{m+j})_{j \geq 1}$$

where $m = \min\{i : x_i > s\}$. Thus X/s is the block sequence that enumerates the tail of X determined by s.

It is clear that Theorem III.1.6 follows from the next lemma.

Lemma III.1.7. *For every $P \subseteq \mathrm{FIN}^{[k]}$ and every infinite block sequence Y of elements of FIN there is an infinite block sequence $X \leq Y$ such that $[X]^{[k]} \subseteq P$ or $[X]^{[k]} \cap P = \emptyset$.*

Proof. The proof is by induction on the integer k. The case $k = 1$ is given by the proof of Hindman's theorem restricted to the semigroup $[Y]$ instead of FIN. So we assume $k > 1$ and that the Lemma is true for $k - 1$.

Recursively on n, we build a block sequence $Z = (z_n)$ in $[Y]$ and a decreasing sequence (Y_n) of members of $\mathrm{FIN}^{[\infty]}$ such that $Y_0 = Y$ and such that:
(1) z_n is the first term of Y_n.
(2) For every n and every $(x_0, \ldots, x_{k-2}) \in [(z_i)_{i=0}^n]^{k-1}$ either $(x_0, \ldots, x_{k-2}, x) \in P$ for all $x \in [Y_{n+1}]$, or $(x_0, \ldots, x_{k-2}, x) \notin P$ for all $x \in [Y_{n+1}]$.
There is no problem in constructing (z_n) and (Y_n) since (2) is provided by the case $k = 1$ of the lemma.

Having constructed $Z = (z_n)$ we note (see (2)) that for $(x_0, \ldots, x_{k-2}, x_{k-1}) \in [Z]^{[k]}$, the sentence

$$(x_0, \ldots, x_{k-2}, x_{k-1}) \in P$$

does not depend on x_{k-1}, or in other words, $P \cap [Z]^{[k]}$ is a cylinder over its projection on $[Z]^{[k-1]}$. So we are done using the inductive hypothesis.   □

III.2   Canonical Equivalence Relations on FIN

Consider the following five equivalence relations on FIN:

$$x E_0 y \text{ iff } x = y$$
$$x E_1 y \text{ iff } x = x$$
$$x E_2 y \text{ iff } \min(x) = \min(y)$$
$$x E_3 y \text{ iff } \max(x) = \max(y)$$
$$x E_4 y \text{ iff } \min(x) = \min(y) \text{ and } \max(x) = \max(y).$$

The purpose of this section is to show that these are all essential equivalence relations on FIN provided one is willing to shrink to a subsemigroup of FIN generated by an infinite block sequence.

Theorem III.2.1 (Taylor). *For every equivalence relation E on FIN there is $i < 5$ such that $E \restriction [X] = E_i \restriction [X]$ for some infinite block sequence $X = (x_i)$.*

Proof. An *equation* (in 4 variables v_0, v_1, v_2, v_3) is a formula $\varphi(a, b, v_0, v_1, v_2, v_3)$ of the form

$$\varphi(a, b, v_0, v_1, v_2, v_3) \equiv a \cup v_{i_0} \cup \cdots \cup v_{i_3} \ E \ b \cup v_{j_0} \cup \cdots \cup v_{j_3}$$

where $i_0, j_0, \ldots, i_3, j_3 \in \{0, 1, 2, 3\}$ and $a, b \in \text{FIN} \cup \{\emptyset\}$.

Given a block sequence Z and an equation $\varphi \equiv \varphi(a, b, v_0, \ldots, v_3)$ we say that φ is *true* in Z iff $\varphi(a_0, \ldots, a_3)$ holds for every $(a_0, \ldots, a_3) \in [Z/(a \cup b)]^4$. We say that φ is *false* in Z iff $\neg\varphi(a_0, \ldots, a_3)$ holds for every $(a_0, \ldots, a_3) \in [Z/(a \cup b)]^4$. We say that φ is *decided* in Z if either φ is true or φ is false in Z.

By Theorem III.1.6, and using a non difficult diagonal procedure, we can find a block sequence $Z = (z_n)_n$ such that all equations $\varphi(a, b, v_0, \ldots, v_3)$ with $a, b \in [Z] \cup \{\emptyset\}$ are decided in Z. Our aim is to show that there is some $i < 5$ such that $E \restriction [Z] = E_i \restriction [Z]$.

Case 1. $v_0 \ E \ v_1$ *is true in* Z. Then $E \restriction [Z] = E_1 \restriction [Z]$: Let $s, t \in [Z]$ and pick $u \in [Z]$ such that $u > s, t$. Then $s \ E \ u$ and $t \ E \ u$, and hence $s \ E \ t$.

Case 2. $v_0 \ E \ v_1$ *is false in* Z, $v_0 \cup v_1 \ E \ v_0$ *is true in* Z *and* $v_0 \cup v_1 \ E \ v_1$ *is false in* Z. Let us check that E is equal to E_2 on $[Z]$. Fix $s, t \in [Z]$. Suppose first that $s \ E_2 \ t$, and write $s = z_n \cup s'$ and $t = z_n \cup t'$, with $s', t' \in [Z]$ and $z_n < s', t'$. Using that $v_0 \cup v_1 \ E \ v_0$ is true in Z, we obtain that $s, t \ E \ z_n$, and hence $s \ E \ t$.

Assume that s E t and suppose now that s $\not{E}_2$ t. Without loss of generality, we may assume that $\min(s) < \min(t)$. Let n be such that $s = z_n \cup s'$, $s' \in [Z]$ and $z_n < s'$. Then s E z_n, $z_n < t$, and hence z_n E t. Since $z_n < t$, we obtain that v_0 E v_1 is true in Z, a contradiction.

Case 3. v_0 E v_1 and $v_0 \cup v_1$ E v_0 are false in Z, and $v_0 \cup v_1$ E v_1 is true in Z. Similar proof than in Case 2 shows that E is equal to E_3 on $[Z]$.

For the last two cases we use the following fact

Claim. If $v_0 \cup v_1$ E v_0 is false in Z, then $E \subseteq E_4$ when restricted to $[Z]$: Suppose that $\max s \neq \max t$ and s E t. We may assume that $\max s < \max t$. Let n be such that $t = t' \cup z_n$ with $t' < z_n$ and $t' \in [Z]$. Since s E t, the equation s E $t' \cup v_0$ is true in Z and hence the equation $t' \cup v_0 \cup v_1$ E $t' \cup v_0$ also is true in Z. This implies that $v_0 \cup v_1$ E v_0 is true in Z.

Case 4. v_0 E v_1, $v_0 \cup v_1$ E v_0 and $v_0 \cup v_1$ E v_1 are false in Z, and $v_0 \cup v_1 \cup v_2$ E $v_0 \cup v_2$ is true in Z. Then E is equal to E_4 on $[Z]$: It is not difficult to prove that $E_4 \subseteq E$ on $[Z]$. For the converse use the previous claim.

Suppose now that $\max s = \max t$ but $\min s \neq \min t$, say $\min s < \min t$. Suppose that s E t and we work for a contradiction. Write $s = z_{n_0} \cup s' \cup z_n$, $t = z_{m_0} \cup t' \cup a_n$ with $z_{n_0} < s' < z_n$, $z_{m_0} < t' < a_n$, $n_0 < m_0$, and all in $[Z]$. Using that the equation $v_0 \cup v_1 \cup v_2$ E $v_0 \cup v_2$ is true in Z, we may assume that $s' = t' = \emptyset$. Since $n_0 < m_0 \leq n$, one of the equations $v_0 \cup v_2$ E $v_1 \cup v_2$ or $v_0 \cup v_1$ E v_1 must be true in Z. The second case is impossible by hypothesis. In the first case we obtain that $v_0 \cup v_3$ E $v_1 \cup v_2 \cup v_3$ and $v_0 \cup v_3$ E $v_2 \cup v_3$ are true in Z and hence so is v_0 E $v_0 \cup v_1$, a contradiction.

Case 5. v_0 E v_1, $v_0 \cup v_1$ E v_0, $v_0 \cup v_1$ E v_1, and $v_0 \cup v_1 \cup v_2$ E $v_0 \cup v_2$ are all false in Z. Then $E \upharpoonright [Z] = E_0 \upharpoonright [Z]$. For suppose that s E t and that $s \neq t$. Since $v_0 \cup v_1$ E v_0 is false in Z, we obtain that $\max s = \max t$ (previous claim). Write $s = \bigcup_{i \in F} z_i$, $t = \bigcup_{i \in G} z_i$ with $\max F = \max G$. Let

$$k = \max(F \triangle G).$$

Without loss of generality we assume that $k \in F \setminus G$. This implies that $s = s' \cup z_k \cup s''$ and $t = t' \cup s''$ with $s', t' < z_k < s''$, $s'' \neq \emptyset$ and all in $[Z]$. Therefore the equation $s' \cup v_0 \cup v_1$ E $t' \cup v_1$ is true in Z, which implies that $s' \cup v_0 \cup v_1 \cup v_2$ E $t' \cup v_2$ and $s' \cup v_0 \cup v_2$ E $t' \cup v_2$ are both true in Z. So, the equation $v_0 \cup v_1 \cup v_2$ E $v_0 \cup v_2$ is true in Z, a contradiction. $\qquad\Box$

III.3   Fronts and Barriers on FIN$^{[<\infty]}$

Let FIN$^{[<\infty]}$ denote the collection of all finite block sequences of elements of FIN and let FIN$^{[\infty]}$ denote the collection of all infinite block sequences of elements of FIN. Recall the ordering $\leq$ on FIN$^{[<\infty]} \cup$ FIN$^{[\infty]}$:

$$X = (x_i) \leq Y = (y_i) \text{ iff } x_i \in [Y] \text{ for all } i.$$

For a family $\mathcal{F}$ of finite block sequences of FIN and an infinite block sequence Y of members of FIN in analogy with the classical case of $\mathbb{N}^{[<\infty]}$ and $\mathbb{N}^{[\infty]}$ it is natural to consider the following form of restriction

$$\mathcal{F} \restriction [Y] = \{s \in \mathcal{F} : s \leq Y\}$$

based on the ordering $\leq$ of being a block subsequence rather than a subset. As in the classical case, there is another natural ordering on $\mathrm{FIN}^{[<\infty]} \cup \mathrm{FIN}^{[\infty]}$:

$$X \sqsubseteq Y \text{ iff } X = Y \restriction n \text{ for some } n,$$

or in other words $X \sqsubseteq Y$ iff the block sequence X is an initial segment of the block sequence Y. By $\sqsubset$ we denote the strict version of this ordering. Using the orderings $\leq$ and $\sqsubseteq$ one can go on and define analogues of Nash–Williams' notions of fronts and barriers in the present context.

   In what follows, we shall reserve the variable $s, t, u, v, \ldots$ for members of $\mathrm{FIN}^{[<\infty]}$, i.e., for finite block sequences and variables $X, Y, Z, \ldots$ for members of $\mathrm{FIN}^{[\infty]}$, i.e., for infinite block sequences of elements of FIN.

Definition III.3.1. (i) A family $\mathcal{F} \subseteq \mathrm{FIN}^{[<\infty]}$ is *thin* if $s \not\sqsubseteq t$ for distinct $s, t \in \mathcal{F}$.
(ii) A family $\mathcal{F} \subseteq \mathrm{FIN}^{[<\infty]}$ is *Sperner* if $s \not\leq t$ for all $s \neq t$ in $\mathcal{F}$.
(iii) A family $\mathcal{F} \subseteq \mathrm{FIN}^{[<\infty]}$ is *Ramsey* if for every finite partition

$$\mathcal{F} = \mathcal{F}_0 \cup \cdots \cup \mathcal{F}_n$$

there is $X \in \mathrm{FIN}^{[\infty]}$ such that at most one of the restrictions

$$\mathcal{F}_0 \restriction [X], \ldots, \mathcal{F}_n \restriction [X]$$

is nonempty.

Example III.3.2. By Theorem III.1.6, for each positive integer k, the family $\mathrm{FIN}^{[k]}$ of all block sequences of length k is thin, Sperner, and Ramsey.

Definition III.3.3. Given $Y \in \mathrm{FIN}^{[\infty]}$ a family $\mathcal{F}$ of finite block subsequences of Y is a *front* on Y if $\mathcal{F}$ is thin and has the property that every infinite block subsequence of Y has an initial segment in $\mathcal{F}$.

Definition III.3.4. For a given $Y \in \mathrm{FIN}^{[\infty]}$, a family $\mathcal{F}$ of finite block sequences of Y is a *barrier on* Y if $\mathcal{F}$ is Sperner and has the property that every infinite block subsequence of Y has an initial segment in $\mathcal{F}$.

   In order to define something analogous to the lexicographical ordering of $\mathbb{N}^{[<\infty]}$ that makes fronts and barriers well-ordered we use the following well ordering on FIN:

$$x <^r_{\mathrm{lex}} y \text{ iff } \max(x \triangle y) \in y.$$

Example III.3.5. Show that $(\mathrm{FIN}, <^r_{\mathrm{lex}})$ is a well ordered set and compute its order type.

Using $<^r_{\text{lex}}$ one can define the corresponding lexicographical ordering on FIN$^{[<\infty]}$ as follows:

$$s <_{\text{lex}} t \text{ iff } s \sqsubset t \text{ or } s(k) <^r_{\text{lex}} t(k),$$

where $k = \min\{i : s(i) \neq t(i)\}$. The following lemma has a proof quite similar to that of Lemma II.2.15 and allows us to prove statements about fronts and barriers by induction on their lexicographical ranks.

Lemma III.3.6. *Every front is lexicographically well ordered.* $\qquad\qquad\square$

We are now ready to state and prove the analogue of Galvin's lemma in this context that will play a crucial role in the further development of the theory of fronts and barriers of FIN$^{[<\infty]}$.

Lemma III.3.7 (Milliken). *For every* $\mathcal{B} \subseteq$ FIN$^{[<\infty]}$ *and every* $Y \in$ FIN$^{[\infty]}$ *there exists* $X \leq Y$ *such that either* $\mathcal{B} \cap [X] = \emptyset$ *or* $\mathcal{B} \cap [X]$ *contains a barrier on* X.

Proof. The proof of this Lemma is quite analogous to the proof of Lemma II.3.8 above. The crucial to the proof is the following notion of combinatorial forcing. First of all we make the analogous definition of a basic set:

$$[s, Y] = \{X \in \text{FIN}^{[\infty]} : s \sqsubseteq X \leq s^\frown(Y/s)\}.^2$$

Definition III.3.8. Fix a family $\mathcal{B} \subseteq$ FIN$^{[<\infty]}$. We say that Y *accepts* s if every $X \in [s, Y]$ has an initial segment in $\mathcal{B}$. We say that Y *rejects* s if there is no $X \leq Y$ accepting s. We say that Y *decides* s if Y either accepts or rejects s.

Definition III.3.9. A subset S of FIN is *small* if it contains no set of the form $[X]$ for some infinite block sequence X of members of FIN.

Note the following immediate properties of the forcing relation:
(i) If Y accepts (rejects) s then every $X \leq Y$ accepts (rejects) s.
(ii) For every Y and s there is $X \leq Y$ which decides s.
(iii) If Y accepts s then Y accepts $s^\frown\langle x\rangle$ for every $x \in [Y/s]$.
(iv) If Y decides of all its finite block subsequences but it rejects s then the set $\{x \in [Y] : Y \text{ does not reject } s^\frown\langle x\rangle\}$ is small.

Starting with $Y \in$ FIN$^{[\infty]}$ and using (i) and (ii) we build a decreasing sequence (Y_n) of infinite block subsequences of Y and an infinite block sequence $Z = (z_n)$ such that for all n, z_n is the first term of Y_n and Y_n decides every $s \leq (z_i)_{i<n}$. If the resulting block sequence Z accepts $\emptyset$ we have the conclusion of Lemma III.3.7, so let us assume that Z rejects $\emptyset$. By (iv) and Hindman's theorem we can find $Z_0 \leq Z$ such that Z rejects all $\langle x\rangle$ for $x \in [Z_0]$. Let x_0 be the first term of Z_0. By (iv) and Hindman's theorem we can find $Z_1 \leq Z/x_0$ such that Z rejects $\langle x_0, x\rangle$ as well as $\langle x\rangle$ for all $x \in [Z_1]$. Let x_1 be the first term of Z_1, and so

---

$^2 s^\frown(Y/s)$ is the infinite block sequence which starts as s and continues with the minimal term of Y above the maximal term of s.

on. At stage n, assuming that Z rejects all $s \leq \langle x_0, \ldots, x_n \rangle$ we find $Z_{n+1} \leq Z_n/x_n$ such that Z rejects $s^\frown \langle x \rangle$ for all $s \leq \langle x_0, \ldots, x_n \rangle$ and $x \in [Z_{n+1}]$. This procedure gives us an infinite block sequence $X = (x_n)$ which rejects all of its finite block subsequences. Clearly, this X satisfies the other alternative of Lemma III.3.7, i.e.,

$$\mathcal{B} \cap [X] = \emptyset.$$

This finishes the proof of Lemma III.3.7.   □

Corollary III.3.10. *Suppose $\mathcal{F}$ is a front on some $Y \in \mathrm{FIN}^{[\infty]}$. Then there exists $Z \leq Y$ in $\mathrm{FIN}^{[\infty]}$ such that $\mathcal{F} \cap [Z]$ is a barrier on Z.*   □

We are now ready to describe the notion of uniformity for fronts and barriers on $\mathrm{FIN}^{[<\infty]}$ that is likely to play a role in applications analogous to that of the classical case.

Definition III.3.11. Let α be a countable ordinal and let $Y \in \mathrm{FIN}^{[\infty]}$. We say that a family $\mathcal{B} \subseteq \mathrm{FIN}^{[<\infty]}$ is α-*uniform on* Y provided that:
(a) $\alpha = 0$ implies $\mathcal{B} = \{\emptyset\}$.
(b) $\alpha = \beta + 1$ implies $\emptyset \notin \mathcal{B}$ and $\mathcal{B}_{(x)}{}^3$ is β-uniform on Y/x for all $x \in [Y]$.
(c) $\alpha > 0$ limit implies that for all $x \in [Y]$ there is (unique) ordinal $\alpha_x < \alpha$ such that $\mathcal{B}_{(x)}$ is α_x-uniform on Y/x and such that the set $\{x \in [Y] : \alpha_x \leq \beta\}$ is small for all $\beta < \alpha$.

Note the following consequences of this definition proved by a simple induction on α.

Lemma III.3.12. *If $\mathcal{B}$ is an uniform family on Y, then $\mathcal{B}$ is a front on Y.*   □

Lemma III.3.13. *If $\mathcal{B}$ is α-uniform on Y, then $\mathcal{B}$ is α-uniform on any $X \leq Y$.*   □

Lemma III.3.14. *If $\mathcal{B}$ is an uniform family on Y (i.e., α-uniform on Y for some α), then $\mathcal{B}$ is a maximal thin family of finite block subsequences of Y.*

Example III.3.15. For every positive integer k the family $\mathrm{FIN}^{[k]}$ of all block sequences of length k is k-uniform on the maximal infinite block sequence $Y = (\{n\})_{n=0}^{\infty}$ of members of FIN.

Exercise III.3.16. Construct an ω-uniform and an $(\omega + 1)$-uniform family on FIN.

One of the main advantages of uniform families over arbitrary fronts is the possibility of proving results about them by typically easy induction arguments on their uniformity indexes α. The following lemma shows that our notion of uniformity is abundant among fronts of $\mathrm{FIN}^{[<\infty]}$ so the induction arguments applicable in the uniform case will typically cover most general cases.

Lemma III.3.17. *For every family $\mathcal{B} \subseteq \mathrm{FIN}^{[<\infty]}$ and for every $Y \in \mathrm{FIN}^{[\infty]}$ there is $X \leq Y$ such that either $\mathcal{B} \cap [X] = \emptyset$, or else $\mathcal{B} \cap [X]$ contains an uniform family on X.*

${}^3\mathcal{B}_{(x)} = \{s \in \mathrm{FIN}^{[<\infty]} : \langle x \rangle^\frown s \in \mathcal{B}\}$.

Proof. By Lemma III.3.7 we may assume to have a $Z \leq Y$ such that $\mathcal{B} \cap [Z]$ contains a barrier on Z. Shrinking $\mathcal{B}$ we assume that $\mathcal{B}$ is actually a barrier on Z. Now the second alternative of the lemma is achieved by induction on the lexicographical rank of $\mathcal{B}$ since clearly,

$$\operatorname{rank}(\mathcal{B}_{(x)}) < \operatorname{rank}(\mathcal{B})$$

for all $x \in [Z]$. $\qquad\qquad\square$

Theorem III.3.18. *The following are equivalent for every family* $\mathcal{F} \subseteq \mathrm{FIN}^{[<\infty]}$:
(a) $\mathcal{F}$ *is Ramsey.*
(b) *There is* $Y \in \mathrm{FIN}^{[\infty]}$ *such that* $\mathcal{F} \cap [Y]$ *is Sperner.*
(c) *There is* $Y \in \mathrm{FIN}^{[\infty]}$ *such that* $\mathcal{F} \cap [Y]$ *is uniform on* Y.
(d) *There is* $Y \in \mathrm{FIN}^{[\infty]}$ *such that* $\mathcal{F} \cap [Y]$ *is thin.*
(e) *There is* $Y \in \mathrm{FIN}^{[\infty]}$ *such that for every* $X \leq Y$ *the restriction* $\mathcal{F} \cap [X]$ *does not contain two disjoint uniform families on* X.

Proof. (a) $\rightarrow$ (b) Apply the Ramsey property of $\mathcal{F}$ to the partition $\mathcal{F} = \mathcal{F}_0 \cup \mathcal{F}_1$ where $\mathcal{F}_0$ is the set of all $\leq$-minimal members of $\mathcal{F}$.
(b) $\rightarrow$ (c) This follows from Lemma III.3.17.
(c) $\rightarrow$ (d) This follows from Lemma III.3.12.
(d) $\rightarrow$ (e) The union of two disjoint fronts on the same $X \in \mathrm{FIN}^{[\infty]}$ cannot be thin.
(e) $\rightarrow$ (a) Consider a finite partition

$$\mathcal{F} = \mathcal{F}_0 \cup \cdots \cup \mathcal{F}_k.$$

By Lemma III.3.17 there is $X \leq Y$ in $\mathrm{FIN}^{[\infty]}$ such that for all $i \leq k$, the restriction $\mathcal{F}_i \cap [X]$ is either empty or it contains a uniform family. Since by our assumption $\mathcal{F} \cap [X]$ contains no two disjoint uniform families at most one of the restrictions $\mathcal{F}_i \cap [X]$ ($i \leq k$) is non-empty. $\qquad\square$

Remark III.3.19. Note that once we have the analogue of Galvin's Lemma for the space $\mathrm{FIN}^{[\infty]}$ (i.e. Lemma III.3.7) the theories of fronts and barriers on $\mathbb{N}^{[<\infty]}$ and $\mathrm{FIN}^{[<\infty]}$ become strikingly similar. The reader is urged to complete the theory of fronts and barriers on $\mathrm{FIN}^{[<\infty]}$ and examine the further analogues from the theory of $\mathbb{N}^{[<\infty]}$. For example, it is quite natural to try to find the descriptions of all canonical equivalence relations on uniform fronts $\mathrm{FIN}^{[k]}$ of finite rank, and more generally to describe all equivalence relations on all uniform fronts of $\mathrm{FIN}^{[<\infty]}$.

III.4   Milliken's Theorem

In this section we prove an infinite-dimensional version of Hindman's theorem, a theorem about coloring of the set $\mathrm{FIN}^{[\infty]}$ of all infinite block sequences of finite nonempty subsets of $\mathbb{N}$.

Definition III.4.1. A set $\mathfrak{X} \subseteq \mathrm{FIN}^{[\infty]}$ is *Ramsey* if for every $[s, Y]$ there is $X \in [s, Y]$ such that $[s, X] \subseteq \mathfrak{X}$ or $[s, X] \cap \mathfrak{X} = \emptyset$. We say that $\mathfrak{X}$ is *Ramsey-null* if the second alternative always holds.

The sets of the form

$$[s, Y] = \{X \in \mathrm{FIN}^{[\infty]} : s \sqsubseteq X \leq s^\frown(Y/s)\}$$

where $s \in \mathrm{FIN}^{[<\infty]}$ and $Y \in \mathrm{FIN}^{[\infty]}$ form a base for a topology on $\mathrm{FIN}^{[\infty]}$ which we call *Milliken's topology* on $\mathrm{FIN}^{[\infty]}$ which is obviously richer than the metrizable topology induced from the Tychonoff power $\mathrm{FIN}^{\mathbb{N}}$ with FIN taken with its discrete topology. It is this topology that we usually refer to when we think of $\mathrm{FIN}^{[\infty]}$ as a topological space. The weaker topology of $\mathrm{FIN}^{[\infty]}$ will be referred to as the *metric topology* of $\mathrm{FIN}^{[\infty]}$.

Lemma III.4.2. *Every open subset of* $\mathrm{FIN}^{[\infty]}$ *is Ramsey.*

Proof. Let $\mathfrak{X}$ be a given open subset of $\mathrm{FIN}^{[\infty]}$ and let $[s, Y]$ be a given basic open set. We give argument only in case $s = \emptyset$ since the general case is a simple relativization of this one.

We first perform an already standard recursive diagonalization procedure to obtain $Z \leq Y$ such that if for some finite block sequence $t \leq Z$ there is $X \in [t, Z]$ such that

$$[t, X] \subseteq \mathfrak{X},$$

then already $[t, Z] \subseteq \mathfrak{X}$. Let

$$\mathcal{F} = \{t \in \mathrm{FIN}^{[<\infty]} : t \leq Z \text{ and } [t, Z] \subseteq \mathfrak{X}\}.$$

Applying Milliken's lemma (Lemma III.3.7) we get an infinite block subsequence X of Z such that either:
(a) $\mathcal{F} \cap [X] = \emptyset$, or
(b) Every infinite block subsequence of X contains an initial segment in $\mathcal{F}$.

Note that if (b) holds, then $[\emptyset, X] \subseteq \mathfrak{X}$. If on the other hand (a) holds, since $\mathfrak{X}$ is open, we must have that $[\emptyset, X] \cap \mathfrak{X} = \emptyset$. $\qquad\square$

A simple diagonalization argument gives us the following fact which shows that the ideal of Ramsey-null subsets of $\mathrm{FIN}^{[\infty]}$ is a σ-ideal.

Lemma III.4.3. *The union of countably many Ramsey-null subsets of* $\mathrm{FIN}^{[\infty]}$ *is Ramsey-null.* $\qquad\square$

Lemma III.4.4. *Every meager subset of* $\mathrm{FIN}^{[\infty]}$ *is Ramsey-null.*

Proof. By Lemma III.4.3 it suffices to show that every nowhere-dense subset of $\mathrm{FIN}^{[\infty]}$ is Ramsey-null. Let $\mathfrak{X}$ be a given nowhere-dense subset of $\mathrm{FIN}^{[\infty]}$ and let $[s, Y]$ be a given basic-open set of $\mathrm{FIN}^{[\infty]}$. Since the closure of a nowhere-dense set is nowhere-dense we may assume $\mathfrak{X}$ is closed. By Lemma III.4.2, $\mathfrak{X}$ is Ramsey, so there is $X \in [s, Y]$ such that

$$[s, X] \subseteq \mathfrak{X} \text{ or } [s, X] \cap \mathfrak{X} = \emptyset.$$

We finish by noting that since $\mathfrak{X}$ is nowhere-dense the first alternative is impossible.
□

Corollary III.4.5. *A subset of* $\mathrm{FIN}^{[\infty]}$ *is meager iff it is Ramsey-null.* □

Theorem III.4.6 (Milliken). *The following are equivalent for an arbitrary subset* $\mathfrak{X}$ *of* $\mathrm{FIN}^{[\infty]}$:
(a) $\mathfrak{X}$ *has the Ramsey property.*
(b) $\mathfrak{X}$ *has the Baire property.*
Moreover, the ideals of meager and Ramsey-null subsets of $\mathrm{FIN}^{[\infty]}$ *coincide.*

Proof. Let $\mathfrak{X} = \mathcal{O} \bigtriangleup \mathfrak{M}$ be a given property of Baire subset of $\mathrm{FIN}^{[\infty]}$ written as symmetric difference of an open set $\mathcal{O}$ and a meager set $\mathfrak{M}$. Let $[s, Y]$ be a given basic-open set. By Corollary III.4.5 there is $Z \in [s, Y]$ such that $[s, Z] \cap \mathfrak{M} = \emptyset$. It follows that

$$\mathfrak{X} \cap [s, Z] = \mathcal{O} \cap [s, Z].$$

By Lemma III.4.2 there is $X \in [s, Z]$ such that $[s, X] \subseteq \mathcal{O}$ or $[s, X] \cap \mathcal{O} = \emptyset$. It follows that $[s, X] \subseteq \mathfrak{X}$ or $[s, X] \cap \mathfrak{X} = \emptyset$ as required. □

Corollary III.4.7. *Every metrically analytic subset of* $\mathrm{FIN}^{[\infty]}$ *is Ramsey.* □

Corollary III.4.8 (Silver). *Every metrically analytic subset of* $\mathbb{N}^{[\infty]}$ *is Ramsey.* □

Proof. Consider the projection $\pi \colon \mathrm{FIN}^{[\infty]} \to \mathbb{N}^{[\infty]}$ defined by

$$\pi((x_i)_{i=0}^{\infty}) = (\min(x_i))_{i=0}^{\infty}.$$
□

Corollary III.4.9 (Parametrized Perfect-Set Theorem). *For every finite Borel coloring of* $\mathbb{N}^{[\infty]} \times \mathbb{R}$ *there is* $M \in \mathbb{N}^{[\infty]}$ *and a perfect* $P \subseteq \mathbb{R}$ *such that* $M^{[\infty]} \times P$ *is monochromatic.*

Proof. Let $\mathfrak{X}$ be a given Borel subset of $\mathbb{N}^{[\infty]} \times \mathcal{P}(\mathbb{N})$. Let

$$\mathfrak{X}^* = \{X = (x_i)_{i=0}^{\infty} \in \mathrm{FIN}^{[\infty]} : (\min(x_{2i+1}))_{i=0}^{\infty}, \bigcup_{i=0}^{\infty} x_{2i}) \in \mathfrak{X}\}.$$

Then $\mathfrak{X}^*$ is a (metrically) Borel subset of $\mathrm{FIN}^{[\infty]}$, so by Corollary III.4.7 there is $Y = (y_i) \in \mathrm{FIN}^{[\infty]}$ such that $[Y]^{[\infty]} \subseteq \mathfrak{X}^*$ or $[Y]^{[\infty]} \cap \mathfrak{X}^* = \emptyset$.
  Choose $\{x_\sigma : \sigma \in 2^{<\infty}\} \subseteq \{y_{2i} : i \in \mathbb{N}\}$ such that:
(1) $x_0 = y_0$ and $x_\sigma \neq x_\tau$ whenever $\sigma \neq \tau$,
(2) $x_\sigma < y_\tau$ whenever $|\sigma| < |\tau|$.

Let $M = (\min(y_{2\varphi(i)+1}))_{i=0}^{\infty}$ where $\varphi \colon \mathbb{N} \to \mathbb{N}$ is a strictly increasing map such that $x_{\sigma} < y_{\varphi(|\sigma|)}$ for all $\sigma \in 2^{<\infty}$, and let

$$P = \{\bigcup_{i=0}^{\infty} x_{a \restriction i} : a \in 2^{\infty}\}$$

Then P is a perfect subset of $\mathcal{P}(\mathbb{N})$ with the topology obtained by identifying $\mathcal{P}(\mathbb{N})$ with $2^{\mathbb{N}}$. Note that for every $N \in M^{[\infty]}$ and $q \in P$ there exists $X = (x_i) \leq Y$ such that

$$N = (\min(x_{2i+1}))_{i=0}^{\infty} \text{ and } q = \bigcup_{i=0}^{\infty} x_{2i}.$$

It follows that $M^{[\infty]} \times P \subseteq \mathfrak{X}$ or $M^{[\infty]} \times P \cap \mathfrak{X} = \emptyset$ depending on whether $[Y]^{[\infty]} \subseteq \mathfrak{X}^*$ or $[Y]^{[\infty]} \cap \mathfrak{X}^* = \emptyset$. $\square$

Corollary III.4.10 (Stern). *For every analytic set $\mathfrak{X}$ of infinite chains of the complete binary tree $2^{<\infty}$ there is a perfect subtree $T \subseteq 2^{<\infty}$ (not necessarily downwards closed) such that either every infinite chain of T is in $\mathfrak{X}$ or every infinite chain of T is not in $\mathfrak{X}$.*

Proof. Let $\mathcal{C}_{\infty}$ be the collection of all infinite chains of $2^{<\infty}$ with the natural topology of pointwise convergence. Consider the map

$$\pi : \mathbb{N}^{[\infty]} \times 2^{\infty} \to \mathcal{C}_{\infty}$$

defined by $\pi(M, a) = \{a \restriction m : m \in M\}$. Clearly, π is a continuous onto map, so for a given analytic set $\mathfrak{X} \subseteq \mathcal{C}_{\infty}$ the set

$$\mathfrak{X}^* = \{(M, a) \in \mathbb{N}^{[\infty]} \times 2^{\infty} : \pi(M, a) \in \mathfrak{X}\}$$

is an analytic subset of $\mathbb{N}^{[\infty]} \times 2^{\infty}$. By Corollary III.4.9 there is $M \in \mathbb{N}^{[\infty]}$ and a perfect set $P \subseteq 2^{\infty}$ such that $M^{[\infty]} \times P \subseteq \mathfrak{X}^*$ or $(M^{[\infty]} \times P) \cap \mathfrak{X}^* = \emptyset$. Let

$$T = \{a \restriction m : m \in M, a \in P\}.$$

Then T is a perfect subtree of $2^{<\infty}$ such that

$$\mathcal{C}_{\infty}(T) \subseteq \pi \left[M^{[\infty]} \times P \right].$$

It follows that

$$\mathcal{C}_{\infty}(T) \subseteq \mathfrak{X} \text{ or } \mathcal{C}_{\infty}(T) \cap \mathfrak{X} = \emptyset.$$ $\square$

Corollary III.4.11 (Parametrized ℓ_1-theorem). *Suppose x_{σ} ($\sigma \in 2^{<\infty}$) is a normalized sequence of elements of some Banach space X indexed by the complete binary tree $2^{<\infty}$. Then there exist $M \in \mathbb{N}^{[\infty]}$ and perfect $P \subseteq 2^{\infty}$ such that either*
(a) For all $a \in P$ the sequence $(x_{a \restriction n})_{n \in M}$ is equivalent to the unit vector basis of ℓ_1, or
(b) For all $a \in P$ the sequence $(x_{a \restriction n})_{n \in M}$ is weakly Cauchy in X.

Proof. Color $\mathbb{N}^{[\infty]} \times P$ according whether for a given pair (M, a) the corresponding sequence $(x_{a\upharpoonright n})_{n\in M}$ is weakly Cauchy, equivalent to the unit vector basis of ℓ_1, or neither of the two. Apply Corollary III.4.9 and Rosenthal's ℓ_1-theorem. $\qquad\square$

III.5 An Approximate Ramsey Theorem

An infinite dimensional Banach space $(X, \|\cdot\|)$ is *distortable* if there is $\lambda > 1$ and an equivalent norm $|\cdot|$ on X such that for every infinite-dimensional subspace Y of X, there exists $x, y \in Y$ such that

$$\|x\| = \|y\| = 1 \text{ and } |x| > \lambda \cdot |y|.$$

When this happens we say that $(X, \|\cdot\|)$ is λ-distortable. We say that $(X, \|\cdot\|)$ is *arbitrarily distortable* if it is λ-distortable for every $\lambda > 1$. The first known example of a distortable Banach space is the Tsirelson space T discussed above in Section 2. However it is still unknown if T is arbitrarily distortable. The following is the first known example of an arbitrarily distortable Banach space.

Example III.5.1 (Schlumprecht). Let $\mathfrak{X}_S$ be the completion of c_{00} under the norm $\|\cdot\|$ defined by the following implicit formula

$$\|x\| = \max\{\|x\|_\infty, \sup_n \sup \frac{1}{\log_2(n+1)} \sum_{i=1}^n \|E_i x\|\}$$

where $E_1 < E_2 < \cdots < E_n$ is a sequence of intervals of non-negative integers and where for $x \in c_{00}$ and an interval E by Ex we denote the restriction of x to E, i.e., $Ex(k) = 0$ if $k \notin E$ and $Ex(k) = x(k)$ if $k \in E$. Then $\mathfrak{X}_S$ is arbitrarily distortable. $\qquad\square$

The distortion problem for classical spaces c_0, ℓ_p $(1 \le p < \infty)$ is solved by the following two results.

Theorem III.5.2 (James). *The spaces c_0 and ℓ_1 are not distortable.* $\qquad\square$

The fact that Schlumprecht space $\mathfrak{X}_S$ is arbitrarily distortable was one of the key ingredients in the proof of the following result which solved a long-standing distortion problem.

Theorem III.5.3 (Odell–Schlumprecht). *The spaces ℓ_p $(1 < p < \infty)$ are all arbitrarily distortable.* $\qquad\square$

Hence, in particular, the Hilbert space ℓ_2 is arbitrarily distortable. The Odell–Schlumprecht method of proof is to transfer distortion from one space to another via uniform homeomorphism between spheres of the spaces. For example the *Mazur map* $M_p\colon S(\ell_1) \to S(\ell_p)$ defined by

$$M_p((x_i)_{i=0}^\infty) = (\text{sgn}(x_i) \cdot |x_i|^{1/p})_{i=0}^\infty$$

is an uniform homeomorphism which preserves block basic sequences as well as asymptotic sets. Recall that a subset A of some Banach space X is *asymptotic* if it intersects an arbitrary closed infinite dimensional subspace of X. For $1 < p < \infty$ the space ℓ_p is distortable iff there exist asymptotic sets $A, B \subseteq S(\ell_p)$ such that $\mathrm{dist}(A, B) > 0$. The Mazur map M_p preserves asymptotic sets so ℓ_p is distortable iff $S(\ell_1)$ contains a pair of asymptotic sets on positive distance from each other. Thus, the distortion problem for the Hilbert space was solved by transferring separated asymptotic sets from Schlumprecht space into $S(\ell_1)$ and then to $S(\ell_2)$.

The distortion problem is related to the problem of oscillation stability of functions $f \colon S(X) \to \mathbb{R}$.

Definition III.5.4. Given a Banach space X, a function $f \colon S(X) \to \mathbb{R}$ is *oscillation stable* on X if for all infinite dimensional closed subspaces Y of X and $\varepsilon > 0$ there exists a closed infinite dimensional subspace Z of Y such that $\sup\{|f(x) - f(y)| : x, y \in S(Z)\} < \varepsilon$.

Lemma III.5.5. *The following are equivalent for every Banach space X:*
(1) *Every Lipschitz function $f \colon S(X) \to \mathbb{R}$ is oscillation stable.*
(2) *Every uniformly continuous function $f \colon S(X) \to \mathbb{R}$ is oscillation stable.*

Proof. Suppose $f \colon S(X) \to \mathbb{R}$ is an uniformly continuous function which is not oscillation stable. Then there exists $\varepsilon < \delta$ and an infinite dimensional closed subspace Y of X such that

$$A = \{x \in S(Y) : f(x) < \varepsilon\} \quad \text{and} \quad B = \{x \in S(Y) : f(x) > \delta\}$$

are two asymptotic sets in Y such that $\mathrm{dist}(A, B) > 0$. Consider $g \colon S(X) \to \mathbb{R}$ defined by $g(x) = \mathrm{dist}(x, A)$. Then g is a Lipschitz function which is not oscillation stable on X. $\qquad\square$

A combination of results of Gowers, Milman and Odell and Schlumprecht gives us the following remarkable synthesis.

Theorem III.5.6. *The following are equivalent for every infinite dimensional Banach space X:*
(1) *Every Lipschitz function $f \colon S(X) \to \mathbb{R}$ is oscillation stable.*
(2) *Every closed infinite dimensional subspace Y contains an isomorph of c_0.*

We shall prove here the implication (2) → (1) of this theorem, or more precisely the following result that is inherently Ramsey theoretic in nature.

Theorem III.5.7 (Gowers). *Every Lipschitz function $f \colon S(c_0) \to \mathbb{R}$ is oscillation stable.*

Let K be the Lipschitz constant of the given Lipschitz map $f \colon S(c_0) \to \mathbb{R}$ and let $\varepsilon > 0$ be a given number. Choose a sufficiently large positive integer k such that if $(1 + \delta_k)^{1-k} = \delta_k$ then $\delta_k \cdot K \le \varepsilon/3$. Let $\Delta_{\pm k}$ be the collection of all finitely supported maps

$$p \colon \mathbb{N} \to \{0, \pm(1 + \delta_k)^{1-k}, \pm(1 + \delta_k)^{2-k}, \dots, \pm(1 + \delta_k)^{-1}, \pm 1\}$$

which achieves at least once one of the values ± 1. Then $\Delta_{\pm k}$ forms an $(\varepsilon/3)$-net on the unit sphere $S(c_0)$ of c_0, and Theorem III.5.7 gets transferred to the following approximate Ramsey theoretic result.

Theorem III.5.8. *For every finite coloring of $\Delta_{\pm k}$ there exists an infinite block sequence $X = (x_i) \subseteq \Delta_{\pm k}$ and a color P such that $[X] \subseteq (P)_\delta$.*

Here $[X]$ denotes the intersection of the subspace of c_0 generated by (x_i) with the set $\Delta_{\pm k}$ and

$$(P)_\delta = \{x \in \Delta_{\pm k} : \exists p \in P \, \|p - x\|_\infty \leq \delta\}.$$

One can make Theorem III.5.8 even more close to Hindman's Theorem by identifying $(1 + \delta_k)^{\ell-k}$ with ℓ for $0 < \ell \leq k$ which amounts to considering the set $\text{FIN}_{\pm k}$ of all finitely supported maps

$$p: \mathbb{N} \to \{0, \pm 1, \pm 2, \ldots, \pm k - 1, \pm k\}$$

that obtain one of the values $\pm k$ at least once. To get the analogous of the subspace $[X]$ generated by a block sequence besides the partial semigroup operation $+$ we need to consider the operation of multiplying by -1 as well as multiplying by the scalar $(1 + \delta_k)^{-1}$. The operation on $\text{FIN}_{\pm k}$ corresponding to the scalar multiplication by $(1 + \delta_k)^{-1}$ is the operation

$$T: \text{FIN}_{\pm k} \to \text{FIN}_{\pm(k-1)}$$

defined by

$$T(p)(n) = \begin{cases} p(n) - 1 & if \quad p(n) > 0 \\ p(n) + 1 & if \quad p(n) < 0 \\ 0 & if \quad p(n) = 0. \end{cases}$$

Then for an (infinite) block sequence $X = (x_i)$ of elements of $\text{FIN}_{\pm k}$ we let[4]

$$[X] = \{ \pm T^{(i_0)}(x_{n_0}) \pm \cdots \pm T^{(i_\ell)}(x_{n_\ell}) : n_0 < \cdots < n_\ell, i_0, \ldots, i_\ell \in \{0, \ldots, k-1\},$$
$$\text{and } i_j = 0 \text{ for some } j \leq \ell \}.$$

Then Theorem III.5.7 has yet another Ramsey-theoretic reformulations as follows.

Theorem III.5.9. *For every positive integer k and every finite coloring of $\text{FIN}_{\pm k}$ there is an infinite block sequence $X = (x_i) \subseteq \text{FIN}_{\pm k}$ and a color P such that $[X] \subseteq (P)_1$.*[5]

Remark III.5.10. Note that the approximate constant 1 in Theorem III.5.9 can be replaced with any other fixed positive integer m (i.e, in the conclusion $[X] \subseteq (P)_1$ can be replaced by $[X] \subseteq (P)_2$ or $[X] \subseteq (P)_3$, etc.) without changing the strength of the result.

---

[4]$T^{(i)}$ denotes the *ith* iterate of T with $T^{(0)} = identity$, $T^{(1)} = T$, $T^{(i+1)} = T(T^{(i)})$, etc.

[5]$(P)_1 = \{x \in \text{FIN}_{\pm k} : \exists p \in P \, \|x - p\|_\infty \leq 1\}$.

Proof of Theorem III.5.9. The proof follows closely the proof of Hindman's theorem applied to the (partial) semigroup

$$\text{FIN}_{[0,\pm k]} = \bigcup_{\ell=0}^{k} \text{FIN}_{\pm\ell}.$$

As before, we start with the closed subspace $\gamma\text{FIN}_{[0,\pm k]}$ of the Stone–Čech compactification $\beta\text{FIN}_{[\pm k]}$ consisting of all ultrafilters $\mathcal{U}$ on $\text{FIN}_{[0,\pm k]}$ such that

$$\{p \in \text{FIN}_{[0,\pm k]} : p(i) = 0 \text{ for all } i \le n\} \in \mathcal{U}$$

for all n. As before we consider $\gamma\text{FIN}_{[0,\pm k]}$ as a semigroup with the operation $+$ defined by

$$A \in \mathcal{U} + \mathcal{V} \text{ iff } \{p : \{q : p < q \text{ and } p+q \in A\} \in \mathcal{V}\} \in \mathcal{U}.^{6}$$

Then $\mathcal{U} \mapsto \mathcal{U} + \mathcal{V}$ is a continuous map from $\gamma\text{FIN}_{[0,\pm k]}$ into $\gamma\text{FIN}_{[0,\pm k]}$ for all $\mathcal{V} \in \gamma\text{FIN}_{[0,\pm k]}$. So, as before, every nonempty closed subsemigroup of $\gamma\text{FIN}_{[0,\pm k]}$ contains an idempotent. We shall need the following general fact about idempotents and homomorphisms in compact semigroups.

Lemma III.5.11. *Let $(S, +)$ be a compact semigroup, let I be a closed subsemigroup of S, and let k be a non negative integer. Suppose $H \colon S \to S$ is a continuous homomorphism that stabilizes after k steps* [7]. *Then there is an $e \in I$ such that*

$$e + H^{(\ell)}(e) = H^{(\ell)}(e) + e = e,$$

for all $\ell = 0, 1, \ldots, k$.

Proof. For $0 \le \ell \le k$ let
$$I_\ell = H^{(k-\ell)}[I].$$

Thus $I_k = I$, $I_{k-1} = H[I]$, $I_{k-2} = H^{(2)}[I]$, etc. Then $I_0, I_1, \ldots, I_k$ are all compact subsemigroups of S_0. By Ellis' lemma we can pick an idempotent e_0 in I_0. Suppose that for some $0 < \ell < k$, $e_i \in I_i$ $(i < \ell)$ have been selected such that:
(1) $H^{(j-i)}(e_j) = e_i$ whenever $i \le j < \ell$,
(2) $e_i + e_j = e_j + e_i = e_j$ whenever $i \le j < \ell$.
Let
$$R_\ell = \{x \in I_\ell : H(x) = e_{\ell-1}\}.$$

Then R_ℓ is a nonempty closed subsemigroup of S. Note also that $R_\ell + e_{\ell-1}$ is also a nonempty closed subsemigroup of S so by Ellis' lemma we can fix an idempotent $b = a + e_{\ell-1} \in R_\ell + e_{\ell-1}$. Let

$$e_\ell = e_{\ell-1} + b + e_{\ell-1}.$$

---

[6] $p < q$ means that $\text{supp}(p) < \text{supp}(q)$ and $p + q$ is defined by taking pointwise addition.
[7] i.e., $H^{(k+\ell)} = H^{(k)}$ for all ℓ, where for a given integer ℓ, $H^{(\ell)}$ is the ℓth iterate of H with $H^{(0)}$ taken as the identity map and $H^{(1)} = H$.

Then (1) and (2) continue to be valid for $i \leq j \leq \ell$. Proceeding this way we get $e_i \in I_i$ $(0 \leq i \leq k)$ satisfying the equations of (1) and (2) for every choice of $0 \leq i \leq j \leq k$. So e_k satisfies the conclusion of the Lemma. $\quad\square$

We apply the lemma to $S = \gamma\mathrm{FIN}_{[0,\pm k]}$, $I = \mathrm{FIN}_{\pm k}$, and

$$H \colon \gamma\mathrm{FIN}_{[0,\pm k]} \to \gamma\mathrm{FIN}_{[0,\pm k]}$$

defined by $H(\mathcal{U}) = -T(\mathcal{U})$. Note that H is indeed a continuous homomorphism being continuous extension of the partial homomorphism[8]

$$H \colon \mathrm{FIN}_{[0,\pm k]} \to \mathrm{FIN}_{[0,\pm k]}$$

defined by $H(p) = -T(p) = T(-p)$. This gives us an ultrafilter $\mathcal{U} \in \gamma\mathrm{FIN}_{\pm k}$ such that
(3) $\mathcal{U} + (-T)^{(\ell)}(\mathcal{U}) = (-T)^{(\ell)}(\mathcal{U}) + \mathcal{U} = \mathcal{U}$ for all $\ell = 0, 1, \ldots, k$.
From now on we fix $\mathcal{U} \in \gamma\mathrm{FIN}_{\pm k}$ satisfying (3).
For an (infinite) block sequence $X = (x_n)$ of members of $\mathrm{FIN}_{\pm k}$, and a homomorphism $H \colon \mathrm{FIN}_{[0,\pm k]} \to \mathrm{FIN}_{[0,\pm k]}$,

$$[X]_H = \{ H^{(\ell_0)}(x_0) + \cdots + H^{(\ell_t)}(x_t) : n_0 < \cdots < n_t, \ell_0, \ldots, \ell_t \leq k,$$
$$\ell_i = 0 \text{ for some } i \leq t \}.$$

Note that the proof of Lemma III.1.4 gives us the following.

Lemma III.5.12. *For every $P \in \mathcal{U}$ there is an infinite block sequence $X = (x_n) \subseteq \mathrm{FIN}_{\pm k}$ such that $[X]_{(-T)} \subseteq P$.* $\quad\square$

We shall also need the following property of the ultrafilter $\mathcal{U}$ satisfying (3).

Lemma III.5.13. $-(P)_1 \in \mathcal{U}$ *for all $P \in \mathcal{U}$.*

Proof. This follows from the following sequence of equivalent statements where we use the fact that $\mathcal{U} = (-T)(\mathcal{U}) + \mathcal{U} = \mathcal{U} + (-T(\mathcal{U}))$:

$$-(P)_1 \in \mathcal{U} + (-T)(\mathcal{U}) = \mathcal{U}$$
$$(\mathcal{U}_p)(\mathcal{U}_q) \, p - T(q) \in -(P)_1$$
$$(\mathcal{U}_p)(\mathcal{U}_q) -p + T(q) \in (P)_1$$
$$(\mathcal{U}_p)(\mathcal{U}_q) -T(p) + q \in P$$
$$P \in -T(\mathcal{U}) + \mathcal{U} = \mathcal{U}.$$

$\quad\square$

---

[8]Partial homomorphism means that the equation $H(p+q) = H(p) + H(q)$ holds for disjointly supported elements p and q of $\gamma\mathrm{FIN}_{[0,\pm k]}$.

Back to the proof of Theorem III.5.9. We shall actually show that there is an infinite block sequence $X = (x_n) \subseteq \mathrm{FIN}_{\pm k}$ and a color P such that

$$[X] \subseteq (P)_2.$$

As noted earlier (see Remark III.5.10) this will complete the proof. The paper of Kanellopoulos [41] from where we have taken some of the above arguments contains a direct combinatorial reductions of the conclusion $[X] \subseteq (P)_1$ from the conclusion $[X] \subseteq (P)_2$.

Pick a color P such that $P \in \mathcal{U}$. By Lemma III.5.13, $-(P)_1 \in \mathcal{U}$. Let $Q = -(P)_1 \cap (P)_1$. Then $Q \in \mathcal{U}$ and $Q = -Q$. By Lemma III.5.12 there is an infinite block sequence $X = (x_n)$ such that $[X]_{(-T)} \subseteq Q$. So the desired conclusion $[X] \subseteq (P)_2$ follows from the following general fact.

Lemma III.5.14. *Suppose $X = (x_n)$ is an infinite block sequence of members of* $\mathrm{FIN}_{\pm k}$ *such that* $[X]_{(-T)} \subseteq Q$ *for some* $Q \subseteq \mathrm{FIN}_{\pm k}$ *such that* $Q = -Q$. *Then* $[X] \subseteq (Q)_1$.

Proof. Consider a $p \in [X]_{(-T)}$. Then there exist $n_0 < \cdots < n_t$, $\ell_0, \ldots, \ell_t \leq k$ with $\ell_i = 0$ for at least one $i \leq t$ and signs $\varepsilon_0, \ldots, \varepsilon_t \in \{\pm\}$ such that

$$p = \varepsilon_0 T^{(\ell_0)}(x_{n_0}) + \cdots + \varepsilon_t T^{(\ell_t)}(x_{n_t}).$$

We need to find a $q \in Q$ such that $\|p - q\|_\infty \leq 1$.
Case 1: There is $i \leq t$ such that $\varepsilon_i = +$ and $\ell_i = 0$. For $i \leq t$, let $y_i = \varepsilon_i T^{(\ell_i)}(x_{n_i})$. We shall find $y_i' \in [X]_{(-T)}$ $(i \leq t)$ such that $y_i' = (-T)^{(\ell_i')}(x_{n_i})$ for some $\ell_i' \in \{\ell_i + 1, \ell_i\}$. Then $\|y_i - y_i'\|_\infty \leq 1$ for all $i \leq t$ and therefore we have that

$$q = (-T)^{(\ell_0')}(x_{n_0}) + \cdots + (-T)^{(\ell_t')}(x_{n_t})$$

is a member of $[X]_{(-T)}$ such that $\|p - q\|_\infty \leq 1$, as required.
(i) If $\varepsilon_i = -$ and ℓ_i is odd, or if $\varepsilon_i = +$ and ℓ_i is even, we put $y_i' = y_i$. Note that in both cases:

$$y_i = y_i' = (-T)^{(\ell_i)}(x_{n_i}).$$

(ii) If $\varepsilon_i = -$ and ℓ_i is even, or if $\varepsilon_i = +$ and ℓ_i is odd, we put $y_i' = T(y_i)$. Note that in both cases

$$y_i' = (-T)^{(\ell_i+1)}(x_{n_i}).$$

Note that these choices and the assumption of Case 1 ensures that the sum $q = y_0' + \cdots + y_t'$ belongs to $[X]_{(-T)}$.
Case 2. For every $i \leq t$, if $\ell_i = 0$ then $\varepsilon_i = -$. Let $\bar{p} = -p$. Then the representation

$$\bar{p} = (-\varepsilon_0)T^{(\ell_0)}(x_{n_0}) + \cdots + (-\varepsilon_t)T^{(\ell_t)}(x_{n_t})$$

falls into Case 1, so working as in Case 1 we find $\bar{q} \in [X]_{(-T)}$ such that $\|\bar{p} - \bar{q}\|_\infty \leq 1$. Since $[X]_{(-T)} \subseteq Q = -Q$ we have that $q = -\bar{q}$ belongs to Q. Finally note that $\|p - q\|_\infty = \|\bar{p} - \bar{q}\|_\infty \leq 1$. This completes the proof of Lemma III.5.14. $\qquad\square$

For a positive integer k, let

$$\text{FIN}_k = \{p \colon \mathbb{N} \to \{0, 1, \ldots, k\} : \text{supp}(p) \text{ is finite and } k \in \text{rang}(p)\}.$$

Thus $(\text{FIN}_k, +)$ is a partial subsemigroup of $\text{FIN}_{\pm k}$ so one can consider the partial homomorphism

$$T \colon \text{FIN}_k \to \text{FIN}_{k-1}$$

defined as before: $T(p)(n) = p(n) - 1$ if $p(n) > 0$; $T(p)(n) = 0$ if $p(n) = 0$. Given an (infinite) block sequence $X = (x_n)$ of members of FIN_k the partial subsemigroup of FIN_k generated by X is defined as before:

$$[X] = \{T^{(\ell_0)}(x_{n_0}) + \cdots + T^{(\ell_t)}(x_{n_t}) : n_0 < \cdots < n_t, \ell_0, \ldots, \ell_t \leq k \text{ and}$$
$$\ell_i = 0 \text{ for some } i \leq t\}.$$

Then we have the following exact Ramsey-theoretic result.

Theorem III.5.15 (Gowers). *For every finite coloring of* FIN_k *there is an infinite block sequence* $X = (x_n) \subseteq \text{FIN}_k$ *such that* $[X]$ *is monochromatic.*

Proof. Let $\text{FIN}_{[0,k]} = \bigcup_{\ell=0}^{k} \text{FIN}_\ell$ and the corresponding compact semigroup $\gamma\text{FIN}_{[0,k]}$ and its continuous homomorphism

$$T \colon \gamma\text{FIN}_{[0,k]} \to \gamma\text{FIN}_{[0,k]}$$

extending T. Applying Lemma III.5.11 to $\gamma\text{FIN}_{[0,k]}$, γFIN_k and T we get an ultrafilter $\mathcal{U} \in \gamma\text{FIN}_k$ such that

$$\mathcal{U} + T^{(\ell)}(\mathcal{U}) = T^{(\ell)}(\mathcal{U}) + \mathcal{U} = \mathcal{U}$$

for all $\ell = 0, 1, \ldots, k$. Choose $P \in \mathcal{U}$ that is monochromatic relative to the given coloring of FIN_k. The proof of Lemma III.1.4 (see also Lemma III.5.12) gives us an infinite block sequence $X = (x_n)$ of members of FIN_k such that $[X] \subseteq P$, as required. $\qquad\square$

Note that FIN_1 can be identified with the space FIN of all finite nonempty subsets of $\mathbb{N}$, so Hindman's theorem (Theorem III.1.1) is an immediate consequence of Theorem III.5.15.

For $\ell = 1, 2, 3, \ldots, \infty$ let $\text{FIN}_k^{[\ell]}$ be the collection of all block sequences of elements of FIN_k of length ℓ. Thus $\text{FIN}_k^{[1]}$ is naturally identified with FIN_k. The proof of Theorem III.1.6, replacing the use there of Hindman's Theorem by the use of Gowers' Theorem, lead us to the following more general result.

Lemma III.5.16. *For every pair of positive integers* k *and* ℓ *and for every finite coloring of* $\text{FIN}_k^{[\ell]}$ *there is an infinite block sequence* $X = (x_n)$ *such that the set* $[X]^{[\ell]}$ *of all block subsequences of* X *of length* ℓ *is monochromatic.*

One can go on and define a notion of a (uniform) barrier on the set $\mathrm{FIN}_k^{[<\infty]}$ of all finite block sequences in FIN_k and prove the following generalization of Lemma III.3.7.

Lemma III.5.17. *For every family $\mathcal{F}$ of finite block sequences of elements of FIN_k there is an infinite block sequence $X = (x_n)$ such that either $\mathcal{F} \cap [X] = \emptyset$, or every infinite block subsequence of X has an initial segment in $\mathcal{F}$.*   ◻

Starting from this lemma one goes on and proves the analogue of the Ellentuck theorem and therefore obtains the following generalization of Milliken's. Theorem.

Theorem III.5.18. *For every analytic set $\mathfrak{X} \subseteq \mathrm{FIN}_k^{[\infty]}$ there is an infinite block sequence $X = (x_n)$ such that either all infinite block-subsequences of X belongs to $\mathfrak{X}$ or all infinite block-subsequences of X fall outside $\mathfrak{X}$.*

Taylor's theorem (Theorem III.2.1) also allows to be generalized.

Theorem III.5.19 (Lopez–Abad). *For each positive integer k there is a finite list $E_i(k)$ $(i < \varphi(k))$ of equivalence relations on FIN_k with the property that for every equivalence relation E on FIN_k there is $i < \varphi(k)$ and an infinite block sequence $X = (x_n)$ such that $E \upharpoonright [X] = E_i(k) \upharpoonright [X]$.*

The number $\varphi(k)$ of the smallest list of equivalence relations on FIN_k satisfying this conclusion allows to be expressed using standard enumerating functions so the first few values of $\varphi(k)$ can be computed. For example $\varphi(1) = 5$, $\varphi(2) = 43$, $\varphi(3) = 619$, etc.

Chapter IV

Approximate and Strategic Ramsey Theory of Banach Spaces

IV.1 Gowers' Dichotomy

Fix a Banach space X with a Schauder basis (c_i). Recall the notion of a *block sequence* of (e_i) is a sequence (x_n) again typically formed of norm 1 vectors, such that for each n there is a finite set $D_n \subseteq \mathbb{N}$ such that

(a) $x_n = \sum_{i \in D_n} a_i e_i$ for some sequence a_i $(i \in D_n)$ of scalars.

(b) $D_m < D_n$[1] whenever $m < n$.

Note that if (x_n) is a (normalized) block sequence of (e_i) then (x_n) is a Schauder basis of its closed linear span $Y = \overline{(x_n)}$.

Theorem IV.1.1 (Bessaga–Pelczynski). *Let X be a given infinite-dimensional Banach space with a Schauder basis (e_i). Then for every $\varepsilon > 0$ and every infinite-dimensional closed subspace Y of X there is a (normalized) block sequence (x_n) and a sequence $(y_n) \subseteq Y$ such that*

$$\sum_{n=0}^{\infty} \|x_n - y_n\| < \varepsilon.$$

Proof. Going to an equivalent norm we may assume that all projections $P_I(x) = \sum_{i \in I} a_i x_i$ on intervals $I \subseteq \mathbb{N}$ have norm 1. Pick a normalized $y_0 \in Y$. Then there is k_0 such that $\|P_{[k_0, \infty)}(y_0)\| < \varepsilon_0$, so if we put $x_0 = P_{k_0}(y_0)$, then $\|x_0 - y_0\| < \varepsilon_0$. Since Y is infinite-dimensional there is a normalized $y_1 \in Y$ such that $P_{k_0}(y_1) = 0$.

---

[1] For subsets D and E of some ordered set, the inequality $D < E$ means that every element of D is smaller than every element of E.

Choose $k_1 > k_0$ such that $\|P_{[k_1,\infty)}(y_1)\| < \varepsilon_1$ and let $x_1 = P_{[k_0,k_1)}(y_1)$. Then $\|x_1 - y_1\| < \varepsilon_1$, and so on. Continuing this way we get the desired block-sequence (x_n). □

In what follows a typical use of Theorem IV.1.1 is in reducing problems about arbitrary infinite-dimensional subspaces of X to subspaces spanned by infinite block-subsequences of (e_i). For example, suppose we are interested whether X contains an infinite-dimensional *hereditarily indecomposable* subspace, i.e., a subspace Y with the property that no infinite-dimensional closed subspace Z of Y can be decomposed as a sum $Z_0 \oplus Z_1$ of two closed-infinite-dimensional subspaces. Then this is same as asking for the existence of a hereditarily indecomposable *block-subspace* of X, i.e., a closed subspace Y of X spanned by a block-subsequence of (e_i). Namely, if $Y \subseteq X$ is hereditarily indecomposable and if (x_n) and (y_n) satisfy the conclusion of Theorem IV.1.1 then the closed linear span of the block-subsequence (x_n) is also hereditarily indecomposable.

There are Banach spaces with no unconditional bases. For example $C[0,1]$ is one such space. However, the problem whether every infinite-dimensional Banach has an unconditional basic sequence remained open for quite some time until was answered negatively by Gowers and Maurey in 1992. The Gowers–Maurey example has turned out to be hereditarily indecomposable and the following subsequent result of Gowers explains why this always must be so.

Theorem IV.1.2 (Gowers). *Let X be a Banach space with a Schauder basis (e_i). Then every block subsequence (x_n) of (e_i) has a block subsequence (x_{n_k}) which is either unconditional or its closed linear span is hereditarily indecomposable.*

The proof of this result uses a combinatorial forcing argument quite analogous to those presented in Sections 4 and 5. Note that the dichotomy is really about normed spaces since completeness plays no role either in the definition of an unconditional basic sequence or a hereditarily indecomposable space. Note that for example an infinite-dimensional space X is hereditarily indecomposable (H.I.) if for any pair Y and Z of infinite-dimensional subspaces of X,

$$\text{dist}(S_Y, S_Z) = \inf\{\|y - z\| : y \in S_Y, z \in S_Z\} = 0.^2$$

Note also that we may restrict ourselves to normed spaces over the *rationals* rather than reals. This will make the forcing argument a bit more natural. So from now on a "space" or a "subspace", unless otherwise specified, refers to infinite dimensional normed spaces over the rationals.

For $\varepsilon > 0$ we say that a space X is ε-*H.I.* if for all subspaces $Y, Z \subseteq X$ there exists $y \in Y, z \in Z$ such that $\|y - z\| < \varepsilon \|y + z\|$. Clearly, X is H.I. iff X is ε-H.I. for all $\varepsilon > 0$. So the Gowers dichotomy will follow once we prove the following.

Theorem IV.1.3. *For a given $\varepsilon > 0$ an infinite-dimensional normed space X either contains a $(2/\varepsilon)$-unconditional basic sequence or an infinite-dimensional subspace Y which is (2ε)-H.I.*

---

[2]For a normed space X, we denote by S_X its unit sphere.

Proof. We assume that $0 < \varepsilon < 1$ and fix an infinite-dimensional normed space X over the rationals, its basis (e_i) and assume that the projections on intervals of $\mathbb{N}$ are uniformly bounded by 1. The variables U, V, Y, Z will run over infinite-dimensional block-spaces of X. For $\varepsilon > 0$ we set

$$P(\varepsilon) = \{(x,y) \in X : \|x - y\| < \varepsilon\|x + y\|\}.$$

Definition IV.1.4 (Combinatorial forcing). We say that $Z \subseteq X$ *accepts* a pair $(x,y) \in X^2$ if for all $U, V \subseteq Z$ there exists $(u,v) \in U \times V$ such that $(x + u, x + v) \in P(\varepsilon)$. We say that Z *rejects* a pair (x,y) if no subspace Y of Z accepts (x,y). We say that Z *decides* (x,y) if Z either accepts or rejects (x,y).

Note the following immediate properties of these notions:
(1) If Z accepts (rejects) (x,y), so does any $Y \subseteq Z$.
(2) For every Z and (x,y) there is $Y \subseteq Z$ deciding (x,y).
(3) If Z accepts $(0,0)$, then Z is (2ε)-H.I.
(4) If Z decides all $(x,y) \in X^2$ but rejects (x_0, y_0), then for all $Y \subseteq Z$ there is $V \subseteq Y$ such that Z rejects all $(x_0 + v, y)$ for $v \in V$.

Only (4) needs some argument. Assume (4) fails and fix a subspace $Y \subseteq Z$ such that for all $V \subseteq Y$ there is $v \in V$ such that Z accepts $(x_0 + v, y_0)$. Choose arbitrary $U, V \subseteq Y$. Pick $v_0 \in V$ such that Z accepts $(x_0 + v_0, y_0)$. Then we can find $(u,v) \in U \times V$ such that

$$(x_0 + v_0 + v, y_0 + u) \subset P(\varepsilon)$$

This checks that Y accepts (x_0, y_0) contradicting (1).

Using (1) and (2) we build an infinite decreasing sequence $X \supseteq X_1 \supseteq \cdots \supseteq X_n \supseteq \ldots$ of block-subspaces of X such that for all $(x,y) \in X^2$ there is some n such that X_n decides (x,y). Let X_∞ be the block subspace which diagonalizes (X_n). Then X_∞ decides all pairs (x,y). By (3) we may assume X_∞ rejects $(0,0)$. For every n, let $\mathcal{N}_n$ be a fixed $(1/4)$-net of the unit ball of ℓ_1^n. For a finite block sequence $(z_i)_{i<n}$ let $[(z_i)_{i<n}]$ be the collection of all vectors of the form

$$\sum_{i=0}^{n-1} a_i z_i,$$

where $(a_i)_{i=0}^{n-1} \in \mathcal{N}_n$. A pair x, y of vectors of $[(z_i)_{i<n}]$ is *disjointly supported* if

$$x = \sum_{i \in I} a_i z_i, \quad y = \sum_{j \in J} a_j z_j$$

and $I \cap J = \emptyset$.

Recursively on i we build an infinite block sequence (z_i) such that for all n, X_∞ rejects all pairs of disjointly supported vectors from $[(z_i)_{i<n}]$. Note that our assumption that X_∞ rejects $(0,0)$ gives us this inductive assumption for $n = 0$.

So, suppose we have constructed a block sequence $(z_i)_{i<n}$ satisfying this inductive hypothesis. Applying (4) successively to all pairs of disjointly supported vectors from $[(z_i)_{i<n}]$ we obtain a block-subspace $Y \subseteq X_\infty$ such that
(5) X_∞ rejects $(x + v, y)$ whenever (x, y) is a pair of disjointly supported vectors from $[(z_i)_{i<n}]$ and $v \in Y$.

Pick arbitrary $z_n \in Y$ with support above the maximum of support of z_{n-1} and of norm between 1 and 2. Then every pair of disjointly supported vectors from $[(z_i)_{i\leq n}]$ has the form

$$(x + az_n, y) \text{ or } (x, y + az_n)$$

for some pair (x, y) of disjointly supported vectors of $[(z_i)_{i<n}]$. By (5) it follows that the inductive hypothesis remains preserved.

The block sequence $(z_i)_{i<n}$ constructed this way has the following properties for all n:
(6) $1 \leq \|z_n\| \leq 2$
(7) $\| \sum_{i=0}^{n-1} a_i z_i \| \leq (1/\varepsilon) \| \sum_{i=0}^{n-1} \varepsilon_i a_i z_i \|$, for every $(\varepsilon_i)_{i<n} \in \{\pm 1\}^n$ and $(a_i)_{i<n} \in \mathcal{N}_n$.

This follows from the fact that (7) can be written in the form $\varepsilon \|x + y\| \geq \|x - y\|$ for (x, y) a pair of disjointly supported vectors from $[(z_i)_{i<n}]$ and the fact that $(x, y) \notin P(\varepsilon)$ whenever X_∞ rejects it. It remains to be checked that a block sequence (z_i) satisfying (6) and (7) is $(2/\varepsilon)$-unconditional: Fix n, $(a_i)_{i<n}$ and $(\varepsilon_i)_{i<n}$, and assume that $\| \sum_{i<n} \varepsilon_i a_i z_i \| = 1$. Choose $(b_i)_{i<n} \in \mathcal{N}_n$ be such that $\|(a_i)_{i<n}, (b_i)_{i<n}\|_\infty \leq 1/4$. Then,

$$\| \sum_{i<n} a_i z_i \| \leq \| \sum_{i<n} (a_i - b_i) z_i \| + \| \sum_{i<n} b_i z_i \| \leq 2 \sum_{i<n} |a_i - b_i| + \frac{1}{\varepsilon} \| \sum_{i<n} \varepsilon_i b_i z_i \| \leq$$

$$\leq \frac{1}{2} + \frac{1}{\varepsilon}(\| \sum_{i<n} \varepsilon_i a_i z_i \| + \| \sum_{i<n} \varepsilon_i (a_i - b_i) z_i \|) \leq \frac{1}{2} + \frac{1}{\varepsilon}(1 + \frac{1}{2}) \leq$$

$$\leq \frac{1}{2\varepsilon} + \frac{3}{2\varepsilon} = \frac{2}{\varepsilon}.$$

$\square$

IV.2 Approximate and Strategic Ramsey Sets

Gowers' dichotomy presented in the previous section suggests the corresponding Ramsey theory of finite and infinite block sequences in Banach spaces with Schauder bases. The purpose of this and the following sections is to give an overview of such a theory. In what follows E is a fixed Banach space with a Schauder basis (e_i). For $k = 0, 1, 2, \ldots, \infty$ by $\mathcal{B}_1^{[k]}(E)$ we denote the collection of all normalized block subsequences of (e_i) of length k. Let

$$\mathcal{B}_1^{[<\infty]}(E) = \bigcup_{k=0}^{\infty} \mathcal{B}_1^{[k]}(E)$$

and

$$\mathcal{B}_1^{[\leq\infty]}(E) = \mathcal{B}_1^{[<\infty]}(E) \cup \mathcal{B}_1^{[\infty]}(E).$$

We shall identify a given block sequence $(z_n) \in \mathcal{B}_1^{[\leq\infty]}(E)$ with the corresponding closed linear span $Z = \overline{\langle z_n \rangle_n}$ and use the capital letter Z to denote also the block sequence itself. When talking about topological properties of a subset $\mathfrak{X}$ of some $\mathcal{B}_1^{[k]}(E)$ we refer to the topology induced from E^k with the product topology.

For $s \in \mathcal{B}_1^{[k]}(E)$, $X \in \mathcal{B}_1^{[l]}(E)$ and $k \leq m \leq l$, let

$$[s, X]^{[m]} = \{Y \in \mathcal{B}_1^{[m]}(E) : s = Y \restriction k \text{ and } Y \leq s \frown (X/s)\}.^3$$

We shall frequently suppress the exponent $[m]$ from $[s, X]^{[m]}$ when it is clear in which dimension we are working at a particular moment. Let $[X]^{[m]} = [\emptyset, X]^{[m]}$. We shall frequently suppress even this exponent $[m]$ especially when it is equal to the length of the block sequence X. The sets $[s, X]$ of block sequences will be frequently called *basic sets* for reason that will be clear later in this chapter.

The lack of a true pigeon-hole principle in this context requires that we consider cubes of the form $[X]^{[m]}$ which are 'approximately monochromatic' rather than 'monochromatic'. It is for this reason that we need to consider the 'errors' of monochromaticity. It turns out that these 'errors' are naturally described by sequences $\Delta = (\delta_n)_{n=0}^\infty \subseteq \mathbb{R}_+$ of nonnegative real numbers. For this it is useful to have the following piece of notation, where $\mathfrak{X} \subseteq \mathcal{B}_1^{[\leq\infty]}(E)$ and $\Delta = (\delta_n)_{n=0}^\infty \subseteq \mathbb{R}_+$,

$$(\mathfrak{X})_\Delta = \{(z_n)_0^k \in \mathcal{B}_1^{[\leq\infty]}(E) : \exists (y_n)_0^k \in \mathfrak{X} \ \forall n \ \|z_n - y_n\| \leq \delta_n\}.$$

For $\Delta = (\delta_n), \Gamma = (\gamma_n) \in \mathbb{R}_+^\infty$, we write $\Gamma < \Delta$ whenever $\delta_n < \gamma_n$ for all n. Unless explicitly specified, we always work with $\Gamma \in \mathbb{R}_+^\infty$ which are decreasing and strictly positive. In this context, we shall frequently perform coordinatewise addition and scalar multiplication among such sequences.

We are now ready to define two analogues of the classical notion of a Ramsey set in the context of sets of infinite normalized block sequences of a given Banach space E with a Schauder basis (e_i).

Definition IV.2.1. A subset $\mathfrak{X}$ of $\mathcal{B}_1^{[\infty]}(E)$ is *approximately Ramsey* if for every $\Delta = (\delta_n)_0^\infty \subseteq \mathbb{R}_+$, every $X \in \mathcal{B}_1^{[\infty]}(E)$ and every $s \in \mathcal{B}_1^{[<\infty]}(E)$ there is $Y \in [s, X]$ such that either $[s, Y] \cap \mathfrak{X} = \emptyset$ or $[s, Y] \subseteq (\mathfrak{X})_\Delta$.

Note that Theorem III.5.6 says that it makes sense to study this notion only when E is a c_0-*saturated space*, i.e., when every closed infinite-dimensional subspace of E contains an isomorphic copy of c_0. In other words, the notion of approximately Ramsey sets of infinite block sequences make sense studying only

---

[3] Recall that $Y \leq X$ denotes the fact that Y is a block subsequence of X and that X/s denotes the maximal tail of X that lies entirely above s.

in the case $E = c_0$. When the given Banach space E is not c_0-saturated one needs to weaken the second alternative

$$[Y] \subseteq (\mathfrak{X})_\Delta$$

in the definition of Ramsey set $\mathfrak{X} \subseteq \mathcal{B}_1^{[\infty]}(E)$ using the terminology of infinite games. As customary in this context in order to avoid repeating the same words over and over again, we reserve the letters X,Y,Z, etc for denoting infinite normalized block sequence and the letters s,t,u, etc for denoting finite normalized block sequences.

Definition IV.2.2. Given $Y \in \mathcal{B}_1^{[\infty]}(E)$ and $\mathfrak{X} \subseteq \mathcal{B}_1^{[\infty]}(E)$, we define an infinite perfect-information game $G_{\mathfrak{X}}(Y)$ between two players I and II as follows:

- I plays an infinite block subsequence Y_0 of Y

- II plays $z_0 \in S_{[Y_0]}$

- I plays another infinite block subsequence Y_1 of Y

- II plays $z_1 \in S_{[Y_1]}$ such that $z_1 > z_0$

- I plays an infinite block subsequence Y_2 of Y, and so on.

We say that II wins the resulting infinite play $Y_0, z_0, Y_1, z_1, \ldots, Y_n, z_n, \ldots$ if the sequence (z_n) belongs to $\mathfrak{X}$. Otherwise I wins the play.

Definition IV.2.3. (1) We say that $\mathfrak{X} \subseteq \mathcal{B}_1^{[\infty]}(E)$ is *large for* $[s, X]$ if for every infinite $Y \in [s, X]$ there is infinite $Z \in [s, Y]$ such that $Z \in \mathfrak{X}$. We say that $\mathfrak{X}$ is *very large for* $[s, X]$ if $[s, X] \subseteq \mathfrak{X}$.

(2) We say that $\mathfrak{X} \subseteq \mathcal{B}_1^{[\infty]}(E)$ is *strategically large for* $[s, X]$ if player II has a winning strategy for the game $G_{(\mathfrak{X}^s)}(X)$, where $\mathfrak{X}^s = \{x \in \mathcal{B}_1^{[\infty]}(E) : s \frown X \in \mathfrak{X}\}$.

(3) We say that $\mathfrak{X} \subseteq \mathcal{B}_1^{[\infty]}(E)$ is *strategically Ramsey* if for every $\Delta = (\delta_n)_n \in \mathbb{R}_+^\infty$ and every $(s, Y) \in \mathcal{B}_1^{[<\infty]}(E) \times \mathcal{B}_1^{[\infty]}(E)$ there is $Z \in [s, Y]$ such that either

(a) $[s, Z] \cap \mathfrak{X} = \emptyset$, or

(b) $\mathfrak{X}_\Delta$ is strategically large for $[s, Z]$.

Remark IV.2.4. (1) Note that what we call here *strategically Ramsey* should more properly be called *strategically approximately Ramsey*. The notion that deserves to be called *strategically Ramsey* would have its second alternative strengthened to the conclusion that player II has a winning strategy in the game $G_{\mathfrak{X}^s}(Z)$ rather than in the game $G_{(\mathfrak{X}^s)_\Delta}(Z)$. While the notion that deserves the name *strategically Ramsey* is both well behaved and well studied in Ramsey theory, it would still be too strong in the context of infinite block basic sequences in Banach spaces. Since we will never deal with the properly named notion of strategically Ramsey sets we have decided, in the context of these lecture notes only, to adopt this simplification of terminology.

(2) Note that in the terminology of various notions of largeness introduced above, a set $\mathfrak{X}$ is approximately Ramsey iff for every basic set $[s, X]$ there is $Y \in [s, X]$ such that either $[s, Y] \cap \mathfrak{X} = \emptyset$ or $(\mathfrak{X})_\Delta$ is *very large* for $[s, Y]$, i.e., $[s, Y] \subseteq (\mathfrak{X})_\Delta$. So the formal difference between these two notions is that *very large* is weakened to *strategically large*. From Theorem III.5.6 we learn however that the actual difference between these two notions is immense.

The terminology 'large' and 'strategically large' is chosen to suggest the corresponding ideals of 'non large' and 'non strategically large' sets. It is therefore quite natural to investigate how much these ideals do resemble σ-ideals in their behavior. In the classical context, that smallness is a σ-complete notion corresponds to the possibility of being able to perform various diagonal arguments with basic sets. In the approximate Ramsey theory the diagonal arguments are typically facilitated by net-approximations.

Definition IV.2.5. Let $X \in \mathcal{B}_1^{[\le\infty]}(E)$, $\Delta = (\delta_i) \in \mathbb{R}_+^\infty$. A Δ-net of $[X]^{[<\infty]}$ is a countable subset $\mathfrak{N} \subseteq [X]^{[<\infty]}$ such that
(a) for every $(y_0, \dots, y_n) \in [X]^{[<\infty]}$ there is some $(z_0, \dots, z_n) \subset \mathfrak{N}$ such that $\|y_i - z_i\| \le \delta_i$ and supp $y_i =$ supp z_i for every $i \le n$.
(b) $\mathfrak{N} \cap [(e_j)_0^i]^{[<\infty]}$ is finite, for every $i \in \mathbb{N}$.

For a single real number $\varepsilon > 0$, an ε-*net* is by definition an $(\varepsilon, \varepsilon^2, \dots)$-net. Finally, for a sequence $(\Delta_k)_k \in \mathbb{R}_+^\infty$, a $(\Delta_k)_k$-*net* of $[X]^{[<\infty]}$ is any countable set $\mathfrak{N}$ of finite block subsequences of X which can be written as the union of a sequence $(\mathfrak{N}_k)_k$ such that $\mathfrak{N}_k$ is a Δ_k-net of $[X]^{[<\infty]}$ for all $k \in \mathbb{N}$.

Clearly, countable Δ-nets in $[X]^{[<\infty]}$ do always exist.

Proposition IV.2.6. *Let $(\mathfrak{X}_n)_n$ be a given sequence of families of block subsequences of E. Let $\Delta = (\delta_n)_n \in \mathbb{R}_+^\infty$, and let $\phi : \mathbb{N} \to \mathbb{N}$. Suppose that the union $\mathfrak{X} = \bigcup_{n=0}^\infty \mathfrak{X}_n$ is large for some basic set $[s, X]$. Then there exist $t \in \mathcal{B}_1^{[<\infty]}(X)$ with $t > s$, an infinite block sequence $Y \in [s, X]$, and an $n \in \mathbb{N}$ such that $|t| \ge \phi(n)$ and such that $(\mathfrak{X}_n)_\Delta$ is large for $[s \frown t, Y]$.*

Proof. Otherwise, for every $Y \le X$, every $n \in \mathbb{N}$ and every $t > s$ with $|t| \ge \phi(n)$ there is some $Z \in [s \frown t, Y]$ such that $[s \frown t, Z] \cap (\mathfrak{X}_n)_\Delta = \emptyset$. Using this, fixing a Δ-net $\mathfrak{N}$ of $[X]^{[<\infty]}$, and then performing a simple diaginalization argument, we find block subsequence $Z = (z_n) \le Y$ and a decreasing sequence Z_n $(n \in \mathbb{N})$ of block subsequences of X such that for every n,
(1) $z_{n+1} \in Z_n^{[1]}$, and
(2) for every $m \le n$ with $\phi(m) \le n$, and every $t \in \mathfrak{N} \cap [(e_i)_0^{\max \text{supp} z_n}]^{[\phi(m)]}$

$$[s \frown t, Z_n] \cap (\mathfrak{X}_m)_\Delta = \emptyset.$$

Having obtained such a block subsequence $Z = (z_n)$, we claim that $\mathfrak{X} \cap [s, Z] = \emptyset$. Otherwise, there is $W \in [s, Z] \cap \mathfrak{X}_m$ for some $m \in \mathbb{N}$. Let (w_n') be such that $W = s \frown (w_n')$. Choose minimal n such that $u = (w_i')_0^{\phi(m)-1} \in [(z_0, \dots, z_n)]$ and pick $t \in \mathfrak{N} \cap [(e_i)_0^{\max \operatorname{supp} z_n}]^{[\phi(m)]}$ such that $d(u,t) \leq \Delta$. By the choice of Z_n we have that

$$[s \frown t, Z_n] \cap (\mathfrak{X}_m)_\Delta = \emptyset.$$

Since $s \frown t \frown (w_i')_{i \geq \phi(m)}$ is a member of this basic set on distance $\leq \Delta$ from W, we obtain that $W \notin \mathfrak{X}_m$, a contradiction. $\qquad \square$

Proposition IV.2.7. *Let $(\mathfrak{X}_n)_n$ be a given sequence of subsets of $\mathcal{B}_1^{[\infty]}(E)$ and let X be a given sequence of $\mathcal{B}_1^{[\infty]}(E)$. Then there is infinite $Z \leq X$ such that for every n, if there is $Y \leq Z$ such that $\mathfrak{X}_n$ is strategically large (large) for $[Y]$, then $\mathfrak{X}_n$ is also strategically large (large) for $[Z]$.*

Proof. We construct recursively on n a decreasing sequence (Z_n) $(n \in \mathbb{N})$ of block subsequences of X such that for every $n \in \mathbb{N}$ if there is some $Y \leq Z_n$ such that $\mathfrak{X}_n$ is strategically large (large) for $[Y]$, then $\mathfrak{X}_n$ is also strategically large (large) for $[Z_{n+1}]$. Let $Z = (z_n)$ be an infinite normalized block sequence such that $(z_{i \geq n}) \leq Z_n$ for all n. Now observe a general fact that if some set $\mathfrak{Y} \subseteq \mathcal{B}_1^{[\infty]}(E)$ is strategically large (large) for some infinite block sequence Y, then $\mathfrak{Y}$ is also strategically large (large) for any infinite normalized block sequence Y' which has a tail that is a block subsequence of Y. $\qquad \square$

IV.3   Combinatorial Forcing on Block Sequences in Banach Spaces

The asymmetric nature of approximate and strategic Ramsey sets makes these notions behave differently from the classical notion of Ramsey sets considered above in Sections II.5 and II.4, though one can still prove some analogous results. For example, as in the classical case one has results that corresponds to Galvin's lemma which deals with sets $\mathfrak{X}$ of finite block sequences. In the strategical Ramsey context, the game $G_{\mathfrak{X}}(Y)$ is then of course slightly modified to the effect that the player II wins a play $Y_0, z_0, Y_1, z_1, \ldots, Y_n, z_n, \ldots$ just in case there is a k such that $(z_n)_0^k \in \mathfrak{X}$.

Lemma IV.3.1 (Gowers). *Let $\mathfrak{X}$ be an arbitrary set of finite normalized block subsequences of a fixed Schauder basis (e_i) of some Banach space E, let X be an arbitrary infinite normalized block subsequence of (e_i), and let $\Delta = (\delta_i)_i \in \mathbb{R}_+^\infty$. Then there is an infinite normalized block subsequence Y of X such that either $[Y]^{[<\infty]} \cap \mathfrak{X} = \emptyset$ or else II has a winning strategy in the game $G_{(\mathfrak{X})_\Delta}(Y)$.*

Proof. We will in fact establish a stronger conclusion. For two (finite or infinite) block sequences s and t of the same length k and $\Delta = (\delta_i) \in \mathbb{R}_+^\infty$, we write $d(s,t) \leq \Delta$ whenever for all $i < k$,

$$\|s(i) - t(i)\| \leq \delta_i \quad \text{and} \quad \operatorname{supp} s(i) = \operatorname{supp} t(i).$$

Fix $\mathfrak{X}$ and $\Delta = (\delta_i)_i$ as in the hypothesis of the Lemma. Given two (finite or infinite) block subsequences $s = (y_i)$, $t = (z_i)$, of the same length let $\Theta_{s,t}$ be the linear isomorphism between the span of (y_i) and the span of (z_i) defined by extending the assignment $\Theta(y_i) = z_i$. Note that $\Theta_{s,t}$ maps in a natural way a given block subsequence u of s to a corresponding block subsequence $\Theta_{s,t}(u)$ of t. The reader can easily verify the following fact of future use.

(1) There is $\Gamma \in \mathbb{R}^\infty_+$ such that $\Gamma < \Delta$ and such that for every two block sequences s and t of the same length and with the property that $d(s,t) \leq \Gamma$ we have that $d(u, \Theta_{s,t}(u)) \leq \Delta$ for every (finite) block subsequence u of s. (Hint: Let $\gamma_n = \min\{\delta_0, \ldots, \delta_n\}/(2^{n+2}C)$, where C is the basis constant of the Schauder basis (e_i) of E.)

Given an infinite normalized block sequence X, a finite normalized block subsequence s and a family $\mathcal{R} \subseteq \mathcal{B}_1^{[<\infty]}(E)$, we say that X $\mathcal{R}$-*rejects* s iff

$$\forall Y \leq X \ \forall n \in \mathbb{N} \ Y \upharpoonright n \in (\mathcal{R})^c.$$

We say that X $\mathcal{R}$-*accepts* s iff no $Y \leq X$ $\mathcal{R}$-rejects s.

Given Δ as in the hypothesis of the Lemma, we let $\Gamma = (\gamma_i)$ be chosen to satisfy the conclusion of (1) for $\Delta/2$. For $n \in \mathbb{N}$, set

$$\Gamma_n = (n\gamma_i/(n+1))_{i=0}^\infty.$$

We are now ready to define the combinatorial forcing that corresponds to that appearing in standard proofs of Galvin's lemma. For this we shall need one more piece of notation,

$$\mathfrak{X}_0 = \{s \in (\mathfrak{X})_{\Delta/2} : [s]^{[<|s|]} \cap (\mathfrak{X})_{\Delta/2} = \emptyset\}.$$

We say that X *rejects* s iff X $(\mathfrak{X}_0)_{\Gamma_{|s|}}$-rejects s. We say that X *accepts* s iff X $(\mathfrak{X}_0)_{\Gamma_{|s|+1}}$-accepts s. We say that X *decides* s if it either accepts or rejects s. We say that $\mathcal{R} \subseteq \mathcal{B}_1^{[<\infty]}(E)$ is *small* if it contains no subset of the form $[Y]^{[<\infty]}$ for any infinite normalized block subsequence Y.

The reader can easily verify the following basic facts about these notions of combinatorial forcing that will be freely used in the arguments that follow.

(a) If X $\mathcal{R}$-accepts ($\mathcal{R}$-rejects) s, then every $Y \leq X$ $\mathcal{R}$-accepts ($\mathcal{R}$-rejects) s.

(b) If X accepts (rejects) s, then every $Y \leq X$ accepts (rejects) s.

(c) If $\mathcal{R} \subseteq \mathcal{R}'$, then if X $\mathcal{R}'$-rejects s, then X $\mathcal{R}$-rejects s.

(d) For every X, s and $\mathcal{R} \subseteq \mathcal{B}_1^{[<\infty]}(E)$ there is some $Y \leq X$ which $\mathcal{R}$-decides s.

(e) For every X and s there is some $Y \leq X$ which decides s.

(f) If X rejects s, then X rejects $s \frown (x)$ for every $x \in X$.

(g) If X accepts s, then $\{x \in X : X \text{ rejects } s \frown (x)\}$ is small.

As in the classical case we are now ready to state and prove the following claim.

(2) There is $Y \leq X$ which decides every $s \in [Y]^{[<\infty]}$.

The infinite normalized block subsequence $Y = (y_i)$ is defined recursively in i along with a decreasing sequence and Y_i $(i \in \mathbb{N})$ of infinite normalized block subsequences of X with the requirement that for every $i \in \mathbb{N}$ and for every $s \in [(y_0, \ldots, y_i)]$, Y_i decides s. Suppose we have already defined $(y_0, \ldots, y_i)$ and Y_i. Choose $y_{i+1} \in Y_i$ such that $y_{i+1} > y_i$. For each $1 \leq k \leq i+1$, let F_k be a finite $\Lambda_k = (\Gamma_{k+1} - \Gamma_k)/3$-net of $[(y_0, \ldots, y_{i+1})]^{[k]}$, i.e.

$$(F_k)_{(\Gamma_{k+1} - \Gamma_k)} = [(y_0, \ldots, y_{i+1})]^{[k]}.$$

Let $Y_{i+1} \in [Y_i]$ be such that for every $t \in \bigcup_{k=1}^{i+1} F_k$ either Y_{i+1} $(\mathfrak{X}_0)_{\Gamma_k + \Lambda_k}$-rejects t or Y_{i+1} $(\mathfrak{X}_0)_{\Gamma_{k+1} - \Lambda_k}$-accepts t, and Y_{i+1} decides $\emptyset$. We claim that Y_{i+1} decides every $s \in [(y_0, \ldots, y_{i+1})]$: Suppose that $s \in [(y_0, \ldots, y_{i+1})]^{[k]}$, and choose $t \in F_k$ be such that $d(s, t) \leq \Lambda_k$ and $\operatorname{supp} s = \operatorname{supp} t$. If Y_{i+1} $(\mathfrak{X}_0)_{\Gamma_k + \Lambda_k}$-rejects t, then Y_{i+1} rejects s. If not, there is some $Z \in [s, Y_{i+1}]$ and some $n \in \mathbb{N}$ such that $Z \restriction n \in (\mathfrak{X}_0)_{\Gamma_k}$. Let $u \in \mathfrak{X}_0$ be such that $d(u, Z \restriction n) \leq \Gamma_k$. Then

$$t \frown u \restriction [|t|, |u| - 1] \in (\mathfrak{X}_0)_{\Gamma_k + \Lambda_k},$$

in contradiction with the fact that Y_{i+1} $(\mathfrak{X}_0)_{\Gamma_k + \Lambda_k}$-rejects t. If Y_{i+1} $(\mathfrak{X}_0)_{\Gamma_{k+1} - \Lambda_k}$-accepts t then one can easily prove that Y_{i+1} accepts s. This describes our recursive construction. It should be clear that the resulting block sequence Y satisfies the conclusion of Claim (2).

From now on we fix Y satisfying the conclusion of Claim (2).
Case 1: There is some $Y' \leq Y$ such that Y' $\mathfrak{X}$-rejects $\emptyset$. Then $[Y']^{[<\infty]} \cap \mathfrak{X} = \emptyset$ and we are done.
Case 2: There is no $Y' \leq Y$ $\mathfrak{X}$-rejecting $\emptyset$. Then we have that

$$\forall Y' \leq Y \exists Y'' \leq Y' \exists n \in \mathbb{N} \ Y'' \restriction n \in \mathfrak{X}. \tag{IV.1}$$

Note that this also gives that Y accepts $\emptyset$. We now claim that player II has a winning strategy for the game $G_{(\mathfrak{X})_\Delta}(Y)$: Suppose that I plays $Z_0 \in [Y]$. By (f) and the fact that Y decides every $s \in [Y]^{[<\infty]}$, II can play $z_0 \in Z_0$ such that Y accepts (z_0). In general, suppose that in the nth-run I plays $Z_n \in [Y]$. Then II can play $z_n > z_{n-1}$ such that Y accepts $(z_0, \ldots, z_n)$. We claim that the payoff of this run (z_i) of the game is such that there must be some $n \in \mathbb{N}$ with the property that $(z_0, \ldots, z_n) \in (\mathfrak{X})_\Delta$: To see this note that by the above displayed property (IV.1) of Y from Case 2, we can find an $n \in \mathbb{N}$ and some $t \in [z_0, \ldots, z_n] \cap \mathfrak{X}$. Since Y accepts $(z_0, \ldots, z_n)$ there is some $W \in [(z_0, \ldots, z_n), Y]$ and some m such that $W \restriction m \in (\mathfrak{X}_0)_{\Gamma_{m+1}}$. If $m \leq n$, then

$$(y_0, \ldots, y_m) \in (\mathfrak{X}_0)_{\Gamma_{m+1}} \subseteq (\mathfrak{X}_0)_\Gamma \subseteq (\mathfrak{X}_0)_{\Delta/2} \subseteq (\mathfrak{X})_\Delta,$$

and we would be done. Otherwise, there is some non-empty u, and some finite block subsequence $(y_0', \ldots, y_n')$ such that $d((y_0, \ldots, y_n), (y_0', \ldots, y_n')) \leq \Gamma_{m+1}$ and such that $(y_0', \ldots, y_n') \frown u \in \mathfrak{X}_0$. Then $d(t, \Theta_{((y_0, \ldots, y_n) \frown u), (y_0', \ldots, y_n') \frown u}(t)) \leq \Delta/2$, which is in a contradiction with the fact that $(y_0', \ldots, y_n') \frown u \in \mathfrak{X}_0$, since $t \in \mathfrak{X}$. $\qquad \square$

Exercise IV.3.2. Supply a proof of Gowers' Banach space dichotomy that uses Lemma IV.3.1.(Hint: Let C be the basis constant of the Schauder basis (e_n) of E and choose summable $\Delta \in \mathbb{R}^\infty$ such that $1 + 2Cd < \sqrt{2}$, where d is the sum of Δ. Let $\mathfrak{X}$ be the set of all all finite block subsequences $(x_i)_0^k$ of (e_n) such that

$$\forall K \ \exists n < m \ \exists \lambda_0, ..., \lambda_{m-n} \ \|\sum_n^m \lambda_{i-n} x_i\| > K \|\sum_n^m (-1)^i \lambda_{i-n} x_i\|.$$

Show that if E contains no infinite-dimensional subspace with an unconditional basis then $\mathfrak{X}$ is large and therefore by Lemma IV.3.1 the corresponding set $(\mathfrak{X})_\Delta$ is strategically large below some infinite block subsequence of (e_n). Use the strategy to build an infinite block sequence whose closed linear span is a hereditarily indecomposable space.)

The corresponding lemma for the approximate Ramsey property for c_0-saturated Banach spaces is also true. In fact, the corresponding notion of combinatorial forcing resembles considerably more the combinatorial forcing from the classical case than the one appearing in the proof of Lemma IV.3.1. It is for this reason that we choose to spell it out in a more explicit manner. So again we simplify the notation by letting the variables $\mathcal{R}$, $\mathcal{S}$, $\mathcal{T}$, etc run over subsets of $\mathcal{B}_1^{[<\infty]}(c_0)$, the variables X, Y, Z, etc run over elements of $\mathcal{B}_1^{[\infty]}(c_0)$, and variables s, t, u, etc run over elements of $\mathcal{B}_1^{[<\infty]}(c_0)$.

Definition IV.3.3. For a given $\mathcal{S}$, s and X as above, we say that X $\mathcal{S}$-*accepts* s iff for every $Y \in [s, X]$ there is some $n \in \mathbb{N}$ such that $Y \restriction n \in \mathcal{S}$. An infinite block subsequence X $\mathcal{S}$-*rejects* s iff no $Y \in [X]$ $\mathcal{S}$-accepts s. An infinite block subsequence X $\mathcal{S}$-*decides* s iff either X $\mathcal{S}$-accepts s or X $\mathcal{S}$-rejects s. More generally, for a pair of sets $\mathcal{R} \subseteq \mathcal{S} \subseteq \mathcal{B}_1^{[<\infty]}(c_0)$, we say that X $(\mathcal{S}, \mathcal{R})$-*decides* s iff either X $\mathcal{S}$-accepts s or X $\mathcal{R}$-rejects s.

The following three propositions summarize the facts about these notions that will be quite useful below.

Proposition IV.3.4. (a) If $\mathcal{R} \subseteq \mathcal{S}$ and X $\mathcal{R}$-accepts s, then X $\mathcal{S}$-accepts s.
(b) If $\mathcal{R} \subseteq \mathcal{S}$ and X $\mathcal{S}$-rejects s, then X $\mathcal{R}$-rejects s. $\qquad\square$

Proposition IV.3.5. Let $\varepsilon \geq 0$ and let $\Gamma \in \mathbb{R}^\infty$ be defined by $\Gamma(i) = \varepsilon$ if $i = |s|$ and 0 otherwise. Suppose that for some X, $\mathcal{R}$, and s we have that the block sequence X $(\mathcal{R})_\Gamma$-rejects s. Then the set

$$(\{x \in X : X \ \mathcal{R}\text{-accepts } s \frown (x)\})_\varepsilon$$

is small in the sense that it contains no set of the form $[Y]^{[1]}$ for Y an infinite block subsequence of X.

Proof. Otherwise, there is infinite $Y \leq X$ such that for every $y \in [Y]^{[1]}$ there is some y' such that $\|y - y'\| \leq \varepsilon$ and X $\mathcal{R}$-accepts $s \frown (y')$. It follows that for every $y \in [Y]^{[1]}$, the block sequence X $(\mathcal{R})_\Gamma$-accepts $s \frown (y)$ contradicting the assumption that X $(\mathcal{R})_\Gamma$-rejects s. $\qquad\square$

Proposition IV.3.6. *Suppose $\Gamma_n < \Delta_n$ for $n \in \mathbb{N}$ is a given sequence of pairs of elements of $\mathbb{R}^\infty_+$. Then for every $X \in \mathcal{B}^{[\infty]}_1(E)$ and $\mathcal{R} \subseteq \mathcal{B}^{[\infty]}_1(E)$ there is infinite $Y \leq X$ such that for every $s \in [Y]^{[<\infty]}$, the block sequence Y $((\mathcal{R})_{\Delta_{|s|}}, (\mathcal{R})_{\Gamma_{|s|}})$-decides s.*

Proof. This is similar to the proof of Claim (2) from the proof of Lemma IV.3.1.

□

Now we are ready to state and prove the analogue of Galvin's Lemma for the approximate Ramsey property of the space c_0.

Lemma IV.3.7 (Gowers). *Let $\mathfrak{X}$ be an arbitrary set of finite normalized block sequences of the Banach space c_0. Let X be an infinite normalized block sequence of c_0 and let $\Delta = (\delta_i)_i \in \mathbb{R}^\infty_+$. Then there is infinite $Y \leq X$ such that either $[Y]^{[<\infty]} \cap \mathfrak{X} = \emptyset$, or else for every infinite $Z \leq Y$ there is $n \in \mathbb{N}$ such that $Z \upharpoonright n \in (\mathfrak{X})_\Delta$.*

Proof. Fix $\mathfrak{X}$ and Δ. For $n \in \mathbb{N}$, we set

$$\Delta_n = (1/(n+1))\Delta.$$

Using these objects, we now define the following notion of combinatorial forcing.

Definition IV.3.8. We say that X *accepts* s iff X $(\mathfrak{X})_{\Delta_{4|s|}}$-accepts s. We say that X *rejects* s iff X $(\mathfrak{X})_{\Delta_{4|s|+1}}$-rejects s. We say that X *decides* s iff either X accepts s or X rejects s.

Note the following properties of this notion of combinatorial forcing.

Proposition IV.3.9. (a) *If X accepts (rejects) s, then for every block subsequence Y of X accepts (rejects) s.*
(b) *For every X and s there is some infinite block subsequence Y of X which decides s.*
(c) *If X accepts s, then for every $x \in X$, the sequence X accepts $s \frown (x)$.*

Proof. (a) This is immediate from the definitions.
(b) If no $Y \leq X$ accepts s, then by the definition X must $(\mathfrak{X})_{\Delta_{4|s|}}$-reject s. Since $(\mathfrak{X})_{\Delta_{4|s|+1}} \subseteq (\mathfrak{X})_{\Delta_{4|s|}}$, we get that X rejects s.
(c) This is immediate from the definition of acceptance.   □

We are now ready to continue with the proof of Lemma IV.3.7. Using Proposition IV.3.6 and a simple diagonalization procedure we obtain $X_0 \in [X]$ which $((\mathfrak{X})_{\Delta_{4|s|+i}}, (\mathfrak{X})_{\Delta_{4|s|+j}})$-decides every $s \in [X_0]^{[<\infty]}$ for $0 \leq i < j \leq 4$. If X_0 accepts $\emptyset$ then the first alternative from the conclusion of the lemma holds. Otherwise, X_0 rejects $\emptyset$. We are going to recursively find $Y = (y_i) \leq X_0$ such that for every $s \in [Y]^{[<\infty]}$ Y rejects s. Suppose we defined $(y_0, \ldots, y_n)$ such that the minimal term of Y_n starts above y_n, $Y_n \leq Y_{n-1}$, and for every $s \in [(y_0, \ldots, y_n)]$ Y_n rejects s. Let

$$\varepsilon = (\Delta_{4n+4}(n) - \Delta_{4n+3}(n))/2.$$

Fix $s \in [(y_0, \ldots, y_n)]$. Consider the coloring $[Y_n]^{[1]} = C_a \cup C_r$ where

$$C_a = \{ y \in [Y_n]^{[1]} : Y_n(\mathfrak{X})_{\Delta_{4|s|+2}}\text{-accepts } s \frown (y) \}$$
$$C_r = \{ y \in [Y_n]^{[1]} : Y_n(\mathfrak{X})_{\Delta_{4|s|+3}}\text{-rejects } s \frown (y) \}.$$

By Gowers' c_0-theorem III.5.7 we obtain that there is some $Y_n(s) \leq Y_n$ such that either $[Y_n(s)]^{[1]} \subseteq (C_a)_\varepsilon$ or $[Y_n(s)]^{[1]} \subseteq (C_r)_\varepsilon$. Notice that $\Delta_{4|s|+2} + \Gamma \leq \Delta_{4|s|+1}$ where $\Gamma \in \mathbb{R}_+^\infty$ is given by $\Gamma(i) = \varepsilon$ if $i = |s|$ and 0 otherwise. So by our recursive assumption, the block sequence Y_n $(\mathfrak{X})_{\Delta_{4|s|+2}+\Gamma}$ rejects s. Then by Proposition IV.3.5 applied to $\mathcal{R} = (\mathfrak{X})_{\Delta_{4|s|+2}}$ we obtain that $(C_a)_\varepsilon$ is small. So, we have the second alternative. This gives us that for every $y \in [Y_n(s)]^{[1]}$, the block sequence $Y_n(s)$ $(\mathfrak{X})_{\Delta_{4|s|+4}}$-rejects $s \frown (y)$.

Now fix a finite ε-net $\mathfrak{N}$ of $[(y_0, \ldots, y_n)]$. Applying the above procedure successively for every $s \in \mathfrak{N}$ we obtain a single $Y_{n+1} \leq Y_n$ such that for every $s \in \mathfrak{N}$ and every $y \in [Y_{n+1}]^{[1]}$, Y_{n+1} $(\mathfrak{X})_{\Delta_{4|s|+4}}$-rejects $s \frown (y)$. Hence for *every* $s \in [(y_0, \ldots, y_n)]$ and every $y \in [Y_{n+1}]^{[1]}$, Y_{n+1} $(\mathfrak{X})_{\Delta_{4(|s|+1)+1}}$-rejects $s \frown (y)$, as desired. Let y_{n+1} be the minimal term of the sequence Y_{n+1} above y_n. This finishes our recursive construction of the infinite block sequence $Y = (y_i)$.

Now we claim that the existence of such sequence $Y = (y_i)$ gives us the first alternative of the Lemma. To see this note that since $\mathfrak{X} \subseteq (\mathfrak{X})_{\Delta_n}$ for every n, by the definition of rejection, we obtain that $[Y]^{[<\infty]} \cap \mathfrak{X} = \emptyset$. $\qquad\square$

IV.4 Coding into Approximate and Strategic Ramsey Sets

In what follows, E is a fixed Banach space with a Schauder basis (e_n). It will be convenient to denote its unit sphere by $S(E)$. More generally, given a finite or infinite normalized block subsequence $(x_n)_n$, we use the notation $S((x_n)_n)$ for the unit sphere of the corresponding closed linear span of $(x_n)_n$. We shall use $\mathcal{N}^\uparrow$ to denote the set of strictly increasing sequences of positive integers as a topological subspace of the Baire space $\mathcal{N} = \mathbb{N}^\mathbb{N}$. Recall that all topological properties of subsets of powers of the form $\mathcal{B}_1^{[k]}(E)$ refer to the topology induced from the Tychonoff cube $S(E)^k$ where $S(E)$ is taken with its separable metric norm-topology.

Definition IV.4.1. A subset A of the sphere $S(E)$ is called *asymptotic* iff $A \cap S((x_n)) \neq \emptyset$ for every infinite block sequence $(x_n)_n$ of E. An *asymptotic pair* of E is a pair $\mathfrak{A} = \langle A_0, A_1 \rangle$ where A_0 and A_1 are disjoint asymptotic sets of E.

Note that asymptotic pairs always exist. For example, the sets

$$A_i = \{ x \in S((e_n)_n) : (-1)^{i+1} e^*_{\min \operatorname{supp} x}(x) > 0 \}$$

for $i = 0, 1$ are clearly asymptotic and they form an asymptotic pair. Since our coding procedure will be based on asymptotic pairs and since we will be interested in calculations of the descriptive complexity of various sets of block sequence it will be quite important to have asymptotic pairs of as low complexity as possible. This motivates our next definition.

Definition IV.4.2. An asymptotic pair $\mathfrak{A} = \langle A_0, A_1 \rangle$ is called *discrete* if A_0 and A_1 are F_σ subsets of $S(E)$. An asymptotic pair $\mathfrak{A} = \langle A_0, A_1 \rangle$ is called *separated* if A_0 and A_1 are closed subsets of the sphere $S(E)$ and if $\delta(\mathfrak{A}) > 0$, where

$$\delta(\mathfrak{A}) = d(A_0, A_1) = \inf\{\|a_0 - a_1\| \; : \; a_0 \in A_0, a_1 \in A_1\}.$$

Remark IV.4.3. (a) The asymptotic pair

$$A_i = \{x \in S((e_n)_n) \; : \; (-1)^{i+1} e^*_{\min \text{supp} x}(x) > 0\}$$

for $i = 0, 1$ is an example of a discrete asymptotic pair since

$$A_i = \bigcup_{(L,N) \in \mathbb{N}^2} \{x \in S(E) \; : \; (-1)^{i+1} e^*_L(x) \geq \frac{1}{N} \text{ and } (\forall l < L) \; e^*_l(x) = 0\}, \quad \text{(IV.2)}$$

gives us a representation of these sets as countable union of closed sets.
(b) By Theorem III.5.6 separated asymptotic pairs exist if the Banach space E is not c_0-saturated. If the Banach space E does not contain an isomorphic copy of c_0 then for every block normalized block sequence X there is a normalized block subsequence Y of X such that $S([Y])$ contains a separated asymptotic pair.

From now on we fix an asymptotic pair $\mathfrak{A} = \langle A_0, A_1 \rangle$ and study the following coding procedure based on it.

Definition IV.4.4. A sequence $(x_k)_k \in \mathcal{B}_1(E)^{[\infty]}$ *codes the pair* $((y_n)_n, (k_n)_n)$, provided that $y_n = x_{2n}$ for all n and provided that $\{k_n\}_n$ is the increasing enumeration of the set $\{k \; : \; x_{2k+1} \in A_1\}$.

Let

$$\mathcal{B}^{[\infty]}_{\mathfrak{A}}(E) = \{(x_n)_n \in \mathcal{B}^{[\infty]}_1(E) \; : \; (\forall n) \; x_{2n+1} \in A_0 \cup A_1 \text{ and } (\exists^\infty n) \; x_{2n+1} \in A_1\}.$$

Then every member $(x_k)_k$ of $\mathcal{B}^{[\infty]}_{\mathfrak{A}}(E)$ codes the pair

$$\Lambda_{\mathfrak{A}}((x_k)_k) = ((x_{2n})_n, (k_n)_n),$$

where $(k_n)_n$ is the increasing enumeration of $\{k \; : \; x_{2k+1} \in A_1\}$. This gives us a mapping

$$\Lambda_{\mathfrak{A}} : \mathcal{B}^{[\infty]}_{\mathfrak{A}}(E) \to \mathcal{B}^{[\infty]}_1(E) \times \mathcal{N}^{\uparrow}$$

which we will study in some detail below. But let us first show that our coding procedure is in fact possible.

Proposition IV.4.5. *For every block sequence X there is some block subsequence Y of X such that every pair from $[Y]^{[\infty]} \times \mathcal{N}^{\uparrow}$ is coded by some block subsequence of X.*

Proof. Fix a block sequence $X = (x_n)_n$. Let n_0 be the first integer n such that $S(\langle x_i \rangle_{i=0}^{n-1}) \cap A_0$ and $S(\langle x_i \rangle_{i=0}^{n-1}) \cap A_1$ are non empty, which is well defined since A_0 and A_1 are asymptotic sets and, by definition, every element of them has finite support. With n_k defined, let n_{k+1} be the first integer $n > n_k$ such that $S(\langle x_i \rangle_{i=n_k+1}^{n-1}) \cap A_0$ and $S(\langle x_i \rangle_{i=n_k+1}^{n-1}) \cap A_1$ are non empty. Set $Y = (x_{n_k})_k$. We are going to show that $[Y] \times \mathcal{N}^\uparrow \subseteq (\Lambda_{\mathfrak{A}}"[X]^{[\infty]}) \cap \mathcal{B}_{\mathfrak{A}}^{[\infty]}(E)$. Fix a block sequence $Z = (z_n)_n \in [Y]$ and $\vec{\varepsilon} = (\varepsilon_n)_n \in \mathcal{N}^\uparrow$. Let k_r be the minimal integer k such that $z_r \in S(\langle x_{n_0}, ..., x_{n_k}\rangle)$ for every r. Choose

$$w_r \in \begin{cases} S(\langle x_j \rangle_{j=n_{k_r}+1}^{n_{k_r+1}-1}) \cap A_1 & \text{if } r \in \{\varepsilon_n\}_n \\ S(\langle x_j \rangle_{j=n_{k_r}+1}^{n_{k_r-1}+1}) \cap A_0 & \text{if } r \notin \{\varepsilon_n\}_n \end{cases} \tag{IV.3}$$

for every r. Then the block sequence $W' = (z_0, w_0, z_1, w_1, ...)$ is in $[X] \cap \mathcal{B}_{\mathfrak{A}}^{[\infty]}(E)$, and clearly $\Lambda_{\mathfrak{A}}(W') = (Z, \vec{\varepsilon})$. $\square$

Recall the notions of large and strategically large for subsets of $\mathcal{B}_1^{[\infty]}(E)$ given above in Definition IV.2.3. Let

$$\pi_0 : \mathcal{B}_1^{[\infty]}(E) \times \mathcal{N}^\uparrow \to \mathcal{B}_1^{[\infty]}(E)$$

be the projection on the first coordinate.

Corollary IV.4.6. *Let C be a subset of $\mathcal{B}_1^{[\infty]}(E) \times \mathcal{N}^\uparrow$ and X be an infinite block sequence. If $\pi_0"C$ is large for $[X]$ then $\Lambda_{\mathfrak{A}}^{-1}(C)$ is also large for $[X]$.*

Proof. Suppose that $\pi_0"C$ is large for $[X]$ and fix some infinite block sequence X' of X. From Proposition IV.4.5 there is infinite block subsequence Y of X' such that $[Y] \times \mathcal{N}^\uparrow \subseteq \Lambda_{\mathfrak{A}}"(\mathcal{B}_1^{[\infty]}(E)) \cap [X']^{[\infty]}$. Since $\pi_0"C$ is large for $[X]$ there is some $Z \in [Y]^{[\infty]} \cap \pi_0"C$. Fix $\vec{\varepsilon} \in \mathcal{N}^\uparrow$ such that $(Z, \vec{\varepsilon}) \in C$ and choose $W \in \mathcal{B}_1^{[\infty]}(E) \cap [s, X']$ such that $\Lambda_{\mathfrak{A}}(W) = (Z, \vec{\varepsilon})$. Clearly $W \in \Lambda_{\mathfrak{A}}^{-1}(C) \cap [X']^{[\infty]}$. $\square$

Proposition IV.4.7. *Let C and X be as in the hypothesis of the previous proposition, and let $\Delta \in \mathbb{R}_+^\infty$. If $(\Lambda_{\mathfrak{A}}^{-1}(C))_\Delta$ is strategically (very) large for $[X]$, then $(\pi_0"C)_\Delta$ is also strategically (very) large for $[X]$.*

Proof. Suppose that there is some X such that Player II has a winning strategy Φ for the game $G_{(\Lambda_{\mathfrak{A}}^{-1}(C))_\Delta}[X]$. Let us describe a winning strategy Φ' for Player II for the game $G_{(\pi_0"C)_\Delta}[X]$: Start the game with Player I choosing $X_0 \in [X]$. Then Player II splits X_0 into two subsequences Y_0 and Z_0 and he picks $y_0 = \Phi'(X_0) = \Phi(Y_0)$. Suppose that the next choice of Player I is $X_1 \in [X]$. Then Player II splits X_1 into two subsequences Y_1 and Z_1 and he chooses $y_1 = \Phi'(X_0, X_1) = \Phi(Y_0, Z_0, Y_1)$, and so on. At the end of the game the block sequence

$$(y_0, y_1, ...) = \Phi' * (Y_n)_n \in (\pi_0"C)_\Delta \tag{IV.4}$$

since

$$(y_0, z_0, y_1, z_1, ...) = \Phi * (Y_0, Z_0, Y_1, Z_1, ...) \in (\Lambda_{\mathfrak{A}}^{-1}(C))_\Delta. \tag{IV.5}$$

The proof for the case of the very large property is similar.   □

Corollary IV.4.8. *Let $\mathcal{SR}_E$ denote the family of strategically Ramsey subsets of $\mathcal{B}_1(E)^{[\infty]}$. Then for every family family $\mathcal{C}$ of subsets of $\mathcal{B}_1(E) \times \mathcal{N}^{\uparrow}$, the inclusion $\{\Lambda_{\mathfrak{A}}^{-1}(C) : C \in \mathcal{C}\} \subseteq \mathcal{SR}_E$ implies the inclusion $\{\pi_0"C : C \in \mathcal{C}\} \subseteq \mathcal{SR}_E$.*   □

A quite analogous argument gives a corresponding result about the approximate Ramsey property in the class of c_0-saturated spaces E.

Proposition IV.4.9. *Let E be a c_0-saturated space, and let $\mathcal{AR}_E$ be the family of approximate Ramsey subsets of $\mathcal{B}_1(E)^{[\infty]}$. Then for every family family $\mathcal{C}$ of subsets of $\mathcal{B}_1(E) \times \mathcal{N}^{\uparrow}$, the inclusion $\{\Lambda_{\mathfrak{A}}^{-1}(C) : C \in \mathcal{C}\} \subseteq \mathcal{AR}_E$ implies the inclusion $\{\pi_0"C : C \in \mathcal{C}\} \subseteq \mathcal{AR}_E$.*   □

Next we discuss the Borel complexity of the set $\mathcal{B}_{\mathfrak{A}}^{[\infty]}(E)$ and the mapping $\Lambda_{\mathfrak{A}}$ depending on the complexity of the asymptotic pair $\mathfrak{A}$. We have the following two simple facts that describe these complexities.

Lemma IV.4.10. *Suppose that $\mathfrak{A}$ is a discrete asymptotic pair. Then $\mathcal{B}_{\mathfrak{A}}^{[\infty]}(E)$ is a $F_{\sigma\delta}$-set and the mapping $\Lambda_{\mathfrak{A}}$ is a Baire-class-1 function. Hence $\Lambda_{\mathfrak{A}}^{-1}(C)$ is an $F_{\sigma\delta}$ set for every closed set $C \subseteq \mathcal{B}_1^{[\infty]}(E) \times \mathcal{N}^{\uparrow}$. More precisely, for every closed set $C \subseteq \mathcal{B}_1^{[\infty]}(E) \times \mathcal{N}^{\uparrow}$, the set $\Lambda_{\mathfrak{A}}^{-1}(C)$ is the intersection of $\mathcal{B}_{\mathfrak{A}}^{[\infty]}(E)$ and a G_δ subset of $\mathcal{B}_1^{[\infty]}(E)$.*   □

Corollary IV.4.11. *Let E be a given Banach space with a Schauder basis (e_n) and let $\mathfrak{A}$ be a discrete asymptotic pair of E. If all relatively G_δ-subsets of $\mathcal{B}_{\mathfrak{A}}^{[\infty]}(E)$ are approximately Ramsey then so are all analytic subsets of $\mathcal{B}_1^{[\infty]}(E)$.*   □

It follows that the study of the approximate Ramsey property in a given c_0-saturated Banach space E reduces to the study of this property on relatively G_δ-subsets of $\mathcal{B}_{\mathfrak{A}}^{[\infty]}(E)$ for conveniently chosen asymptotic pairs $\mathfrak{A}$ of E. For example, in the case of the Banach space c_0 itself, we may restrict ourselves to G_δ-subsets of $\mathcal{B}_{\mathfrak{A}}^{[\infty]}(c_0)$, where $\mathfrak{A}$ is the asymptotic pair given in Remark IV.4.3(a) above.

Lemma IV.4.12. *Suppose that $\mathfrak{A}$ is a separated asymptotic pair. Then $\mathcal{B}_{\mathfrak{A}}^{[\infty]}(E)$ is a G_δ subset of $\mathcal{B}_1^{[\infty]}(E)$ and $\Lambda_{\mathfrak{A}}$ is a continuous mapping on this domain. Hence $\Lambda_{\mathfrak{A}}^{-1}(C)$ is a G_δ set for every closed set $C \subseteq \mathcal{B}_1^{[\infty]}(E) \times \mathcal{N}^{\uparrow}$. More precisely, for every closed set $C \subseteq \mathcal{B}_1^{[\infty]}(E) \times \mathcal{N}^{\uparrow}$, the set $\Lambda_{\mathfrak{A}}^{-1}(C)$ is the intersection of $\mathcal{B}_{\mathfrak{A}}^{[\infty]}(E)$ and a closed subset of $\mathcal{B}_1^{[\infty]}(E)$.*   □

Corollary IV.4.13. *If a Banach space E does not contain an isomorphic copy of c_0 and if all G_δ subsets of $\mathcal{B}_1^{[\infty]}(E)$ are strategically Ramsey, then so are all analytic subsets of $\mathcal{B}_1^{[\infty]}(E)$.*   □

It follows that the notions of approximate and strategic Ramsey sets behave strikingly different from the corresponding classical notion of Ramsey sets where results about analytic sets are typically considerably deeper.

Recall that subsets of Polish spaces can be classified according to their topological complexity. This yields the so-called projective (or Lusin) hierarchy of pointclasses. We shall use the following standard notation(see [42]): Σ_1^1 is the class of analytic sets, i.e., the continuous images of Borel sets; Π_1^1 is the class of coanalytic sets, i.e., the complements of analytic sets; Σ_{n+1}^1 is the class of the continuous images of Π_n^1 sets, and Π_{n+1}^1 is the class of complements of Σ_{n+1}^1 sets. Then we have proved also the following more general result.

Theorem IV.4.14 (Lopez–Abad). *Let E be a given Banach space with a Schauder basis (e_n). If every coanalytic subset of $\mathcal{B}_1^{[\infty]}(E)$ is approximately (strategically) Ramsey, then every Σ_2^1 subset of $\mathcal{B}_1^{[\infty]}(E)$ is also approximately (strategically) Ramsey. More generally, for every $n \geq 1$, if every Π_n^1 subset of $\mathcal{B}_1^{[\infty]}(E)$ is approximately (strategically) Ramsey, then every Σ_{n+1}^1 subset of $\mathcal{B}_1^{[\infty]}(E)$ is approximately (strategically) Ramsey .*

IV.5 Topological Ramsey Theory of Block Sequences in Banach Spaces

We start again with a Banach space E with a Schauder basis (e_n) and study the corresponding family $\mathcal{B}_1^{[\infty]}(E)$ of infinite normalized block sequences. The separable metric topology of $\mathcal{B}_1^{[\infty]}(E)$ is refined by letting the sets of the form

$$[s, X] \quad (s \in \mathcal{B}_1^{[<\infty]}(E), X \in \mathcal{B}_1^{[\infty]}(E))$$

be open. This is what we are going to call the *Gowers topology* of $\mathcal{B}_1^{[\infty]}(E)$. The following result relates this topology to the notions of approximately and strategically Ramsey sets introduced above.

Lemma IV.5.1. *Every Gowers-open subset $\mathfrak{O}$ of $\mathcal{B}_1^{[\infty]}(E)$ is strategically Ramsey.*

Proof. Let $[s, X]$ and $\Delta \in \mathbb{R}^\infty$ be given. Clearly we may assume that s is empty.
Claim. *There is $X_0 \leq X$ be such that*

$$\forall s \leq X_0 \forall Y \leq X_0 \quad ([s, Y] \subseteq (\mathfrak{O})_{\Delta/2} \to [s, X_0] \subseteq (\mathfrak{O})_\Delta). \qquad (IV.6)$$

Proof of the claim. Fix a $\Delta/4$-net $\mathfrak{N}$ of $[X]^{[<\infty]}$. The required block sequence $X_0 = (x_i) \leq X$ will be defined recursively on i together with a descending sequence (Y_i) of block subsequences of X such that for every i,
(a) $x_i < Y_i$ and $x_{i+1} \in [Y_i]^{[1]}$,
(c) $\forall s \in \mathfrak{N} \cap [(e_i)_0^{\max \operatorname{supp} x_i}] \forall Y \leq Y_i \quad ([s, Y] \subseteq (\mathfrak{O})_{3\Delta/4} \longrightarrow [s, Y_i] \subseteq (\mathfrak{O})_{3\Delta/4})$.
We claim that the resulting $X_0 = (x_i)_i$ has the desired property. For suppose that we have $s \in [X_0]^{[<\infty]}$ and $s < Y \leq X_0$ with the property that $[s, Y] \subseteq (\mathfrak{O})_{\Delta/2}$.

Let i be the minimal j such that $s \in [(x_0, \ldots, x_j)]$. Choose $t \in \mathfrak{N}$ such that $d(s,t) \leq \Delta/4$ and $\operatorname{supp} s(j) = \operatorname{supp} t(j)$ for every $j < |s|$. (Note that t can always be found by the definition of a $\Delta/4$-net). Hence $[t,Y] \subseteq (\mathfrak{O})_{3\Delta/4}$, and $Y \leq Y_i$. So we obtain that $[t,Y_i] \subseteq (\mathfrak{O})_{3\Delta/4}$, and hence $[t,X_0] \subseteq (\mathfrak{O})_{3\Delta/4}$, which implies that $[s,X_0] \subseteq (\mathfrak{O})_\Delta$, as promised. $\qquad \square$

From now on we fix $X_0 \leq X$ as in the claim. Let

$$\mathfrak{X} = \{s \in [X_0]^{[<\infty]} : \forall t \in [X_0]^{[<\infty]} (d(s,t) \leq \Delta/2 \to [t,X_0] \subseteq (\mathfrak{O})_\Delta)\}.$$

Applying Lemma IV.3.1 we obtain $X_1 \leq X_0$ that falls into one of the following two cases:

Case 1: $[X_1]^{[<\infty]} \cap \mathfrak{X} = \emptyset$. We claim that in this case, $[X_1]^{[\infty]} \cap \mathfrak{O} = \emptyset$. Otherwise, fix $Y \in [X_1]^{[\infty]} \cap \mathfrak{O}$. Since $\mathfrak{O}$ is Gowers-open there is some $n \in \mathbb{N}$ such that $[Y \upharpoonright n, Y] \subseteq \mathfrak{O}$. Set $s = Y \upharpoonright n$. Consider an arbitrary finite block sequence t such that $d(s,t) \leq \Delta/2$. Let $m \in \mathbb{N}$ be such that $t < Y/m$. Then $[t, Y/m] \subseteq (\mathfrak{O})_{\Delta/2}$ and by (IV.6) we obtain that $[t,X_0] \subseteq (\mathfrak{O})_\Delta$. This checks that $s \in \mathfrak{X} \cap [X_1]^{[<\infty]}$ and this is in direct contradiction with the alternative of Case 1.

Case 2: The player II has a winning strategy for the game $G_{(\mathfrak{X})_{\Delta/2}}(X_1)$. We claim that in this case $(\mathfrak{O})_\Delta$ is strategically large for $[X_1]$. The winning strategy for player II in the game $G_{(\mathfrak{O})_\Delta}(X_1)$ is described as follows. He first chooses a winning strategy σ for the game $G_{(\mathfrak{X})_{\Delta/2}}(X_1)$. Then he plays according to σ until he reaches some $t \in (\mathfrak{X})_{\Delta/2}$. After that he plays arbitrarily $z_0 < z_1 < \ldots$ above t. We claim that after any such infinite run, $t ^\frown (z_i) \in (\mathfrak{O})_\Delta$. To see this note that since $t \in (\mathfrak{X})_{\Delta/2}$ there is some $s \in \mathfrak{X}$ such that $d(t,s) \leq \Delta/2$. By the definition of $\mathfrak{X}$, this means that $[t,X_0] \in (\mathfrak{O})_\Delta$. Hence we obtain that $t ^\frown (z_i) \in (\mathfrak{O})_\Delta$ as promised. $\qquad \square$

An almost identical net-approximation argument based on Lemma IV.3.7 will give us the following.

Lemma IV.5.2. *Every Gowers-open subset $\mathfrak{O}$ of $\mathcal{B}_1^{[\infty]}(c_0)$ is approximately Ramsey.*
$\qquad \square$

Unfortunately, this is as much as we can do with the Gowers topology in this context. For example, we will not be able to mimic the classical setup and relate the notion of, say, Gowers-nowhere-dense sets with the notion of strategic or approximate Ramsey property. Recall, however, that one always implicitly assumes (unless otherwise stated) the topology on $\mathcal{B}_1^{[\infty]}(E)$ to be the separable metric topology induced from $(S_E)^{\mathbb{N}}$ where the sphere S_E is taken with its norm topology. Since this topology is weaker than the Gowers topology, the two previous lemmas say also that open sets relative to the metric topology are strategically and approximately Ramsey, respectively. We can indeed proceed to generalize these two lemmas to more complex subsets of $\mathcal{B}_1^{[\infty]}(E)$ but the complexity is expressed relative to this metric topology. Recall that in the previous Section we have learned that crucial results will be about G_δ-sets relative to the metric topology of $\mathcal{B}_1^{[\infty]}(E)$.

We start with the following technical result that will be used in the proof that G_δ-sets are strategically Ramsey.

Proposition IV.5.3. *Let $(\mathfrak{X}_n)_n$ be a given sequence of families of block subsequences of E. Let $\Delta = (\delta_n)_n \in \mathbb{R}_+^\infty$, and let $\phi : \mathbb{N} \to \mathbb{N}$. If the union $\mathfrak{X} = \bigcup_{n=0}^\infty \mathfrak{X}_n$ is large for some $[s, X]$, then there exists $Y \in [s, X]$ such that the set*

$$\mathfrak{X}(\Delta, s, Y, \phi) = \{(z_m)_m : \exists n \; \exists k \geq \phi(n) \; (\mathfrak{X}_n)_\Delta \text{ is large already for}$$
$$[s \frown (z_0, \ldots, z_k), Y]\},$$

is strategically (very) large for $[s, Y]$.

Proof. Fix all data as in the statement, and suppose that the union $\mathfrak{X} = \bigcup_{n=0}^\infty \mathfrak{X}_n$ is large for some $[s, X]$. Notice that the sets $\mathfrak{X}(\Delta, s, Y, \phi)$ are always open. For a set of block subsequences $\mathfrak{Z}$, set $(\mathfrak{Z})^s = \{W : s \frown W \in \mathfrak{Z}\}$. Let $\mathfrak{N}$ be a countable $\Delta/4$-net of $\mathcal{B}_1^{[<\infty]}(E)$. Apply Proposition IV.2.7 to the family $((\mathfrak{X}_n)_{\Delta/2})^t$ $(t \in \mathfrak{N}$, $n \in \mathbb{N})$ and block sequence X and obtain X_0 such that, if for some $t \in \mathfrak{N}$, some $n \in \mathbb{N}$, and some $Y \leq X_0$, the family $(\mathfrak{X}_n)_{\Delta/2}$ is large for $[s \frown t, Y]$, then $(\mathfrak{X}_n)_{\Delta/2}$ is large already for $[s \frown t, X_0]$. Since $\mathfrak{N}$ was a $\Delta/4$-net, for every t and n, we obtain that

$$\text{if } (\mathfrak{X}_n)_{\Delta/4} \text{ is large for } [s \frown t, Y] \text{ for some } Y \leq X_0,$$
$$\text{then } (\mathfrak{X}_n)_{3\Delta/4} \text{ is large already for } [s \frown t, X_0]. \tag{IV.7}$$

Applying Proposition IV.2.6, we obtain that for every $Y \in [s, X_0]$ there is some $n \in \mathbb{N}$ some $t \in [Y]^{[<\infty]}$ such that $|t| \geq \phi(n)$ and some $Z \leq X_0$ such that $(\mathfrak{X}_n)_{\Delta/4}$ is large for $[s \frown t, Z]$. Now by (IV.7) one concludes that the set $\mathfrak{X}(3\Delta/4, s, X_0, \phi)$ is large for $[X_0]$. Since this is an open set, by Lemma IV.5.1, we obtain that there is some $Y \leq X_0$ such that $(\mathfrak{X}(3\Delta/4, s, X_0, \phi))_{\Delta/4}$ is strategically large for $[Y]$. This easily implies the desired result. $\square$

We are now ready for the key result about the strategic Ramsey property.

Lemma IV.5.4. *All G_δ-subsets of $\mathcal{B}_1^{[\infty]}(E)$ are strategically Ramsey.*

Proof. Suppose that $\mathfrak{X} = \bigcap_{n=0}^\infty \mathfrak{X}_n$ is a given G_δ-subset of $\mathcal{B}_1^{[\infty]}(E)$ written as the intersection of a decreasing sequence of open sets. We define $\Phi : \mathfrak{X} \to \mathbb{N}^\mathbb{N}$ by letting

$$\Phi(X)(n) = \min\{k : B(X \restriction k, 1/k)^4 \subseteq \mathfrak{X}_n\}.$$

The mapping Φ is well defined by our assumption that every $\mathfrak{X}_n$ is an open subset of $\mathcal{B}_1^{[\infty]}(E)$. For $\sigma \in \mathbb{N}^{<\infty}$, let $[\sigma] = \{a \in \mathbb{N}^\mathbb{N} : a \restriction |\sigma| = \sigma\}$, and let $\mathfrak{Y}_\sigma = \Phi^{-1}[\sigma]$. It is clear that for every $\sigma \in \mathbb{N}^{<\infty}$,

$$\mathfrak{Y}_\sigma = \bigcup_{n=0}^\infty \mathfrak{Y}_{\sigma \frown \langle n \rangle}. \tag{IV.8}$$

---

[4]$B((x_1, \ldots, x_n), \varepsilon) = \{(y_n)_n : \forall i \leq k \; \|x_i - y_i\| < \varepsilon\}.$

Fix also a bijection $\psi : \mathbb{N} \to \mathbb{N}^{<\infty}$ such that $n \geq \max \psi(n)$ for every n. For $\sigma \in \mathbb{N}^{<\infty}$, let $\phi_\sigma : \mathbb{N} \to \mathbb{N}$ be defined by $\phi_\sigma(n) = \psi^{-1}(\sigma \frown (n))$.

Suppose that $\mathfrak{X}$ is large for $[s, X]$. We may assume that $s = \emptyset$ and that $\Delta = (\delta_n)_n$ is decreasing. We assume also that for every n, $\delta_n < 1/C$, where C is the basic constant of the fixed Schauder basis (e_i) of E. For $m \in \mathbb{N}$, let

$$\Delta_m = \Delta(2^m - 1)/2^m \text{ and } \Gamma_m = (\Delta_m - \Delta_{m-1})/4.$$

Let $\{s_n\}_n$ be a countable (Γ_m)-net of $\mathcal{B}_1^{[<\infty]}(X)$ closed under concatenation. Let X_0 be the result of the application of Proposition IV.2.7 to X and the family

$$\{((\mathfrak{Y}_\sigma)_{\Delta_m + 2\Gamma_{m+1}})^{s_n} : \sigma \in \mathbb{N}^{<\infty}, n, m \in \mathbb{N}\}.$$

Now let X_1 be the result of the application of Proposition IV.2.7 to X_0 and the family

$$\{(\mathfrak{Y}_\sigma)_{\Delta_k}(3\Gamma_{m+1}, s_n, X_0, \phi_\tau) : \sigma, \tau \in \mathbb{N}^{<\infty}, k, n, m \in \mathbb{N}\}.$$

Claim. *For every $\sigma, \tau \in \mathbb{N}^{<\infty}$, $k, n, m \in \mathbb{N}$, $s \in [X]^{[<\infty]}$ and every $Y \leq X_1$, if $(\mathfrak{Y}_\sigma)_{\Delta_k}(\Gamma_{m+1}, s, Y, \phi_\tau)$ is strategically large for $[Y]$, then $(\mathfrak{Y}_\sigma)_{\Delta_k}(4\Gamma_{m+1}, s, X_1, \phi_\tau)$ is strategically large for $[X_1]$.*

Proof of the claim. Suppose that $(\mathfrak{Y}_\sigma)_{\Delta_k}(\Gamma_{m+1}, s, Y, \phi_\sigma)$ is strategically large for $[Y]$. This means that player II has a strategy to produce some t, such that for some $r \in \mathbb{N}$ with $|t| \geq \phi_\sigma(r)$ we have that

$$(\mathfrak{Y}_{\sigma \frown \langle r \rangle})_{\Delta_k + \Gamma_{m+1}} \text{ is large for } [s \frown t, Y].$$

Let $s_n = s_{n_1} \frown s_{n_2} \in \mathfrak{N}$ be such that $d(s_n, s \frown t) \leq \Gamma_m$. Then

$$(\mathfrak{Y}_{\sigma \frown \langle r \rangle})_{\Delta_k + 2\Gamma_{m+1}} \text{ is large for } [s_n, Y].$$

Hence by the property of X_0, we get

$$(\mathfrak{Y}_{\sigma \frown \langle r \rangle})_{\Delta_k + 2\Gamma_{m+1}} \text{ is large for } [s_{n_1} \frown s_{n_2}, X_0],$$

which implies that

$$(\mathfrak{Y}_{\sigma \frown \langle r \rangle})_{\Delta_k + 3\Gamma_{m+1}} \text{ is large for } [s_{n_1} \frown t, X_0]. \tag{IV.9}$$

So we have just shown that indeed player II has a winning strategy to produce t such that (IV.9) holds, or in other words,

$$(\mathfrak{Y}_{\sigma \frown \langle r \rangle})_{\Delta_k}(3\Gamma_{m+1}, s_{n_1}, X_0, \phi_\sigma) \text{ is strategically large for } [Y] \tag{IV.10}$$

By the property of X_1, we obtain from (IV.10), and the relationship $X_1 \leq X_0$, that

$$(\mathfrak{Y}_{\sigma \frown \langle r \rangle})_{\Delta_k}(3\Gamma_{m+1}, s_{n_1}, X_1, \phi_\sigma) \text{ is strategically large for } [X_1]. \tag{IV.11}$$

So,

$$(\mathfrak{Y}_{\sigma \frown \langle r \rangle})_{\Delta_k})(4\Gamma_{m+1}, s, X_1, \phi_\sigma) \text{ is strategically large for } [X_1], \qquad \text{(IV.12)}$$

as desired, thus finishing the proof of the claim. $\qquad\qquad\qquad\qquad \Box$

We now describe a winning strategy for player II in the game $G_{\mathfrak{X}_\Delta}(X_1)$. It will consist of concatenations of infinitely many strategies of different games, all of them played in $[X_1]$. We begin by noting that since $\mathfrak{X} = \bigcup_n \mathfrak{Y}_{\langle n \rangle}$ is large for $[X_1]^{[\infty]}$, by Proposition IV.5.3, and the claim, we get that

$$(\mathfrak{Y}_\emptyset)_{\Delta_0}(4\Gamma_1, \emptyset, X_1, \phi_\emptyset) \text{ is strategically large for } [X_1].$$

So the player II can play according to a winning strategy τ_0 for the game

$$G_{(\mathfrak{Y}_\emptyset)_{\Delta_0}(4\Gamma_1, \emptyset, X_1, \phi_\emptyset)}(X_1)$$

until he produces a finite block sequence t_0 such that for some integer n_0 the set $(\mathfrak{Y}_{\langle n_0 \rangle})_{\Delta_1}$ is large for $[t_0, X_0]$ and $|t_0| \geq \phi_\emptyset(n_0)$. Since

$$(\mathfrak{Y}_{\langle n_0 \rangle})_{\Delta_1} = \bigcup_{n=0}^{\infty} (\mathfrak{Y}_{\langle n_0, n \rangle})_{\Delta_1},$$

by Proposition IV.5.3 and the claim, we know that

$$(\mathfrak{Y}_{\langle n_0 \rangle})_{\Delta_1}(4\Gamma_2, t_0, X_1, \phi_{\langle n_0 \rangle}) \text{ is strategically large for } [X_1].$$

The player II now continues playing according to a winning strategy τ_1 for the game

$$G_{(\mathfrak{Y}_{\langle n_0 \rangle})_{\Delta_1}(4\Gamma_2, t_0, X_1, \phi_{\langle n_0 \rangle})}(X_1)$$

until he produces a finite block sequence $t_1 > t_0$ such that for some integer n_1, the set $(\mathfrak{Y}_{\langle n_0, n_1 \rangle})_{\Delta_2}$ is large for $[t_0 \frown t_1, X_1]$ and $|t_1| \geq \phi_{\langle n_0 \rangle}(n_1)$, and so on.

We have to show that any sequence $(y_n) = t_0 \frown t_1 \frown \cdots$ produced by player II by playing according to this strategy belongs to $\mathfrak{X}_\Delta$. Fix $(n_k) \in \mathcal{N}^\uparrow$ such that for every k, we have that $|t_k| \geq \psi^{-1}(\langle n_0, \dots, n_k \rangle)$ and

$$\mathfrak{Y}_{\langle n_0, \dots, n_k \rangle})_{\Delta_{k+1}} \text{ is large for } [t_0 \frown \cdots \frown t_k, X_1].$$

For every k fix $Z_k \in [t_0 \frown \cdots \frown t_k, X_1]$ and

$$W_k \in \mathfrak{Y}_{\langle n_0, \dots, n_k \rangle} \qquad \text{(IV.13)}$$

such that $d(Z_k, W_k) \leq \Delta_{k+1}$. Note that (IV.13) is equivalent to saying that

$$(\forall i \leq k)\, B(Z_k \restriction n_i, 1/n_i) \subseteq \mathfrak{X}_i. \qquad \text{(IV.14)}$$

Set $r_k = |t_0 \frown \dots \frown t_k|$ $(k \in \mathbb{N})$. Note that for every $k \in \mathbb{N}$ and every $k' \geq k$

$$d(W_k \restriction r_k, t_0 \frown \dots \frown t_k) \leq \Delta_{k+1} \qquad \text{(IV.15)}$$

Set $W_k = (w_{k,i})_i$ $(k \in \mathbb{N})$. We claim that (IV.15) implies that

$$\text{for every } l \geq k+1 \; w_{l,k} \in [e_i]_0^{\max \text{supp } y_{k+1}}. \tag{IV.16}$$

Otherwise, fix $l \geq k+1$ such that for some $i \in \text{supp } w_{l,k}$, we have that $i > \max \text{supp } y_{k+1}$. Hence $w_{l,k+1} > y_{k+1}$, which implies that $\|w_{l,k+1} - y_{k+1}\| \geq 1/C$, where C is the basic constant of (e_i). But then $\|w_{l,k+1} - y_{k+1}\| > \delta_{k+1}$, contradicting (IV.15).

Fix a nonprincipal ultrafilter $\mathcal{U}$ on $\mathbb{N}$. Note that the condition (IV.16) implies that the $\mathcal{U}$-limit of the sequence (W_k) exists. Set $W_\infty = \mathcal{U} - \lim W_k$. It is clear also that $d((y_n), W_\infty) \leq \Delta$. Let us check that $W_\infty = (w_n) \in \mathfrak{X}$: Fix k, and let $k' \geq k$ be such that

$$\max\{\|w_i - w_{k',i}\| : i < n_k\} < 1/n_k. \tag{IV.17}$$

Since $B(W_k \restriction n_k, 1/n_k) \subseteq \mathfrak{X}_k$, (IV.17) implies that $W_\infty \in \mathfrak{X}_k$. $\qquad \square$

Our next goal is to show the corresponding result for the approximate Ramsey property for the Banach space c_0. As pointed out before, in this case there are no closed asymptotic pairs $\mathfrak{A}$ but only discrete ones. Hence the mapping $\Lambda_{\mathfrak{A}}$ is not continuous but only of Baire-class-1, for a closed set $C \subseteq \mathcal{B}_{\mathfrak{A}}^{[\infty]}(c_0) \times \mathcal{N}^\uparrow$ the corresponding set $\Lambda_{\mathfrak{A}}^{-1}(C)$ is not a G_δ-set but an $F_{\sigma\delta}$-set, or more precisely, the intersection of a G_δ-set $\mathfrak{X}$ and $\mathcal{B}_{\mathfrak{A}}^{[\infty]}(c_0)$. Hence the key result in this context is the approximate Ramsey property for all relatively G_δ-subsets of $\mathcal{B}_{\mathfrak{A}}^{[\infty]}(c_0)$ for some conveniently chosen discrete asymptotic pair $\mathfrak{A}$. This explains our next lemma.

Lemma IV.5.5. *Let $\mathfrak{A} = \langle A_0, A_1 \rangle$ be the discrete asymptotic pair:*

$$A_i = \{x \in S(c_0) : (-1)^{i+1} e_{\min \text{supp} x}^*(x) > 0\}$$

($i = 0, 1$) of the Banach space c_0. Then for every G_δ set $\mathfrak{X}$ of block subsequences of c_0, the intersection $\mathfrak{X} \cap \mathcal{B}_{\mathfrak{A}}^{[\infty]}(c_0)$ is approximately Ramsey. In particular, for every closed set $C \subseteq \mathcal{B}_{\mathfrak{A}}^{[\infty]}(c_0) \times \mathcal{N}^\uparrow$ the set $\Lambda_{\mathfrak{A}}^{-1}(C)$ is approximately Ramsey.

Proof. The proof will follow the proof Lemma IV.5.4 except that at places where the Proposition IV.2.7 (which fails if 'large' is replaced by 'very large') is used, we use instead the Claim A below. Let $\mathfrak{X} = \bigcap_n \mathfrak{X}_n$ be a G_δ set of block sequences of c_0 written as a decreasing intersection of open sets $\mathfrak{X}_n$. Suppose that $\mathfrak{X}$ is large for some basic set $[s, X]$ and that $\Delta = (\delta_n)_n \in \mathbb{R}_+^\infty$ is decreasing. We assume that $s = \emptyset$. For $m \in \mathbb{N}$, let

$$\Delta_m = \Delta(2^m - 1)/2^m \text{ and } \Gamma_m = (\Delta_m - \Delta_{m-1})/4.$$

Let $\varphi : \mathbb{N} \to \mathbb{N}^2$ be any bijection and let φ_i, $i = 0, 1$, be its two projections. Redefine the mapping $\Phi : \sigma \to \mathcal{N}$ from the proof of Lemma IV.5.4 by letting $\Phi((x_i)_i)(n) = k_n$ iff

(a) $B(X \upharpoonright \varphi_0(k_n), 1/\varphi_0(k_n)) \subseteq \mathfrak{X}_n$.

(b) $e^*_{\min \operatorname{supp} x_i}(x_i) \geq 1/\varphi_1(k_n)$ for every i in the set of the first n-many j such that $x_{2j+1} \in A_1$

The new condition (b) allows us to read some information out of a given block vector $(x_n)_n \in \mathcal{B}_{\mathcal{U}}^{[\infty]}(c_0)$, such as, for example, which of its odd positions x_{2n+1} have the first non zero coordinate positive (i.e., for which n do we have $x_{2n+1} \in A_1$), and how large these coordinates are. It is clear that Φ is well defined. For $\sigma \in \mathbb{N}^{<\infty}$, let $[\sigma] = \{a \in \mathbb{N}^{\mathbb{N}} : a \upharpoonright |\sigma| = \sigma\}$, and let $\mathfrak{Y}_\sigma = \Phi^{-1}[\sigma]$. Fix also a bijection $\psi : \mathbb{N} \to \mathbb{N}^{<\infty}$ as in the proof of Lemma IV.5.4. Choose also a countable Δ-net $\{s_n\}_n$ of $\mathcal{B}_1^{[<\infty]}(X)$ as in the proof of Lemma IV.5.4 with the additional property that $\{s_n\}_n \cup [(e_i)_0^k]^{[\leq k]}$ is finite for every k. Let X_0 be the result of the application of Proposition IV.2.7 for the property of being *large* to X and the family

$$\{((\mathfrak{Y}_\sigma)_{\Delta_{m+2\Gamma_{m+1}}})^{s_n} : \sigma \in \mathbb{N}^{<\infty}, n, m \in \mathbb{N}\}.$$

Claim A. *There is $X_1 = (x_n) \leq X_0$ such that for every $\sigma \in \mathbb{N}^{<\infty}$ and $n \in \mathbb{N}$ with properties $|s_n| \geq \psi^{-1}\sigma$ and $s_n \in [(e_i)_0^{\min \operatorname{supp} x_{|s_n|}-1}]$, and every $Y \leq (x_i)_{i \geq |s_n|}$, if*

$$(\mathfrak{Y}_\sigma)_{\Delta_{|\sigma|}}(3\Gamma_{|\sigma|+1}, s_n, X_0, \phi_\sigma)$$

is very large for $[Y]$, then

$$(\mathfrak{Y}_\sigma)_{\Delta_{|\sigma|}}(3\Gamma_{|\sigma|+1}, s_n, X_0, \psi_\sigma)$$

is very large for $[X_1]$.

Proof of the claim. Set $\mathfrak{Z}_k = \mathfrak{Y}_{\psi(k)}$ $(k \in \mathbb{N})$. Define a decreasing sequence $Y_i = (y_j^i)$ of infinite normalized block subsequences of X_0 such that for every i, every n such that $s_n \in [(e_k)_0^{\min \operatorname{supp} y_0^i}]^{[<\infty]}$, every $k \leq i$, and every $Y \leq Y_i$,

if $(\mathfrak{Z}_k)_{\Delta_{|\psi(k)|}}(3\Gamma_{|\psi(k)|+1}, s_n, X_0, \phi_{\psi(k)})$ is very large for $[Y]$, then
$(\mathfrak{Y}_\sigma)_{\Delta_{|\psi(k)|}}(3\Gamma_{|\psi(k)|+1}, s_n, X_0, \phi_{\psi(k)})$ is very large for $[Y_i]$.

It is not difficult to show that the diagonal sequence $X_1 = (y_i^i)$ satisfies the conclusions. $\square$

Fix $X_1 = (x_n)$ as in Claim A. Following the proof of the claim from the proof of Lemma IV.5.4, we get the following result.

Claim B. *For every $\sigma \in \mathbb{N}^{<\infty}$, every $s \in [X]^{[<\infty]}$ such that $|s| \geq \psi^{-1}\sigma$ and every $Y \leq (x_i)_{i \geq |s_n|}$, if*

$$(\mathfrak{Y}_\sigma)_{\Delta_{|\sigma|}}(\Gamma_{|\sigma|+1}, s, Y, \phi_\sigma)$$

is very large for $[Y]$, then also

$$(\mathfrak{Y}_\sigma)_{\Delta_{|\sigma|}}(4\Gamma_{|\sigma|+1}, s, X_1, \phi_\sigma)$$

is very large for $[X_1]$. $\square$

Using again arguments analogous to those used in the proof of Lemma IV.5.4, we can show that for every $Y \leq X_1$ there are two sequences $(n_k), (m_k) \in \mathcal{N}^\uparrow$ such that $m_k \geq \psi^{-1}(\langle n_0, \ldots, n_k \rangle)$ for every k, and such that

$$(\mathfrak{Y}_{\langle n_0, \ldots, n_k \rangle})_{\Delta_{k+1}} \text{ is large for } [Y \upharpoonright (m_0 + \cdots + m_k), X_1]. \tag{IV.18}$$

An argument analogous to the one used by the end of the proof of Lemma IV.5.4 shows that (IV.18) implies that there is some $Y_\infty \in \mathfrak{X}$ such that $d(Y, Y_\infty) \leq \Delta$. Moreover, by the new condition (b) on Φ, we have $Y_\infty \in \mathcal{B}_\mathfrak{A}^{[\infty]}(c_0)$. This shows that $[X_1]^{[\infty]} \subseteq (\mathfrak{X} \cap \mathcal{B}_\mathfrak{A}^{[\infty]}(c_0))_\Delta$, as desired. □

Combining Lemmas IV.5.4 and IV.5.5 with Corollaries IV.4.11 and IV.4.13, we obtain the following result.

Theorem IV.5.6 (Gowers). (a) *For every Banach space E with a Schauder basis (e_n), all analytic subsets of $\mathcal{B}_1^{[\infty]}(E)$ are strategically Ramsey.*
(b) *Analytic subsets of $\mathcal{B}_1^{[\infty]}(c_0)$ are approximately Ramsey.* □

Exercise IV.5.7. Find a more direct proof of the previous theorem. (Hint: Given an analytic set $\mathfrak{X}$, fix a continuous function $f : \mathcal{N} \to \mathcal{B}_1^{[\infty]}(E)$ such that $f''\mathcal{N} = \mathfrak{X}$, declare $\mathfrak{Y}_\sigma = f''[\sigma]$ for $\sigma \in \mathbb{N}^{<\infty}$, and follow the proof of Lemma IV.5.4.)

IV.6 An Application to Rough Classification of Banach Spaces

The most famous consequence of Theorem IV.5.6 is of course Gowers' dichotomy presented above in the first section of this chapter. This theorem however has some other interesting applications and the purpose of this section is to present a typical one. It involves a notion of minimality in the class of Banach spaces. Recall that an infinite-dimensional Banach space E is *minimal* if every infinite-dimensional subspace of E has a further subspace isomorphic to E. Not every infinite-dimensional Banach space has a minimal infinite-dimensional subspace. A typical counterexample is the Tsirelson space. However, Tsirelson space has the following weaker minimality property so one can go on and investigate when a given Banach space contains a subspace with this weaker property.

Definition IV.6.1. We say that two infinite-dimensional Banach spaces E and F are *totally incomparable* if no infinite-dimensional subspace of E is isomorphic to a subspace of F. A Banach space E is said to be *quasi-minimal* if it contains no pair of totally incomparable subspaces.

Note that every hereditarily indecomposable Banach space is also an example of a quasi-minimal space. Since every infinite-dimensional Banach space has an infinite-dimensional subspace with a Schauder basis, in studying this notion we may restrict ourselves to Banach spaces E with a Schauder basis (e_n).

Proposition IV.6.2. *Let E be a given Banach space with a Schauder basis (e_n).*
(1) Let X and Y be two infinite normalized block sequences of E. Then their closed linear spans are totally incomparable iff there are no infinite normalized block subsequences $X' \leq X$ and $Y' \leq Y$ such that the closed linear spans of X' and Y' are isomorphic.
(2) The space E is quasi-minimal iff for every pair of infinite normalized block sequences X and Y of E there exist infinite normalized block subsequences $X' \leq X$ and $Y' \leq Y$ such that the closed linear spans of X' and Y' are isomorphic.

Proof. By Theorem IV.1.1 for every infinite-dimensional closed linear subspace F of E there is an infinite normalized block sequence X and a closed subspace F' of F such that the closed linear span of X is isomorphic to F'. When F is spanned by an infinite normalized sequence Y the subspace F' can be chosen to be spanned by an infinite normalized block subsequence of Y. From these facts, the conclusions (1) and (2) are immediate. $\square$

Let us say that two block subspaces E_0 and E_1 of E are *disjointly supported* if the support (relative the fixed Schauder basis (e_n) of E) of every vector of E_0 is disjoint from the support of every vector of E_1. The following lemma hints towards an alternative of quasi-minimality which involves this notion.

Lemma IV.6.3 (Casazza). *Suppose that a given space E with a Schauder basis is isomorphic to a proper subspace of itself. Then E contains two equivalent infinite block sequences (x_n) and (y_n) such that $x_n < y_n < x_{n+1}$ for all n.*

Proof. Let $(e_n)_{n=0}^\infty$ be a fixed Schauder basis of E. Let $T : E \to E$ be an isomorphism between E and its proper subspace. Perturbing T if necessary we may assume that $T(e_n)$ is a finitely supported vector for all n. Since the range of T is contained in some subspace of E of codimension one and since all such subspaces of E are isomorphic, we may assume that the range of T is contained in the closed linear span of $e_1, e_2, \dots$,i.e., the subspace that does not involve the basis vector e_0. For $k \in \mathbb{N}$, let E_n denote the linear span of $e_0, \dots, e_k$. Note that by our assumption, $\dim(E_k \cap T''E_k) < \dim(T''E_k)$ for all k. Let $x_0 = e_0$. Note that $x_0 < T(x_0)$. Since $T(x_0)$ is finitely supported there is an integer k_0 such that $T(x_0) \in E_{k_0}$. Since

$$\dim(E_{k_0} \cap T''E_{k_0}) < \dim(T''E_{k_0})$$

there is a norm-one vector $x_1 \in E_{k_0}$ such that $T(x_1) > k_0$ and therefore $x_1 < T(x_1)$. Clearly we can continue and produce an infinite sequence (x_n) such that $x_n < T(x_n) < T(x_{n+1})$ for all n. By the Bessaga–Pelczynski argument (see Theorem IV.1.1 above), the sequence (x_n) can be refined to a subsequence (x_{n_i}) which is equivalent to a block basic sequence (x'_{n_i}) such that $x'_{n_i} < T(x_{n_i})$ for all i. This gives the conclusion of the lemma. $\square$

Theorem IV.6.4 (Gowers). *Let E be a Banach space with a Schauder basis (e_n). Then either E has a quasi-minimal subspace or E has a block subspace F such* *that no two disjointly supported subspaces of F are isomorphic, and therefore, every pair of two disjointly supported subspaces of F are totaly incomparable.*

Proof. Let us assume that that any block subspace F of E contains a pair of disjointly supported block subspaces that are isomorphic.

For the notational convenience during the course of this proof, we shall identify the family $\mathbb{N}^{[\infty]}$ of infinite subsets of $\mathbb{N}$ with the family of all infinite increasing sequences $\sigma = (\sigma_n)_n$ of non-negative integers such that $\sigma_0 = 0$. Given a pair of finite or infinite normalized block sequences $X = (x_n)_n$ and $Y = (y_n)_n$ of E and a positive constant C, we write $X \sim_C Y$ if there is an isomorphism T from the closed linear span of X onto the closed linear span of Y such that $T(x_n) = y_n$ for all n and such that $\| T \| . \| T^{-1} \| \leq C$. We write $X \sim Y$ whenever $X \sim_C Y$ for some constant $C > 0$.

Consider the following set of normalized block sequences of E:

$$\mathfrak{X} = \{(x_n)_n \in \mathcal{B}_1^{[\infty]}(E) : \exists \sigma \in \mathbb{N}^{[\infty]} \ \exists M \in \mathbb{R}_+ \ \exists (y_n)_n \sim_M (z_n)_n$$
$$\forall n \ \ y_n \in [(x_{2k+1})_{k=\sigma_n}^{\sigma_{n+1}}]^{[1]}, z_n \in [(x_{2k})_{k=\sigma_n}^{\sigma_{n+1}}]^{[1]}\}.$$

Note that $\mathfrak{X}$ is an analytic subset of $\mathcal{B}_1^{[\infty]}(E)$ and therefore strategically Ramsey by Theorem IV.5.6. Using our assumption that every block subspace of E contains two disjointly supported block subspaces that are isomorphic, it is not difficult to show that $\mathfrak{X}$ is large. Let C be the basis constant of (e_n) and chose a summable $\Delta \in \mathbb{R}_+^\infty$, let d be its sum and let $K = (1 + 2Cd)^2$. Then $X \sim_K Y$ for every pair $X, Y \in \mathcal{B}_1^{[\infty]}(E)$ such that $d(X, Y) \leq \Delta$. Since $\mathfrak{X}$ is large, by Theorem IV.5.6, there is some Y such that $(\mathfrak{X})_\Delta$ is strategically large for $[Y]^{[\infty]}$. We claim that the closed linear span of Y is quasi-minimal. We shall use the criterion of quasi-minimality given in Proposition IV.6.2. Consider a pair of infinite block sequences $Z_0, Z_1 \leq Y$. Let Z be an infinite block subsequence of Y given by the strategy of player II in the game $G_{(\mathfrak{X})_\Delta}(Y)$, when player I plays alternatively between Z_0 and Z_1. Then $Z \in (\mathfrak{X})_\Delta$. So, there exist $\widetilde{Z} \in \mathfrak{X}$ such that $d(Z, \widetilde{Z}) \leq \Delta$. Then $Z \sim_K \widetilde{Z}$ and so we let T denote an operator witnessing this equivalence. Chose a sequence $(\sigma_n)_n \in \mathbb{N}^{[\infty]}$, a block sequences $\widetilde{W} = (\widetilde{w}_n), \widetilde{U} = (\widetilde{u}_n) \leq \widetilde{Z}$, and a positive constant M witnessing the membership $\widetilde{Z} \in \mathfrak{X}$. Note that, in particular, we have the equivalence $\widetilde{W} \sim_M \widetilde{U}$. Let $W, U \leq Z$ be infinite block sequences given by $W = T(\widetilde{W})$ and $U = T(\widetilde{U})$. Then $W \sim \widetilde{W}$ and $U \sim \widetilde{U}$ since $Z \sim \widetilde{Z}$. Note also that $w_n \in Z_1$ and $u_n \in Z_0$ for all n. But $(w_n)_n \sim (\widetilde{w}_n)_n \sim (\widetilde{u}_n)_n \sim (u_n)_n$, and we are done. $\square$

Corollary IV.6.5 (Gowers). *Every infinite-dimensional Banach space E has an infinite-dimensional subspace F such that F is either quasi-minimal or no subspace of F is isomorphic to any proper subspace of itself.* $\square$

We finish this section by mentioning the following interesting result whose proof involves several ideas exposited above including Gowers' result about quasi-minimal spaces.

Theorem IV.6.6 (Rosendal). *Let E be an infinite-dimensional Banach space. Then E contains either a minimal subspace or continuum many pairwise incomparable subspaces.*   □

IV.7   An Analytic Set whose Complement is not Approximately Ramsey

It is well-known that using the axiom of choice one is likely to find a set that fails to have a given regularity property. The classical such use is the construction of a set of reals that is not Lebesgue measurable. After the discovery of Gödel's universe of constructible sets it has been realized that the corresponding pathological sets can be constructed even on a very low level of Luzin's hierarchy of projective sets provided of course one is willing to confine oneself to the constructible subuniverse. All these constructions are based on a particularly nice well-ordering of the reals.

Definition IV.7.1. A Σ_2^1-*good well ordering* of $\mathbb{N}^{\mathbb{N}}$ is a Σ_2^1-relation $<_w \subseteq \mathbb{N}^{\mathbb{N}} \times \mathbb{N}^{\mathbb{N}}$ that well-orders $\mathbb{N}^{\mathbb{N}}$ in order type ω_1 in such a way that the relation $R \subseteq \mathbb{N}^{\mathbb{N}} \times \mathbb{N}^{\mathbb{N}}$ defined by letting

$$(x,y) \in R \text{ iff } \{z \in \mathbb{N}^{\mathbb{N}} : z <_w y\} = \{(x)_i : i \in \mathbb{N}\}^5$$

is also Σ_2^1.

Recall the classical fact essentially due to Gödel (see [40]) that in the constructible subuniverse L there is a Σ_2^1-good well-ordering of $\mathbb{N}^{\mathbb{N}}$. The point of this definition is that most of the known constructions of irregular sets (such as for example a non Lebesgue measurable set) that use a well-ordering of $\mathbb{R}$ would give us a Σ_2^1 irregular set provided we start with a Σ_2^1-good well-ordering. We shall now see that this is indeed also the case with a natural approach of constructing a set of normalized block-sequences that is not strategically Ramsey by diagonalizing over a well-ordering of all strategies for one of the players. There is however a slight problem with this approach since clearly there are more than continuum many strategies. It is for this reason that we need to reformulate the game $G_{\mathfrak{X}}(Y)$ so that the set of strategies is in a natural way equinumerous with $\mathbb{N}^{\mathbb{N}}$ allowing us to transfer the Σ_2^1-good well ordering.

Definition IV.7.2. Given $Y \in \mathcal{B}_1^{[\infty]}(E)$ and $\mathfrak{X} \subseteq \mathcal{B}_1^{[\infty]}$, let $G_{\mathfrak{X}}^f(Y)$ be the infinite perfect-information game between two players I and II played as follows:

- I plays a norm-one vector y_0 of the linear span of Y.

- II responds by playing either a norm-one vector z_0 in the linear span of $\{y_0\}$, or by playing $z_0 = 0$ meaning that II chooses not to play any vector at the moment.

---

[5]For $x \in \mathbb{N}^{\mathbb{N}}$ and $i \in \mathbb{N}$ we let $(x)_i$ be defined by $(x)_i(j) = x(2^i(2j+1))$.

- I plays another norm-one vector z_1 of the linear span of Y.

- II responds by playing either a norm-one vector z_1 in the linear span of $\{y_0, y_1\}$, or by playing $z_1 = 0$ meaning that II chooses not to play any vector at the moment, and so on.

- At stage k player I plays a norm-one vector y_k in the linear span of Y and II responds either by finding a maximal place $l < k$ where he played a vector and then choosing a norm-one vector z_k in the linear span of $\{y_{l+1}, ..., y_k\}$, or by playing $z_k = 0$.

We require that, if in a given interval $[l, m]$ of integers the player II chooses never to play a vector, then the vectors y_k $(l \leq k \leq m)$ picked by I must form a block subsequence of Y. We also require that if z_l and z_k are vectors played by II and if $l < k$ then $z_l < z_k$. After infinitely many steps, we let $(z_{k_i})_i$ be the increasing enumeration of vectors played by II and we say that II wins the play if this block subsequence of Y belongs to $\mathfrak{X}$. Otherwise (i.e., if II does not produce an infinite sequence, or if $(z_{k_i})_i \notin \mathfrak{X}$) we say that I wins the infinite play of the game.

Thus, while at the first step of the new game $G_{\mathfrak{X}}^{\mathrm{f}}(Y)$ the player I cannot play an infinite block subsequence $Y_0 = (y_n^0)$ of Y as he could in $G_{\mathfrak{X}}(Y)$, he can play instead $y_0^0, y_1^0, \ldots, y_{k_0}^0$ until the first place k_0 when II plays a norm-one vector z_{k_0} belonging to the linear span of $\{y_0^0, y_1^0, \ldots, y_{k_0}^0\}$ (and therefore to the linear span of Y_0). After that, while I cannot play an infinite block subsequence $Y_1 = (y_n^1)$ of Y as he could have done this in the game $G_{\mathfrak{X}}(Y)$, he can instead play vectors $y_0^1, y_1^1, \ldots, y_{k_1}^1$ as long as II allows him to do this , i.e., until the place k_1 when II plays a vector, and so on. In other words, while the two games $G_{\mathfrak{X}}(Y)$ and $G_{\mathfrak{X}}^{\mathrm{f}}(Y)$ look different, they are really just two different formulations of the same game, and so we have the following result.

Proposition IV.7.3. *The player* I *has a winning strategy in the game* $G_{\mathfrak{X}}^f(Y)$ *iff the player* I *has a winning strategy in the game* $G_{\mathfrak{X}}(Y)$. *Similarly, the player* II *has a winning strategy in the game* $G_{\mathfrak{X}}^f(Y)$ *iff the player* II *has a winning strategy in the game* $G_{\mathfrak{X}}(Y)$. $\qquad \square$

The point of replacing the game $G_{\mathfrak{X}}(Y)$ by the new game $G_{\mathfrak{X}}^{\mathrm{f}}(Y)$ is that the sets of strategies for any of the players in the new game can naturally be injected into $\mathbb{N}^{\mathbb{N}}$, so well-orderings of $\mathbb{N}^{\mathbb{N}}$ naturally transfer to well-orderings of the sets of strategies. So we are in a situation to prove the following result.

Theorem IV.7.4 (Bagaria-Lopez Abad, Gowers). *If there is a Σ_2^1-good well ordering of $\mathbb{N}^{\mathbb{N}}$, then for every infinite dimensional Banach space E there is a Σ_2^1-subset $\mathfrak{X}$ of $\mathcal{B}_1^{[\infty]}(E)$ which is not strategically Ramsey. Under the same assumption, there is also a Σ_2^1-subset $\mathfrak{X}$ of $\mathcal{B}_1^{[\infty]}(c_0)$ which is not approximately Ramsey.* $\qquad \square$

Proof. We shall in fact prove a stronger result: For every $\Delta = (\delta_n)_n \in \mathbb{R}_+^\infty$ such that $\delta_n < 1$ for all n there is a Σ_2^1 set of normalized block sequences $\mathfrak{X}$ such that for every block subsequence Y, the player I has no winning strategy in the game $G_{\mathfrak{X}}^{f}(Y)$ and the player II has no winning strategy in the game $G_{(\mathfrak{X})_{\Delta}}^{f}(Y)$.

Fix a Σ_2^1-good well-ordering $<_w$ of $\mathbb{N}^{\mathbb{N}}$. We have already observed that the union of $\mathcal{B}_1^{[\infty]}(E)$ and the two sets

$$\mathcal{S}_I = \{(Y,\sigma) : \sigma \text{ is a strategy of I in } G_{\mathfrak{Y}}^{f}(Y) \text{ for some } \mathfrak{Y}\}$$

and

$$\mathcal{S}_{II} = \{(Y,\sigma) : \sigma \text{ is a strategy of I in } G_{\mathfrak{Y}}^{f}(Y) \text{ for some } \mathfrak{Y}\}$$

can naturally be injected into $\mathbb{N}^{\mathbb{N}}$, so the well-ordering $<_w$ induces the corresponding well-ordering $<_s$ on the union. Recursively on $<_s$ we chose two assignments

$$\{Z_I(Y,\sigma) : (Y,\sigma) \in \mathcal{S}_I\} \text{ and } \{Z_{II}(Y,\sigma) : (Y,\sigma) \in \mathcal{S}_{II}\}$$

such that:

(1) If $(Y,\sigma) \in \mathcal{S}_I$, i.e., σ is a strategy for the player I in the game $G_{\mathfrak{Y}}^{f}(Y)$ for some set $\mathfrak{Y}$ of infinite block sequences, then $Z_I(Y,\sigma)$ is the $<_s$-least block subsequence $Z = (z_n)$ of Y with the following two properties:

(1.a) There is an infinite play of $G_{\mathfrak{Y}}^{f}(Y)$ in which I plays according to σ and the player II produces the infinite block sequence $Z = (z_n)$, i.e.,

$$z_n \in [y_0^n, \dots, y_{k_n}^n],$$

where

$$y_i^n = \sigma(\overbrace{0,\dots,0}^{(k_0)}, z_0, \dots, \overbrace{0,\dots,0}^{(k_{n-1})}, z_{n-1}, \overbrace{0,\dots,0}^{(i)}).$$

(1.b) $d(Z_I(Y',\sigma'), X) \not\leq \Delta$ and $d(Z_{II}(Y',\sigma'), X) \not\leq \Delta$ for every $(Y',\sigma') <_s (Y,\sigma)$.

(2) If $(Y,\sigma) \in \mathcal{S}_{II}$, i.e., if σ is a strategy for player II in the game $G_{\mathfrak{Y}}^{f}(Y)$ for some set $\mathfrak{Y}$ of block sequences, then $Z_{II}(Y,\sigma)$ is the $<_s$-least infinite block sequence $Z = (z_n)$ with the following two properties:

(2.a) There is an infinite run of $G_{\mathfrak{Y}}^{f}(Y)$ in which player II plays according to σ and at the end produces the block sequence $Z = (z_n)$.

(2.b) $d(Z_{II}(Y',\sigma'), Z) \not\leq \Delta$ and $d(Z_I(Y',\sigma'), X) \not\leq \Delta$ for every $(Y',\sigma') <_s (Y,\sigma)$.

Since every (Y,σ) has only countably many predecessors, a simple diagonalisation procedure shows that there are always block sequences satisfying (1) or (2) depending whether $(Y,\sigma) \in \mathcal{S}_I$ or $(Y,\sigma) \in \mathcal{S}_{II}$, respectively. Let

$$\mathfrak{X} = \{Z_I(Y,\sigma) : (Y,\sigma) \in \mathcal{S}_I\}.$$

The fact that $<_s$ is obtained in a natural way from $<_w$ which is a Σ_2^1-good well-ordering guarantees that $\mathfrak{X}$ is a Σ_2^1 set.

Consider a strategy σ for player I in the game $G_{\mathfrak{X}}^{f}(Y)$ for some infinite block sequence Y. Then by (1) the Player II can play against σ and win by producing $Z_I(Y,\sigma) \in \mathfrak{X}$. So, the strategy σ is not winning for I in the game $G_{\mathfrak{X}}^{f}(Y)$.

Suppose now that σ is a strategy for II in the game $G^{\mathrm{f}}_{(\mathfrak{X})_{\Delta}}(Y)$ for some infinite block sequence Y. Then, by definition of $Z_{II}(Y,\sigma)$ in case (2), there is a run of the game in which II uses σ and produces the set $Z_{II}(Y,\sigma)$. We show that $Z = Z_{II}(Y,\sigma)$ is not in $\mathfrak{X}_{\Delta}$. Consider an arbitrary $Z' \in \mathfrak{X}$ and choose $(Y',\sigma') \in \mathcal{S}_I$ such that $Z' = Z_I(Y',S')$. If $(Y',\sigma') <_s (Y,\sigma)$ then Z was chosen so that $d(Z,Z') \not\leq \Delta$. Otherwise, $(Y,\sigma) <_s (Y',\sigma')$, and hence Z' was chosen so that $d(Z',Z) \not\leq \Delta$. Since Z' was an arbitrary member of $\mathfrak{X}$ this shows that the set $Z_{II}(Y,S)$ does not belong to $\mathfrak{X}_{\Delta}$. This proves that σ is not a winning strategy for player II in the game $G^{\mathrm{f}}_{(\mathfrak{X})_{\Delta}}(Y)$.

In conclusion, we have just produced a Σ^1_2 set $\mathfrak{X}$ of infinite normalized block sequences of the Banach space E such that there is no infinite normalized block sequence Y such that the player I has a winning strategy in the game $G^{\mathrm{f}}_{(\mathfrak{X})}(Y)$ (and so in particular $[Y]^{[\infty]} \cap \mathfrak{X}$ is nonempty for every $Y \in \mathcal{B}^{[\infty]}_1(E)$) and also there is no such Y for which the player II has a winning strategy in $G^{\mathrm{f}}_{(\mathfrak{X})_{\Delta}}(Y)$ (and so in particular $[Y]^{[\infty]} \not\subseteq \mathfrak{X}_{\Delta}$ for every $Y \in \mathcal{B}^{[\infty]}_1(E)$). It follows that $\mathfrak{X}$ is neither approximately nor strategically Ramsey. $\square$

Combining this and Theorem IV.4.14 we obtain the following result.

Theorem IV.7.5 (Lopez–Abad). *If there is a Σ^1_2-good well ordering of $\mathbb{N}^{\mathbb{N}}$, then for every infinite-dimensional Banach space E there is a coanalytic subset $\mathfrak{X}$ of $\mathcal{B}^{[\infty]}_1(E)$ which is not strategically Ramsey. Under the same assumption, there is also a coanalytic subset of $\mathcal{B}^{[\infty]}_1(c_0)$ which is not approximately Ramsey.*

So, in particular, the classes of approximately and strategically Ramsey sets are in general not closed under the operation of complementation, and so that these notion behave quite differently from the classical notion of Ramsey sets. Going to Gödel's constructible subuniverse to obtain this non-complementarity result is in some sense necessary as it is shown in [9] (see also [49]) that under some large cardinal assumptions all projective (or more generally 'reasonably' definable) families of infinite normalized block sequences of an arbitrary Banach space E with a Schauder basis (e_n) are strategically Ramsey.

We finish this section and this chapter with the following problem that asks for yet another picture of the class of approximate and strategic Ramsey sets different from the one given by Gödel's constructible universe.

Problem IV.7.6 (Gowers). *Does the axiom of determinacy imply that all sets of infinite normalized block sequences are strategically Ramsey?*

Remark IV.7.7. The material of this chapter is based on sources [35], [36], [56], [9], [10], and [49] where some alternative proofs of these results can be found. The reader is referred to these sources also for a wealth of additional information about the approximate and strategic Ramsey theory that might lie outside the scope of the present notes.

Bibliography

[1] A. Alexiewicz, *On sequences of operations. II.* Studia Math. **11** (1950), 200–236.

[2] D. E. Alspach, S. A. Argyros, *Complexity of weakly null sequences.* Dissertationes Math. (Rozprawy Mat.) **321** (1992), 44 pp.

[3] D. E. Alspach, E. Odell, *Averaging weakly null sequences.* Functional analysis (Austin, TX, 1986–87), 126–144, Lecture Notes in Math., 1332, Springer, Berlin, 1988.

[4] S. A. Argyros, I. Deliyanni, *Examples of asymptotic l_1 Banach spaces.* Trans. Amer. Math. Soc. **349** (1997), no. 3, 973–995.

[5] S. A. Argyros, I. Gasparis, *Unconditional structures of weakly null sequences.* Trans. Amer. Math. Soc. **353** (2001), no. 5, 2019–2058

[6] S. A. Argyros, G. Godefroy, H. P. Rosenthal, *Descriptive set theory and Banach spaces.* Handbook of the geometry of Banach spaces, Vol. 2, 1007–1069, North-Holland, Amsterdam, 2003.

[7] S. A. Argyros, S. Mercourakis, A. Tsarpalias, *Convex unconditionality and summability of weakly null sequences.* Israel J. Math. **107** (1998), 157–193.

[8] A. D. Arvantakis, *Weakly null sequences with an unconditional subsequence.* Preprint 2004.

[9] J. Bagaria, J. Lopez-Abad, *Weakly Ramsey sets in Banach spaces.* Advances in Mathematics **160** (2001), 133–174.

[10] J. Bagaria, J. Lopez-Abad, *Determinacy and weakly Ramsey sets in Banach spaces.* Trans. Amer. Math. Soc. **354** (2002), no. 4, 1327–1349

[11] B. Beauzamy, *Banach-Saks properties and spreading models.* Math. Scand. **44** (1979), no. 2, 357–384.

[12] S. Bellenot, *The Banach space T and the fast growing hierarchy from logic,* Israel J. Math. **47** (1984), 305–313.

[13] Y. Benyamini, J. Lindenstrauss, *Geometric nonlinear functional analysis.*
Vol. 1. A.M.S. Colloquium Publications, 48. American Mathematical Society, Providence, RI, 2000.

[14] C. Bessaga and A. Pelczynski, *On bases and unconditional convergence of series in Banach spaces*, Studia Math., **42** (1958), 151–164.

[15] A. Borichev, R. Deville, E. Matheron, *Strongly sequentially continuous functions.* Quaest. Math. **24** (2001), no. 4, 535–548.

[16] A. Brunel, L. Sucheston, *On B-convex Banach spaces*, Math. System Theory, **7** (1974), 294–299.

[17] J. Burzyk, C. Kliś, Z. Lipecki, *On metrizable abelian groups with a completeness-type property.* Colloq. Math. **49** (1984), no. 1, 33–39.

[18] P. G. Casazza, T. J. Shura, *Tsirelson's space.* Lecture Notes in Mathematics, 1363. Springer-Verlag, Berlin, 1989.

[19] R. Deville, E. Matheron, *Pyramidal vectors and smooth functions on Banach spaces.* Proc. Amer. Math. Soc. **128** (2000), no. 12, 3601–3608

[20] J. Diestel, *Sequences and series in Banach spaces.* Graduate Texts in Mathematics, **92**. Springer-Verlag, New York, 1984.

[21] A. Dvoretzky, *Some results on convex bodies and Banach spaces.* 1961 Proc. Internat. Sympos. Linear Spaces (Jerusalem, 1960), 123–160. Jerusalem Academic Press, Jerusalem; Pergamon, Oxford.

[22] E. Ellentuck, *A new proof that analytic sets are Ramsey*, J. Symbolic Logic, **39** (1974), 163–165.

[23] J. Elton, *Thesis*, Yale University, New Haven, CT.

[24] J. Elton, E. Odell, *The unit ball of every infinite-dimensional normed linear space contains a $(1 + \varepsilon)$-separated sequence.* Colloq. Math. **44** (1981), no. 1, 105–109.

[25] P. Erdös and M. Magidor, *A note on regular methods of summability and the Banach-Saks property*, Proc. Amer. Math. Soc., **59** (1970), 232–234.

[26] P. Erdös and R. Rado, *A combinatorial theorem*, J. London Math. Soc., **25** (1950), 249–255.

[27] J. Farahat, *Espaces de Banach contenant l^1, d'après H. P. Rosenthal.* Espaces L^p, applications radonifiantes et géométrie des espaces de Banach, Exp. No. 26, 6 pp. Centre de Math., École Polytech., Paris, 1974.

[28] V. Farmaki, *Systems of Ramsey families.* Atti Sem. Mat. Fis. Univ. Modena **50** (2002), no. 2, 363–379.

[29] V. Ferenczi and C. Rosendal, *On the number of non-isomorphic subspaces of a Banach space*, Studia Math., to appear.

[30] T. Figiel , R. Frankiewicz , R. Komorowski and C. Ryll-Nardzewski, *On hereditarily indecomposable Banach spaces* Ann. Pure Appl. Logic **126** (2004), 293–299.

[31] F. Galvin, *A generalization of Ramsey's Theorem*, Notices Amer. Math. Soc., **15** (1968), Abstract 68T-368.

[32] F. Galvin, K. Prikry, *Borel sets and Ramsey's Theorem*, J. Symbolic Logic, **38** (1973), 193–198.

[33] I. Gasparis, E. Odell, and B. Wahl, *Weakly null sequences in the Banach space* $C(K)$. ArXiv 2004.

[34] W. T. Gowers, *Lipschitz Function on Classical Spaces*, European J. Combinatorics, **13** (1992), 141–151.

[35] W. T. Gowers, *An Infinite Ramsey Theorem and Some Banach-Space Dichotomies*, Ann. Math. **156** (2002), 797–833.

[36] W. T. Gowers, *Ramsey methods in Banach spaces*, Handbook of the geometry of Banach spaces, Vol. 2, 1071–1097, North-Holland, Amsterdam, 2003.

[37] S. Guerre-Delabrière, *Classical sequences in Banach spaces*. Monographs and Textbooks in Pure and Applied Mathematics, **166**. Marcel Dekker, Inc., New York, 1992.

[38] N. Hindman, *Finite sums from sequences within cells on a partition of* $\mathbb{N}$, J. Combinatorial Theory(A), **17** (1974), 1–11.

[39] W. Just, A. R. D. Mathias, K. Prikry, P. Simon, *On the existence of large p-ideals*. J. Symbolic Logic **55** (1990), no. 2, 457–465.

[40] A. Kanamori, *The higher infinite. Large cardinals in set theory from their beginnings.* Second edition. Springer Monographs in Mathematics. Springer-Verlag, Berlin, 2003.

[41] V. Kanellopoulos, *A proof of Gowers' c_0 theorem*, Proc. Amer. Math. Soc., to appear.

[42] A.S. Kechris *Classical descriptive set theory*, Springer-Verlag 1995.

[43] O. Klein and O. Spinas, *Canonical form of Borel functions on the Milliken's space*, preprint 2002.

[44] J. L. Krivine, *Sous-espaces de dimension finie des espaces de Banach réticulés*. Ann. of Math. (2) **104** (1976), no. 1, 1–29.

[45] R. Laver, *On Fraïssé's order type conjecture*, Ann. Math. **93** (1971), 89–111.

[46] R. Laver, *Well-quasi-orderings and sets of finite sequences*. Math. Proc. Cambridge Philos. Soc. **79** (1976), no. 1, 1–10.

[47] R. Laver, *Better-quasi-orderings and a class of trees*. Studies in foundations and combinatorics, 31–48, Adv. in Math. Suppl. Stud., 1, Academic Press, New York-London, 1978.

[48] J. Lindenstrauss and L. Tzafriri, *Classical Banach spaces I*, Springer-Verlag, 1977.

[49] J. Lopez-Abad, *Coding into Ramsey sets*, CRM Preprint **557**.

[50] J. Lopez-Abad, *Canonical equivalence relations on nets of PS_{c_0}*, CRM Prepint **558**.

[51] J. Lopez-Abad, A. Manoussakis, *On Tsirelson type spaces*. In preparation 2004.

[52] J. Lopez-Abad, S. Todorcevic, *Pre-compact families of finite sets and unconditional basic sequences*. Preprint 2004.

[53] E. Matheron, *A useful lemma concerning subseries convergence*. Bull. Austral. Math. Soc. **63** (2001), no. 2, 273–277.

[54] A. R. D. Mathias, *A remark on rare filters*. Colloq. Math. Soc. Janos Bolyai, Vol. 10, North-Holland, Amsterdam, 1975.

[55] A. R. D. Mathias, *Happy families*. Ann. Math. Logic **12** (1977), no. 1, 59–111.

[56] B. Maurey, *A note on Gowers' dichotomy theorem*, Convex Geom. Analysis, MSRI Publications, Vol **34** (1998), 149–157.

[57] B. Maurey, *Type, cotype and K-convexity*. Handbook of the geometry of Banach spaces, Vol. 2, 1299–1332, North-Holland, Amsterdam, 2003.

[58] B. Maurey, H.P. Rosenthal, *Normalized weakly null sequence with no unconditional subsequence*, Studia Math., **61** (1977), 77–98.

[59] S. Mazur, W. Orlicz, *Sur les espaces métriques linéaires. II*. Studia Math. **13**, (1953). 137–179.

[60] K. Milliken, *Ramsey's theorem with Sums and Unions*, J. Combinatorial Theory (A), **18** (1975), 276–290.

[61] V. D. Milman, G. Schechtman, *Asymptotic theory of finite-dimensional normed spaces. With an appendix by M. Gromov*. Lecture Notes in Mathematics, 1200. Springer-Verlag, Berlin, 1986.

[62] E. C. Milner, *Basic wqo- and bqo-theory. Graphs and order (Banff, Alta., 1984)*, 487–502, NATO Adv. Sci. Inst. Ser. C Math. Phys. Sci., 147, Reidel, Dordrecht, 1985.

[63] C. St. J. A. Nash-Williams, *On well quasi-ordering transfinite sequences*, Proc. Cambridge Philos. Soc. **61** (1965), 33–39.

[64] C. St. J. A. Nash-Williams, *On better-quasi ordering transfinite sequences*, Proc. Cambridge Philos. Soc. **64** (1968), 273–290.

[65] C. St. J. A. Nash-Williams, *On well quasi-ordering infinite trees*, Proc. Cambridge Philos. Soc. **64** (1968).

[66] E. Odell, *Applications of Ramsey theorems to Banach space theory*. Notes in Banach spaces, 379–404, Univ. Texas Press, Austin, Tex., 1980.

[67] E. Odell, *On Schreier unconditional sequences*. Banach spaces (Mérida, 1992), 197–201, Contemp. Math., 144, Amer. Math. Soc., Providence, RI, 1993.

[68] E. Odell, *On subspaces, asymptotic structure, and distortion of Banach spaces; connections with logic*. Analysis and logic (Mons, 1997), 189–267, London Math. Soc. Lecture Note Ser., 262, Cambridge Univ. Press, Cambridge, 2002.

[69] E. Odell, T. Schlumprecht, *The distortion problem*, Acta Math. **173** (1994), 259–281.

[70] E. Odell, N. Tomczak-Jaegermann, R. Wagner, *Proximity to l_1 and distortion in asymptotic l_1 spaces*. J. Funct. Anal. **150** (1997), no. 1, 101–145.

[71] J. Pawlikowski, *Parametrized Ellentuck theorem*. Topology Appl. **37** (1990), no. 1, 65–73.

[72] M. Pouzet, *Sur les prémeillieurordres*, Ann. Inst. Fourier (Grenoble) **22** (1972), 1–10.

[73] H. J. Prömel; B. Voigt, *Canonical forms of Borel-measurable mappings $\Delta \colon [\omega]^\omega \to \mathbb{R}$*. J. Combin. Theory Ser. A **40** (1985), no. 2, 409–417.

[74] V. Pták, *A combinatorial theorem on system of inequalities and its applications to analysis*, Czech. Math. J., **84** (1959), 629–630.

[75] P. Pudlak and V. Rödl, *Partition theorems for systems of finite subsets of integers*, Discrete Math., **39** (1982), 67–73.

[76] F. P. Ramsey, *On a problem of formal logic*, Proc. London Math. Soc., **30** (1929), 264–286.

[77] C. Rosendal, *Incomparable, non isomorphic and minimal Banach spaces*, preprint 2004.

[78] H. P. Rosenthal, *A characterization of Banach spaces containing l^1*. Proc. Nat. Acad. Sci. U.S.A. **71** (1974), 2411–2413.

[79] H. P. Rosenthal, *On a theorem of J. L. Krivine concerning block finite representability of l^p in general Banach spaces*. J. Funct. Anal. **28** (1978), no. 2, 197–225.

[80] H. P. Rosenthal, *A characterization of Banach spaces containing c_0*. J. Amer. Math. Soc. **7** (1994), no. 3, 707–748.

[81] H. P. Rosenthal, *The Banach spaces C(K)*. Handbook of the geometry of Banach spaces, Vol. 2, 1547–1602, North-Holland, Amsterdam, 2003.

[82] T. Schlumprecht, *An arbitrarily distortable Banach space*, Israel J. Math., **76** (1991), 81–95.

[83] J. Schreier, *Ein Gegenbeispiel zur Theorie der Schwachen Konvergenz*, Studia Math., **2** (1930), 58–62.

[84] J. Silver, *Every analytic set is Ramsey*, J. Symbolic Logic, **35** (1970), 60–64.

[85] S. G. Simpson, *Bqo-theory and Fraïsé's conjecture*, in Recursive aspects of descriptive set theory, R. Mansfield and G. Weikkamp, Oxford Univ. Press. 1985.

[86] J. Stern, *A Ramsey theorem for trees, with an application to Banach spaces*, Israel. J. Math. **29** (1978), 179–188.

[87] A. D. Taylor, *A canonical Partition Relation for finite Subsets of ω*, J. Combinatorial Theory (A), **21** (1970), 137–146.

[88] S. Todorcevic, *Topics in topology*. Lecture Notes in Mathematics, **1652**. Springer-Verlag, Berlin, 1997.

[89] S. Todorcevic, *An Introduction to Ramsey Spaces*, in preparation.

[90] N. Tomczak-Jaegermann, *Banach spaces of type p have arbitrarily distortable subspaces*. Geom. Funct. Anal. **6** (1996), no. 6, 1074–1082

[91] B. S. Tsirelson, *Not every Banach space contains ℓ_p or c_0*, Functional Anal. Appl., **8** (1974), 81–95.

Index

Advanced Courses in Mathematics

Your Specialized Publisher in Mathematics

Birkhäuser

Since 1995 the Centre de Recerca Matemàtica (CRM) in Barcelona has conducted a number of annual Summer Schools at the post-doctoral or advanced graduate level. Sponsored mainly by the European Community, these Advanced Courses have usually been held at the CRM in Bellaterra.
The books in this series consist essentially of the expanded and embellished material presented by the authors in their lectures.

For orders originating from all over the world except USA/Canada/Latin America:

Birkhäuser Verlag AG
c/o Springer GmbH & Co
Haberstrasse 7
D-69126 Heidelberg
Fax: +49 / 6221 / 345 4 229
e-mail: birkhauser@springer.de
http://www.birkhauser.ch

For orders originating in the USA/Canada/Latin America:

Birkhäuser
333 Meadowland Parkway
USA-Secaucus
NJ 07094-2491
Fax: +1 201 348 4505
e-mail: orders@birkhauser.com

■ **Da Prato, G.**, Scuola Normale Superiore, Pisa, Italy
Kolmogorov Equations for Stochastic PDEs
2004. 192 pages. Softcover
ISBN 3-7643-7216-8

The subject of this book is stochastic partial differential equations, in particular, reaction-diffusion equations, Burgers and Navier-Stokes equations and the corresponding Kolmogorov equations. For each case the transition semigroup is considered and irreducibility, the strong Feller property, and invariant measures are investigated. Moreover, it is proved that the exponential functions provide a core for the infinitesimal generator. As a consequence, it is possible to study Sobolev spaces with respect to invariant measures and to prove a basic formula of integration by parts (the so-called "carré du champs identity").
Several results were proved by the author and his collaborators and appear in book form for the first time. Presenting the basic elements of the theory in a simple and compact way, the book covers a one-year course directed to graduate students in mathematics or physics. The only prerequisites are basic probability (including finite dimensional stochastic differential equations), basic functional analysis and some elements of the theory of partial differential equations.

■ **Drensky, V.**, Bulgarian Academy of Sciences, Sofia, Bulgaria / **Formanek, E.**, Pennsylvania State University, University Park, PA, USA
Polynomial Identity Rings

2004. 208 pages. Softcover
ISBN 3-7643-7126-9

A ring R satisfies a polynomial identity if there is a polynomial f in noncommuting variables which vanishes under substitutions from R. For example, commutative rings satisfy the polynomial

$$f(x,y) = xy - yx$$

and exterior algebras satisfy the polynomial

$$f(x,y,z) = (xy - yx)z - z(xy - yx).$$

"Satisfying a polynomial identity" is often regarded as a generalization of commutativity.
These lecture notes treat polynomial identity rings from both the combinatorial and structural points of view. The former studies the ideal of polynomial identities satisfied by a ring R. The latter studies the properties of rings which satisfy a polynomial identity.
The greater part of recent research in polynomial identity rings is about combinatorial questions, and the combinatorial part of the lecture notes gives an up-to-date account of recent research. On the other hand, the main structural results have been known for some time, and the emphasis there is on a presentation accessible to newcomers to the subject.
The intended audience is graduate students in algebra, and researchers in algebra, combinatorics and invariant theory.

Advanced Courses in Mathematics

Your Specialized Publisher in Mathematics

Birkhäuser

Available:

Argyros, S. / Todorcevic, S.
Ramsey Methods in Analysis (2005)
ISBN 3-7643-7264-8

Audin, M. / Cannas da Silva, A. / Lerman, E.
Symplectic Geometry of Integrable Hamiltonian Systems (2003)
ISBN 3-7643-2167-9

Brown, K.A. / Goodearl, K.R.
Lectures on Algebraic Quantum Groups (2002)
ISBN 3-7643-6714-8

Da Prato, G.
Kolmogorov Equations for Stochastic PDEs
(2004). ISBN 3-7643-7216 8

Drensky, V. / Formanek, E.
Polynomial Identity Rings (2004)
ISBN 3-7643-7126-9

Dwyer, W.G. / Henn, H.-W.
Homotopy Theoretic Methods in Group Cohomology (2001)
ISBN 3-7643-6605-2

Hess, K. / Sullivan, D. / Voronov, A.A.
String Topology and Hochschild Homology
(2004). ISBN 3-7643-2182-2

Markvorsen, S. / Min-Oo, M.
Global Riemannian Geometry: Curvature and Topology (2003)
ISBN 3-7643-2170-9

Mislin, G. / Valette, A.
Proper Group Actions and the Baum-Connes Conjecture (2003)
ISBN 3-7643-0408-1

Forthcoming:

Catalano, D. / Cramer, R. / Damgård, I. / Di Creszenso, G. / Pointcheval, D. / Takagi, T.
Contemporary Cryptology (2005)
ISBN 3-7643-7294-X

The aim of this text is to treat selected topics of the subject, structured in five quite independent but related themes:

Efficient distributed computation modula a shared secret

Multiparty computation

Modern Cryptography

Provable security for public key schemes

Efficient and secure public-key cryptosystems.